Fractional Order Motion Controls

Fractional Order Motion Controls

Ying Luo

Department of Automation Science and Engineering
South China University of Technology
Guangzhou, China

YangQuan Chen

School of Engineering, University of California – Merced
California, USA

A John Wiley & Sons, Ltd., Publication

Registered office
John Wiley & Sons Ltd, The Atrium, Southern Gate, Chichester, West Sussex, PO19 8SQ, United Kingdom

For details of our global editorial offices, for customer services and for information about how to apply for permission to reuse the copyright material in this book please see our website at www.wiley.com.

Library of Congress Cataloging-in-Publication Data

Luo, Ying, 1973–
  Fractional order motion controls / Ying Luo, YangQuan Chen.
    pages cm
  Includes bibliographical references and index.
  ISBN 978-1-119-94455-3 (cloth)
  1. Motion control devices.   2. Incremental motion control.   I. Chen, YangQuan, 1966–   II. Title.
  TJ214.5.L86 2012
  629.8–dc23

                                                                                    2012020586

A catalogue record for this book is available from the British Library.

Print ISBN: 9781119944553

Typeset in 10/12.5 Palatino by Aptara DELHI, India
Printed and bound in Malaysia by Vivar Printing Sdn Bhd

1   2012

To my father XianShu Luo and my mother Gui'E Xiong
– Ying Luo

To my family, my mentors and my colleagues
– YangQuan Chen

Contents

PART III   FRACTIONAL ORDER POSITION CONTROLS

PART V   FRACTIONAL ORDER DISTURBANCE COMPENSATIONS

Foreword

I am pleased that Professor Ying Luo and Professor YangQuan Chen have completed their book, *Fractional Order Motion Controls*, and it will be published by John
Wiley & Sons Ltd. The mathematical backbone of fractional order control is fractional
order differential equations and fractional order transfer functions. Fractional order
transfer functions were proposed by Bode in the analysis and feedback amplifier
design as early as in the mid-1940s. Tustin applied Bode's idea in the late 1950s to
motion control, and he suggested that the open loop transfer function of the motion
control system may be set $G(s) = (\frac{w_c}{s})^k$ with $k = 1.5$ over a certain frequency range,
say, $(0.2\ 1.2)w_c$ It is easy to check that for this transfer function the phase margin is
45 degree, and the closed loop system remains robust if this phase margin can be
maintained over a finite frequency range. This example may be a strong motivation
for motion control engineers to perform fractional order design. In tracing references
on fractional order control, I noted that a number of interesting successful applications of fractional order control have been reported, and that excellent papers were
published by control practitioners in industry such as Dr. Manabe at Mitsubishi. We
may wonder then why fractional order control has not been more widely used as
yet. One reason may be that fractional order controllers must be approximated by a
combination of standard (integer order) differentiators and integrators for implementation and that implementable controllers may become high order. This is no longer
a major obstacle since high order controllers may be digitally implemented. I think
that fractional order control should not be regarded as a new control theory to replace
standard control such as PID control based on integer order transfer functions. Good
ideas that we learned in the standard (integer order) control theory all remain as
valid design guidelines. Fractional order control adds something more to what we
know and practice. It may be regarded as a means that allows us to design control
systems, for example perform loop shaping, in a wider design domain. When we are
freed from the constraint that each element of a controller must be of integer order,
we may move only in the direction to have better control systems. Thus, fractional
order control is of interest for those who design controllers for physical systems:

for example motion control systems, automotive suspension systems, robots, and so on. This book by Luo and Chen covers the range from fundamental fractional calculus and early development of fractional order control to the most recent developments by the authors themselves as well as by others. It is a fun book to read, and it will surely motivate more motion control engineers to plunge into the world of fractional thinking.

Masayoshi Tomizuka
Cherly and John Neerhout, Jr. Distinguished Professor
University of California – Berkeley, USA

Preface

There is increasing interest in dynamic systems and controls of non-integer orders or fractional orders. Traditional calculus is based on integer order differentiation and integration. The concept of fractional calculus has tremendous potential to change the way we model and control the world around us. Rejecting fractional derivatives is like saying that zero, fractional, or irrational numbers do not exist. Fractional calculus has a firm and enduring theoretical foundation. However, the fractional calculus concept was not widely applied in control engineering for hundreds of years, because the idea was unfamiliar and the fractional operators were limited in their realization. In the past few decades, with the rapid development of computer technology and better understanding of the potential of fractional calculus, the realization of fractional order control systems became much easier and fractional calculus is becoming more and more useful in various science and engineering areas. The present book focuses on fractional order control of motion systems.

Motion control is a sub-field of automation, in which the velocity and position of machines are controlled using certain types of actuation devices such as a hydraulic actuator, a linear actuator, or an electric motor, generally called a servo. Motion control is an important part of robotics and Computerized Numerical Control (CNC) machine tools, and is widely used in packaging, printing, textile, semiconductor production, and the assembly industries. In motion control systems, the control strategies should be stabilizing, fast and precise. In real-time applications, high performance motion control systems must be immune to any kind of disturbance. Thus, motion control research explores enhancement of both performance following command and disturbance rejection. The aim of this book is to introduce fractional calculus-based control methods in motion control applications and to illustrate the advantages and importance of using fractional order controls.

In order to improve the performance following command of motion control, fractional order PID controllers are proposed and designed in a systematic way for integer/fractional order velocity and position systems in this work. With the "flat phase" tuning constraint and other specifications, the motion control systems based on fractional calculus can achieve better robust performances with respect to loop gain variations or time constant variations than using traditionally optimized integer order controllers. From our extensive simulation and experimental efforts, we demonstrated the desirable control performance with faster response and smaller overshoot using properly designed fractional order controllers over those using optimized integer order controllers. In terms of systematic design schemes for fractional order PID controllers satisfying the desired specifications, stability is the minimum requirement for the controller design, and it is better to obtain a feasible region to check the complete set of specifications before the controller is designed and tuned. Therefore, the complete stability regions of the fractional order PID controller parameters, and the achievable regions of the specifications to obtain stabilizing and the desired fractional order PID controllers are discussed in detail in this book. Impressively, the achievable regions of specifications using fractional order PID controllers are significantly larger than those using an integer order PID controller for certain types of systems.

Motion control systems are usually influenced by various disturbances. In high performance motion control systems, maintaining a stable and robust operation by attenuating the influence of disturbances is required. A fractional order disturbance observer (DOB) based on the fractional order Q-filter is presented. A nice feature of this is that the traditional DOB is extended to the fractional order DOB (FO-DOB) with the advantage that the FO-DOB design will no longer be conservative nor aggressive. In addition, a fractional order adaptive feedforward cancellation (FO-AFC) scheme is proposed to cancel periodic disturbances. This FO-AFC method is much more flexible than the integer order AFC in preventing periodic disturbance and suppressing the harmonics or the noise. Meanwhile, a fractional order robust control method is devised for cogging effect compensation on the permanent magnetic synchronous motor position and the velocity systems. Also presented is a fractional order periodic adaptive learning compensation method to reject general state-dependent periodic disturbances.

In this book, nonlinear motion control systems are also considered for fractional calculus applications. A fractional order PID controller design scheme is presented for a DC motor control system with an elastic shaft. Under the same optimization conditions, the best fractional order PID controller outperforms the best integer order PID controller for the motion control system with nonlinearities of backlash and dead zone. Applying the systematic design of fractional order PD (FOPD) controller for ultra-low speed position tracking with a significant nonlinear friction effect, the experimental tracking performance using the designed FOPD controller is much better than that using the optimized integer order PI controller. This advantage of the designed FOPD is explained by the describing function analysis. Furthermore, an optimized fractional order conditional integrator (OFOCI) is proposed. By tuning the fractional order and the other tuning parameter following the analytical optimal design specifications, this proposed OFOCI can achieve an optimized performance not achievable by integer order conditional integrators.

In order to further validate and demonstrate some of the presented fractional order controller design schemes in this work, two real-world applications of fractional order control are included: an unmanned aerial vehicle (UAV) flight control system and an industrial hard-disk-drive (HDD) servo system. These are really exciting real-world applications that clearly show the advantages of using fractional calculus for motion controls.

This book is organized as follows. Part I contains only Chapter 1, introducing fundamentals of fractional order systems and controls followed by research motivations and book contributions. Part II is dedicated to the fractional order velocity controls, which includes Chapters 2–5. Part III focuses on the fractional order position controls, including Chapters 6–10. The feasible regions of the specifications for integer and fractional order controller designs based on the stability analysis are studied in Part IV, which includes Chapters 11 and 12. Part V explains how to design a fractional order disturbance observer, a fractional order adaptive feed-forward controller, a fractional order adaptive controller, and a fractional order periodic adaptive learning controller to compensate for the external disturbances in motion control systems, shown in Chapters 13–16, respectively. Part VI is devoted to the fractional order controls on nonlinear control systems in Chapters 17–19. Applications of fractional order controls in UAV flight control system and the HDD servo system are presented in Part VII including Chapters 20 and 21.

It is our sincere hope that this book can well serve two purposes. For motion control researchers and engineers, this book offers some new schemes that can present further improved performance not achievable before. For researchers and students interested in fractional calculus, this book is a demonstration that fractional calculus is indeed useful in real-world applications, not just a pure math game. Given the pervasive and ubiquitous nature of fractional calculus, we do believe that, as demonstrated in this book, even for simple motion control problems, there are ample opportunities to apply fractional calculus-based control methods. For more complex engineering and non-engineering systems, the opportunities and beneficial consequences of applying fractional calculus are limited only by our imagination.

Ying Luo
YangQuan Chen
California, USA

Acknowledgments

This book provides a comprehensive summary of our research efforts during the past few years in fractional order control theory and its applications in motion systems. This book contains material from papers and articles that have been previously published as well as the Ph.D. dissertation of the first author. We are grateful and would like to acknowledge the copyright permissions from the following publishers who have released our works.

Acknowledgement is given to the Institute of Electrical and Electronics Engineers (IEEE) to reproduce material from the following papers:

© 2009 IEEE. Reprinted, with permission, from YangQuan Chen, I. Petras, and Dingyu Xue, "Fractional order control: A tutorial," in *Proceedings of American Control Conference*, 10–12 June 2009, St. Louis, MO, pages 1397–1411 (material found in Chapter 1). DOI: 10.1109/ACC.2009.5160719.

© 2002 IEEE. Reprinted, with permission, from Dingyu Xue, and YangQuan Chen, "A comparative introduction of four fractional order controllers," in *Proceedings of the 4th World Congress on Intelligent Control and Automation*, 2002, pages 3228–3235 (material found in Chapter 1). DOI: 10.1109/WCICA.2002.1020131.

© 2009 IEEE. Reprinted, with permission, from Chunyang Wang, Ying Luo, and YangQuan Chen, "Fractional order proportional integral (FOPI) and [proportional integral] (FO[PI]) controller designs for first order plus time delay (FOPTD) systems," in *Proceedings of the 21th IEEE Conference on Chinese Control and Decision*, Guilin, China, June 17–19, 2009, pages 329–334 (material found in Chapter 2). DOI: 10.1109/CCDC.2009.5195105.

© 2005 IEEE. Reprinted, with permission, from YangQuan Chen and K. L. Moore, "Relay feedback tuning of robust PID controllers with iso-damping property," *IEEE Transactions on Systems, Man, and Cybernetics, Part B: Cybernetics*, volume 35, issue 1, 2005, pages 23–31 (material found in Chapter 4). DOI: 10.1109/TSMCB.2004.837950.

© 2009 IEEE. Reprinted, with permission, from ChunYang Wang, YongShun Jin, and YangQuan Chen, "Auto-tuning of FOPI and FO[PI] controllers with iso-damping property," in *Proceedings of the 48th IEEE Conference on Decision and Control, 2009*

Held Jointly with the 2009 28th Chinese Control Conference, 15–18 Dec. 2009, pages 7309–7314 (material found in Chapter 5). DOI: 10.1109/CDC.2009.5400057.

© 2010 IEEE. Reprinted, with permission, from HongSheng Li, Ying Luo and YangQuan Chen, "A fractional order proportional and derivative (FOPD) motion controller: Tuning rule and experiments," *IEEE Transactions on Control Systems Technology*, vol. 18, no. 2, March 2010, pages 516–520 (material found in Chapter 6). DOI: 11.1109/TCST.2009.2019120.

© 2009 IEEE. Reprinted, with permission, from Ying Luo and YangQuan Chen, "Fractional-order [proportional derivative] controller for robust motion control: Tuning procedure and validation," in *Proceedings of the 2009 American Control Conference*, St. Louis, Missouri, June 10–12 2009, pages 1412–1417 (material found in Chapter 7). DOI: 10.1109/ACC.2009.5160284.

© 2011 IEEE. Reprinted, with permission, from Yongshun Jin, YangQuan Chen, and Dingyu Xue, "Time-constant robust analysis of a fractional order [proportional derivative] controller," *IET Control Theory and Applications*, volume 5, issue 1, January 2011, pages 164–172 (material found in Chapter 8). DOI: 10.1049/iet-cta.2009.0543.

© 2011 IEEE. Reprinted, with permission, from Ying Luo and YangQuan Chen, "Synthesis of robust PID controllers design with complete information on pre-specifications for the FOPTD systems," in *Proceedings of 2011 American Control Conference*, San Francisco, CA, June 29–July 1, 2011 (material found in Chapter 11).

© 2011 IEEE. Reprinted, with permission, from Ying Luo and YangQuan Chen, "Stabilizing and robust FOPI controller synthesis for first order plus time delay systems," in *Proceedings of the 50th IEEE Conference on Decision and Control and European Control Conference*, Orlando, FL, USA, December 12–15, 2011 (material found in Chapter 12).

© 2011 IEEE. Reprinted, with permission, from Ying Luo, YangQuan Chen and YouGuo Pi, "Fractional order adaptive feedforward cancelation," in *Proceedings of 2011 American Control Conference*, San Francisco, CA, June 29–July 1, 2011 (material found in Chapter 14).

© 2011 IEEE. Reprinted, with permission, from Ying Luo, YangQuan Chen, Hyo-sung Ahn, and Youguo Pi, "Fractional order periodic adaptive learning compensation for the state-dependent periodic disturbance," *IEEE Transactions on Control Systems Technology*, volume 20, issue 2, 2012, pages 465–472 (material found in Chapter 16). DOI: 10.1109/TCST.2011.2117426.

© 2006 IEEE. Reprinted, with permission, from Dingyu Xue, Chunna Zhao, and YangQuan Chen, "Fractional order PID control of a DC-motor with elastic shaft: A case study," in *Proceedings of American Control Conference*, 14–16 June 2006, Minneapolis, MN (material found in Chapter 17). DOI: 10.1109/ACC.2006.1657207.

© 2010 IEEE. Reprinted, with permission, from Ying Luo, Haiyang Chao, Long Di, and YangQuan Chen, "Fractional order [proportional integral] roll channel flight control for small fixed-wing UAV," in *Proceedings of the 8th World Congress on Intelligent Control and Automation*, Jinan, China, July 2010 (material found in Chapter 20).

Acknowledgement is given to the Elsevier B.V. to reproduce material from the following papers:

© 2010 Elsevier B.V. Reprinted, with permission, from Ying Luo, Chunyang Wang, YangQuan Chen and YouGuo Pi, "Tuning fractional order proportional integral controllers for fractional order systems," *Journal of Process Control*, volume 20, issue 7, August 2010, pages 823–831 (material found in Chapter 3). DOI: 10.1016/j.jprocont.2010.04.011.

© 2011 Elsevier B.V. Reprinted, with permission, from Ying Luo, YangQuan Chen, and Youguo Pi, "Experimental study of fractional order proportional derivative controller synthesis for fractional order systems," Mechatronics, volume 21, 2011, pages 204–214 (material found in Chapter 9). DOI: 10.1016/j.mechatronics.2010.10.004.

© 2009 Elsevier B.V. Reprinted, with permission, from Ying Luo and YangQuan Chen, "Fractional-order [proportional derivative] controller for a class of fractional order systems," *Automatica*, volume 45, issue 10, 2009, pages 2446–2450 (material found in Chapter 10). DOI: 10.1016/j.automatica.2009.06.022.

© 2010 Elsevier B.V. Reprinted, with permission, from Ying Luo, YangQuan Chen, Hyo-Sung Ahn and YouGuo Pi, "Fractional order robust control for cogging effect compensation in PMSM position servo systems: Stability analysis and experiments," *Control Engineering Practice*, volume 18, issue 9, September 2010, pages 1022–1036 (material found in Chapter 15). DOI: 10.1016/j.conengprac.2010.05.005.

© 2011 Elsevier B.V. Reprinted, with permission, from Ying Luo, YangQuan Chen, and Youguo Pi, "Fractional order ultra low-speed position servo: Improved performance via describing function analysis," *ISA Transactions*, volume 50, 2011, pages 53–60 (material found in Chapter 18). DOI: 10.1016/j.isatra.2010.09.003.

© 2011 Elsevier B.V. Reprinted, with permission, from Ying Luo, YangQuan Chen, Youguo Pi, Concepción A. Monje, and Blas M. Vinagre, "Optimized fractional order conditional integrator," *Journal of Process Control*, volume 21, issue 6, July 2011, pages 960–966 (material found in Chapter 19). DOI: 10.1016/j.jprocont.2011.02.002.

Acknowledgement is given to the American Society of Mechanical Engineers (ASME) to reproduce material from the following paper:

© 2003 ASME. Reprinted, with permission, from YangQuan Chen, Blas M. Vinagre, and Igor Podlubny, "On fractional order disturbance observer," in *Proceedings of 2003 Design Engineering Technical Conferences and Computers and Information in Engineering Conference*, Chicago, Illinois, USA, September 2–6, 2003 (material found in Chapter 14).

The research described in this book would not have been possible without the inspiration and help from the work of individuals in the research community, and we would like to acknowledge them. We would like to express our thanks to Dr. YouGuo Pi for his efforts in some chapters of this book, and all his great support and help to the first author of this book. Thanks go to Dr. DingYu Xue for his support on the robust analysis and discussion on the nonlinearities effect of fractional order PID controls (Chapters 8 and 17). Our thanks are directed to Dr. Haiyang Chao, Long Di, and Jinlu Han for their joint efforts on fractional order flight control on unmanned aerial vehicles (Chapter 20), to Dr. Yongshun Jin and Dr. Chunyang Wang for their joint work on fractional order controllers' design and tuning (Chapters 2, 5, and 8), and to Dr. HuiFang Dou for her help on linear motor modeling and control. We would like to thank Dr. Tao Zhang, Dr. C. I. Kang, and BongJin Lee for their help and guidance on fractional order control in the industrial hard-disk-drive servo system (Chapter 21), and Dr. Hyo-Sung Ahn for his efforts in our joint research on iterative and repetitive learning controls (Chapters 15 and 16). Our thanks go to Dr. Concepción A. Monje and Dr. Blas M. Vinagre for their work and guidance on the fractional order conditional integrator (Chapter 19).

Ying Luo would like to express his sincere thanks to his parents, XianShu Luo and Gui'E Xiong, for their constant and great support. He would also like to thank former and current CSOIS members: Dr. Yan Li, Yiding Han, Austin Jensen, Calvin Coopmans, Shayok Mukhopadhyay and Dr. Hu Sheng for their support during their studies in CSOIS at Utah State University.

YangQuan Chen would like to thank his wife, Dr. Huifang Dou, and his sons, Duyun, David and Daniel, for their patience, understanding and complete support throughout this work. He is thankful to Utah State University for the support and academic freedom he received where the main work of this book was performed while the final proof was completed during his move to University of California, Merced.

We wish to express our appreciation to five anonymous book proposal reviewers whose comments improved our presentation. In particular, we are thankful to Prof. Tomizuka for preparing an insightful Foreword for this book. Last but not least, we thank Sophia Travis (John Wiley & Sons – Chichester) and Paul Petralia (John Wiley & Sons – Hoboken) for their excellent professional support during the whole cycle of this book project.

Acronyms

AFC	adaptive feed-forward cancellation
CCI	Clegg conditional integrator
CNC	computerized numerical control
CRB	complex root boundary
DC	direct current
DF	describing function
DOB	disturbance observer
ESF	error sensitivity function
FO	fractional order
FO-AFC	fractional order adaptive feed-forward cancellation
FOC	fractional order control
FOCI	fractional order conditional integrator
FODOB	fractional order disturbance observer
FOIMP	fractional order internal model principle
FOLPF	fractional order low-pass filter
FOPI	fractional order proportional integral
FOPID	fractional order proportional integral derivative
FOPTD	first order plus time delay
FOPD	fractional order proportional derivative
FORC	fractional order robust control
FOS	fractional order system
FOVS	fractional order velocity systems
FO[PD]	fractional order [proportional derivative]
FO[PI]	fractional order [proportional integral]
FO-PALC	fractional order periodic adaptive learning compensation
HDD	hard disk drive
HIL	hardware in the loop
ICI	intelligent conditional integrator
IMP	internal model principle
IMU	inertial measurement units

IO	integer order
IOAFC	integer order adaptive feed-forward cancellation
IOCI	integer order conditional integrator
IOPID	integer order proportional integral derivative
IOPI	integer order proportional integral
IORC	integer order robust control
IOS	integer order system
IO-PALC	integer order periodic adaptive learning compensation
IRB	infinity root boundary
IRID	impulse response invariant discretization
ISE	integral of squared error
ITAE	integral time absolute error
LTI	linear time invariant
MICI	modified intelligent conditional integrator
MIMO	multiple input and multiple output
MZNPI	modified Ziegler-Nichols proportional integral
OFOCI	optimized fractional order conditional integrator
PID	proportional integral derivative
PMSM	permanent magnetic synchronous motors
RRB	real root boundary
RTW	realtime workshop
SDPD	State-dependent periodic disturbance
SISO	single input and single output
TID	tilted integral derivative
UAV	unmanned aerial vehicle
VCM	voice coil motor
ZNPID	Ziegler-Nichols proportional integral derivative
w.r.t	with respect to

Part I
Fundamentals of Fractional Order Controls

Part I

1

Introduction

It is known that the nth order derivative of a function $f(t)$ can be mathematically described by $d^n y/dx^n$. With this notation, one may ask "What does $n = 1/2$ mean in the notation?" Actually, this was the question asked in a letter by the French mathematician Guillaume François Antoine L'Hôpital to one of the inventors of calculus, the French mathematician Gottfried Wilhelm Leibnitz more than 300 years ago. In answering to the letter, Leibnitz said: "It will lead to a paradox, from which one day useful consequences will be drawn." This marks the beginning of fractional calculus. However, earlier research concentrated on theoretical math issues. Fractional calculus is now being widely used in many areas. For instance, in the discipline of automatic control, fractional order control is a promising new topic [161].

1.1 Fractional Calculus

The idea of Fractional Calculus has been known since the development of the regular (integer order) calculus, with the first reference probably being associated with Leibniz and L'Hôpital in 1695 where the half-order derivative was mentioned.

Fractional calculus is a generalization of integration and differentiation to the non-integer order fundamental operator $_a D_t^r$, where a and t are the limits of the operation. The continuous integro-differential operator is defined as

$$_a D_t^r = \begin{cases} d^r/dt^r & \Re(r) > 0, \\ 1 & \Re(r) = 0, \\ \int_a^t (d\tau)^{-r} & \Re(r) < 0, \end{cases}$$

where r is the order of the operation, generally $r \in R$ but r could also be a complex number [179].

Fractional Order Motion Controls, First Edition. Ying Luo and YangQuan Chen.
© 2013 John Wiley & Sons, Ltd. Published 2013 by John Wiley & Sons, Ltd.

1.1.1 Definitions and Properties

Various definitions have appeared in the development and studies of fractional calculus. Some of the definitions are directly extended from the conventional integer order calculus. The commonly used definitions are summarized as follows [161]:

A. **Fractional order Cauchy integral formula**
   The formula is extended from integer order calculus

$$D^\alpha f(t) = \frac{\Gamma(\alpha+1)}{j2\pi} \int_C \frac{f(\tau)}{(\tau-t)^{\alpha+1}} \, d\tau, \tag{1.1}$$

   where C is the closed-path that encircles the poles of the function $f(t)$.
   The integrals and derivatives for sinusoidal and cosine functions can be expressed by

$$\frac{d^k}{dt^k}\left[\sin at\right] = a^k \sin\left(at + \frac{k\pi}{2}\right), \quad \frac{d^k}{dt^k}\left[\cos at\right] = a^k \cos\left(at + \frac{k\pi}{2}\right). \tag{1.2}$$

   It can also be shown with Cauchy's formula that, if k is not an integer, the above formula is still valid.

B. **Grünwald-Letnikov definition**
   The fractional order differentiation and integral can be defined in a unified way such that

$$_a D_t^\alpha f(t) = \lim_{h \to 0} \frac{1}{h^\alpha} \sum_{j=0}^{[(t-a)/h]} (-1)^j \binom{\alpha}{j} f(t-jh), \tag{1.3}$$

   where $\binom{\alpha}{j}$ are the binomial coefficients; the subscripts to the left and right of D are the lower- and upper-bounds in the integral. The value of α can be positive or negative, corresponding to differentiation and integration, respectively and the α is non-integer.

C. **Riemann-Liouville definition**
   The fractional order integral is defined as

$$_a D_t^{-\alpha} f(t) = \frac{1}{\Gamma(\alpha)} \int_a^t (t-\tau)^{\alpha-1} f(\tau) d\tau, \tag{1.4}$$

   where $0 < \alpha < 1$, and a is the initial value. Let $a = 0$, the notation of integral can be simplified to $D_t^{-\alpha} f(t)$. The Riemann-Liouville definition is a widely used definition for fractional order differentiation and integral. Similarly, fractional order differentiation is defined as

$$_a D_t^\beta f(t) = \frac{d^n}{dt^n} \left[_a D_t^{-(n-\beta)} f(t) \right] = \frac{1}{\Gamma(n-\beta)} \frac{d^n}{dt^n} \left[\int_a^t \frac{f(\tau)}{(t-\tau)^{\beta-n+1}} d\tau \right], \quad (1.5)$$

where $n - 1 < \beta \leqslant n$.

D. **Caputo definition**

The Caputo fractional order differentiation is defined by

$$_0 D_t^\alpha f(t) = \frac{1}{\Gamma(1-\gamma)} \int_0^t \frac{f^{(m+1)}(\tau)}{(t-\tau)^\gamma} d\tau, \quad (1.6)$$

where $\alpha = m + \gamma$, m is an integer and $0 < \gamma \leqslant 1$. Similarly, by the Caputo definition, the integral is described by

$$_0 D_t^{-\gamma} f(t) = \frac{1}{\Gamma(\gamma)} \int_0^t \frac{f(\tau)}{(t-\tau)^{1-\gamma}} d\tau, \gamma > 0. \quad (1.7)$$

It can be shown that for a great varieties of functions, the Grünwald-Letnikov and the Riemann-Liouville definitions are equivalent [188].

The properties of fractional calculus are summarized as below [188]:

A. If $f(t)$ is an analytical function of t, its fractional derivative $_0 D_t^\alpha f(t)$ is an analytical function of t and α.

B. For $\alpha = n$, where n is an integer, the operation $_0 D_t^\alpha f(t)$ gives the same result as classical differentiation of integer order n.

C. For $\alpha = 0$, the operation $_0 D_t^\alpha f(t)$ is the identity operator:

$$_0 D_t^0 f(t) = f(t).$$

D. Fractional differentiation and fractional integration are linear operations:

$$_0 D_t^\alpha (a f(t) + b g(t)) = a \, _0 D_t^\alpha f(t) + b \, _0 D_t^\alpha g(t).$$

E. The additive index law (semigroup property)

$$_0 D_t^\alpha \, _0 D_t^\beta f(t) = \, _0 D_t^\beta \, _0 D_t^\alpha f(t) = \, _0 D_t^{\alpha+\beta} f(t),$$

holds under some reasonable constraints on the function $f(t)$.

The fractional order derivative commutes with integer order derivative

$$\frac{d^n}{dt^n} (_a D_t^r f(t)) = \, _a D_t^r \left(\frac{d^n f(t)}{dt^n} \right) = \, _a D_t^{r+n} f(t),$$

under the condition $t = a$, we have $f^{(k)}(a) = 0$, $(k = 0, 1, 2, \ldots, n - 1)$. The relationship above says the operators $\frac{d^n}{dt^n}$ and $_a D_t^r$ commute (see [188, Chapter 2] for other commute properties).

1.1.2 Laplace Transform

Consider the linear fractional order differential equation given by

$$a_1 D^{\eta_1} y(t) + a_2 D^{\eta_2} y(t) + \cdots + a_{n-1} D^{\eta_{n-1}} y(t) + a_n D^{\eta_n} y(t)$$
$$= b_1 D^{\gamma_1} u(t) + b_2 D^{\gamma_2} u(t) + \cdots + b_m D^{\gamma_m} u(t). \tag{1.8}$$

If all the initial values of the input and output are zero, the Laplace transform can be applied such that the differential equation can be mapped into an algebraic equation, from which the fractional order transfer function can be defined

$$G(s) = \frac{\mathcal{L}[y(t)]}{\mathcal{L}[u(t)]} = \frac{b_1 s^{\gamma_1} + b_2 s^{\gamma_2} + \cdots + b_m s^{\gamma_m}}{a_1 s^{\eta_1} + a_2 s^{\eta_2} + \cdots + a_{n-1} s^{\eta_{n-1}} + a_n s^{\eta_n}}. \tag{1.9}$$

The Fourier transform for the fractional order derivative and integral can be defined in a unified way as

$$\mathcal{F}\left[_{-\infty} D_t^\alpha f(t)\right] = (j\omega)^\alpha \mathcal{F}[f(t)], \tag{1.10}$$

where α can be either a positive or negative real number. The Laplace transform of a fractional order integral can be expressed by

$$\mathcal{L}\left[D_t^{-\gamma} f(t)\right] = s^{-\gamma} \mathcal{L}[f(t)], \tag{1.11}$$

and the transform for a fractional order derivative (Riemann-Liouville definition) can be evaluated from

$$\mathcal{L}\left[_a D_t^\alpha f(t)\right] = s^\alpha \mathcal{L}[f(t)] - \sum_{k=0}^{n-1} s^k \left[_a D_t^{\alpha-k-1} f(t)\right]_{t=a}, \tag{1.12}$$

where, $n - 1 \leqslant \alpha < n$. In particular, if the derivatives of the function $f(t)$ at $t = a$ are all equal to 0, one simply has $\mathcal{L}\left[_a D_t^\alpha f(t)\right] = s^\alpha \mathcal{L}[f(t)]$.

1.1.3 Fractional Order Dynamic Systems

Many real dynamic systems are better characterized using a non-integer order dynamic model based on fractional calculus or differentiation or integration of non-integer order. Traditional calculus is based on integer order differentiation and integration. The concept of fractional calculus has tremendous potential to change the way we see, model, and control the world around us.

Fractional calculus is a topic more than 300 years old. The number of applications where fractional calculus has been used is rapidly growing. These mathematical phenomena describe a real object more accurately than the classical integer order methods. The real objects are generally fractional [163], [178], [189], 232], however, for many of them the fractionality is very low. A typical example of a non-integer (fractional) order system is the voltage-current relation of a semi-infinite lossy transmission line [228] or the diffusion of the heat into a semi-infinite solid, where, under coefficients normalized to unity, the heat flow $q(t)$ in nature is equal to the semi-derivative of the temperature $T(t)$ [188], [190]

$$\frac{d^{0.5}T(t)}{dt^{0.5}} = q(t).$$

Clearly, using an integer order ordinary differential equation (ODE) description for the above system may differ significantly to the actual situation. However, the fact that the integer order dynamic models are more welcome is probably due to the absence of solutions for fractional order differential equations (FODEs). At present, there are lots of methods for the approximation of fractional derivative and integral equations, therefore, some progress in the analysis of dynamic systems modeled by FODEs has been made in [3], [15], [19], [69], [152], [169], [2], [187], [188], [233], [246]. Recently, fractional calculus can be easily used in wide areas of applications [49], for example, new fractional order system models, electrical circuits theory – fractances, capacitor theory, etc.

A fractional order dynamic system can be described by a fractional differential equation of the following form [187], [188], [220]:

$$a_n D^{\alpha_n} y(t) + a_{n-1} D^{\alpha_{n-1}} y(t) + \cdots + a_0 D^{\alpha_0} y(t)$$
$$= b_m D^{\beta_m} u(t) + b_{m-1} D^{\beta_{m-1}} u(t) + \cdots + b_0 D^{\beta_0} u(t), \tag{1.13}$$

where $D^\gamma \equiv {}_0 D_t^\gamma$; a_k $(k = 0, \cdots n)$, b_k $(k = 0, \cdots m)$ are constants; and α_k $(k = 0, \cdots n)$, β_k $(k = 0, \cdots m)$ are arbitrary real numbers.

Without loss of generality, we can assume that $\alpha_n > \alpha_{n-1} > \cdots > \alpha_0$, and $\beta_m > \beta_{m-1} > \cdots > \beta_0$.

To obtain a discrete model of the fractional order system (1.13), we have to use discrete approximations of the fractional order integro-differential operators and then we obtain a general expression for the discrete transfer function of the controlled system [220]

$$G(z) = \frac{b_m(w(z^{-1}))^{\beta_m} + \cdots + b_0(w(z^{-1}))^{\beta_0}}{a_n (w(z^{-1}))^{\alpha_n} + \cdots + a_0 (w(z^{-1}))^{\alpha_0}}, \tag{1.14}$$

where $(\omega(z^{-1}))$ denotes the discrete equivalent of the Laplace operator s, expressed as a function of the complex variable z or the backward shift operator z^{-1}.

The fractional order linear time-invariant system can also be represented by the following state-space model

$$
\begin{aligned}
{}_0D_t^q x(t) &= Ax(t) + Bu(t) \\
y(t) &= Cx(t),
\end{aligned}
\tag{1.15}
$$

where $x \in R^n, u \in R^r$ and $y \in R^p$ are the state, input and output vectors of the system and $A \in R^{n \times n}, B \in R^{n \times r}, C \in R^{p \times n}, q$ is the fractional commensurate order [161].

1.1.4 Stability of LTI Fractional Order Systems

It is known from the theory of stability that a linear time-invariant (LTI) system is stable if the roots of the characteristic polynomial are negative or have negative real parts if they are complex conjugate. It means that they are located on the left half of the complex plane. In the fractional order LTI case, the stability is different from the integer one. An interesting notion is that a stable fractional order system may have roots in the right half of complex plane (see Figure 1.1). It has been shown that system (1.15) is stable if the following condition is satisfied [153]

$$
|\arg(eig(A))| > q\frac{\pi}{2},
\tag{1.16}
$$

where $0 < q < 1$ and eig(A) denotes the eigenvalues of matrix A.

Matignon's stability theorem says [153]: *The fractional transfer function $G(s) = Z(s)/P(s)$ is stable if and only if the following condition is satisfied in σ-plane:*

$$
|\angle(\sigma)| > q\frac{\pi}{2}, \quad \forall \sigma \in C, \quad P(\sigma) = 0,
\tag{1.17}
$$

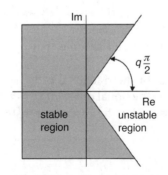

Figure 1.1   Stability region of LTI fractional order systems with order $0 < q \leq 1$

where $\sigma := s^q$. When $\sigma = 0$ is a single root of $P(s)$, the system cannot be stable. For $q = 1$, this is the classical theorem of pole location in the complex plane: no pole is in the closed right half plane of the first Riemann sheet.

Generally, consider the following commensurate fractional order system in the form:

$$D^q w = f(w), \tag{1.18}$$

where $0 < q < 1$ and $w \in R^n$. The equilibrium points of system (1.18) are calculated via solving the following equation

$$f(w) = 0. \tag{1.19}$$

The equilibrium points are asymptotically stable if all the eigenvalues λ_j, ($j = 1, 2, \ldots, n$) of the Jacobian matrix $J = \partial f / \partial w$, evaluated at the equilibrium, satisfy the following condition:

$$|\angle(eig(J))| = |\angle(\lambda_j)| > q\frac{\pi}{2}, \quad j = 1, 2, \ldots, n. \tag{1.20}$$

Figure 1.1 shows the stable and unstable regions of the complex plane for such a case.

1.2 Fractional Order Controls

1.2.1 Why Fractional Order Control?

Using the notion of fractional order may be a step closer to the real world because the real processes are generally or most likely *fractional* [18]. However, for many of them, the fractionality may be very small. As said, a typical example of a non-integer (fractional) order system is the voltage-current relation of a semi-infinite lossy RC line. In theory, the control systems can include both the fractional order dynamic system or plant to be controlled and the fractional order controller. However, in control practice, it is more common to consider the fractional order controller. This is due to the fact that the plant model may have already been obtained as an integer order model in classical sense. In most cases, our objective is to apply the fractional order control (FOC) to enhance the system control performance. For example, the proportional integral derivative (PID) controllers, which have been dominating industrial controllers, have been modified using the notion of fractional order integrator and differentiator. It is shown that the extra degree of freedom from the use of fractional order integrator and differentiator made it possible to further improve the performance of traditional PID controllers [161]. Therefore, in this section, we will concentrate on this scenario–the controller being fractional order.

1.2.2 Basic Fractional Order Control Actions

The applications using fractional calculus have been attracting more and more attentions in the past few decades. These mathematical phenomena describe a real object more accurately than the classical "integer order" methods. As pointed out in [50], clearly, for closed-loop control systems, there are four situations. They are (1) IO (integer order) plant with IO controller; (2) IO plant with FO (fractional order) controller; (3) FO plant with IO controller and (4) FO plant with FO controller. From a control engineering point of view, doing something better is the major concern. Existing evidence confirm that the best fractional order controller can outperform the best integer order controller. It has also been answered in the literature why one should consider fractional order control even when the integer (high) order control works comparatively well [157, 160]. Fractional order PID controller tuning has reached a mature state of practical use. Since (integer order) PID control dominates the industry, we believe FOPID will gain increasing impact and wide acceptance. Furthermore, we also believe that based on some real-world examples, fractional order control is ubiquitous when the dynamic system is of distributed parameter nature [50].

1.2.3 A Historical Review of Fractional Order Controls

In [29], Bode mentioned, maybe the first time and in a comprehensive way, the interest of considering a fractional integro-differential operator in a feedback loop without using the term "fractional". After that, Manabe [150] introduced the frequency and transient responses of the fractional order integral and its application to control systems. As a further step in automatic control, the tilted integral derivative (TID) controller was proposed in a patent [145] by Lurie, and Oustaloup proposed the CRONE (Commande Robuste d'Ordre Non Entier) method with three generations [176], [178], [182], over the traditional PID controller for the control of dynamic systems to achieve better performance. A generalization of the traditional integral order PID controller was proposed by Podlubny, e.g. the $PI^\lambda D^\mu$ controller with an integrator of order λ and a differentiator of order μ [190]. Extending the classical lead-lag compensator to the fractional order case was studied in [193].

More early attempts to apply fractional calculus to systems control can be found in [15], [18], [71], [152], [175], [191], [199]. In this section, four representative fractional order controllers in the literature will be briefly introduced, namely, TID controller, CRONE controller, the $PI^\lambda D^\mu$ controller and fractional lead-lag compensator.

1.2.3.1 TID Controller

In [144], a feedback control system using a PID controller is provided, wherein the proportional component of the PID controller is replaced with a tilted component having a transfer function $s^{-\frac{1}{n}}$. The resulting transfer function of the entire system with this TID more closely approximates an optimal transfer function, thereby achieving improved feedback controller. Further, as compared to conventional PID controllers, this TID controller allows for simpler tuning, better disturbance rejection ratio, and smaller effects of plant parameter variations on closed-loop response.

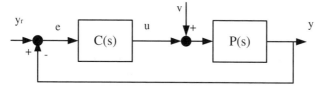

(a) Block diagram of the classic feedback control system with disturbance

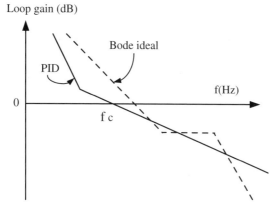

(b) Bode plots for PID controlled plant and the ideal loop response

Figure 1.2 Classic control system and its ideal Bode plot

A. Basic Motivations

The motivation for this TID control is from the consideration of the so-called theoretically optimal loop response due to Bode. Consider the conventional feedback control system block diagram in Figure 1.2(a) where C is the feedback controller, y_r is the reference input signal, e is the control error signal, u and y are input and output signals respectively. In Figure 1.2(a), the additive disturbance is denoted by v. The major goals of the feedback control system are to minimize the effect of disturbances at the output of the system, and to minimize sensitivity of the closed-loop response to plant parameter variations. To satisfy these requirements, the feedback of the system, properly weighted in frequency, must be maximized. These constraints uniquely define the optimal transfer function for the feedback loop. The purpose of the controller of the feedback system is to implement a loop response reasonably close to the optimal one. A commonly used controller employed in feedback control systems is a PID controller. In fact, a PID controller provides varying degrees of gain and phase shift of the signal according to the frequency contents. The conventional PID controller transfer function typically has two real zeros. Typically, the P-term dominates near f_c, the D-term dominates at frequencies over $4f_c$, and the I-term dominates at frequencies up to $f_c/4$, where f_c is the crossover frequency at which the loop gain is $0 \, dB$ as shown in Figure 1.2(b).

Referring to Figure 1.2(b), a theoretically optimal loop response has been determined by Bode. For the purpose of industrial control, a simplified suboptimal Bode loop response can be employed. The suboptimal response is illustrated in Figure 1.2(b) by a dashed line. The slope of this suboptimal gain response is about -10 dB/octave. The transcendental loop transfer function which characterizes the suboptimal response can be closely approximated by a rational function. As can be seen from Figure 1.2(b), rather sharp corners occur at the sides of the Bode step. Any smoothing of the corners, especially the left one, caused by an improper or inaccurate rational function approximation, reduces the available feedback, resulting in reduced performance. A typical loop gain Bode plot of the system with a PID controller is also shown in Figure 1.2(b). When provided with the same stability margin and the same average loop gain as an optimal Bode controller, the crossover frequency f_c of the PID controller is about one-half that of the optimal Bode loop response. The feedback at frequency $f_c/4$ is about 10 dB lower than that of a simplified suboptimal Bode loop response. The conventional PID controllers in common use when applied to a great variety of plants, are easy to tune to provide robust and fairly good performance. However, the performance is not optimal as explained above.

The aim of TID is to provide an improved feedback loop controller having the advantages of the conventional PID controller, but providing a response which is closer to the theoretically optimal response.

B. **Brief Introduction to TID Control**

Similar to PID control, the TID control scheme is shown in Figure 1.3. where the the proportional compensating unit is replaced with a compensator having a transfer function characterized by $1/s^{\frac{1}{n}}$ or $s^{-1/n}$. This compensator is herein referred to as a "Tilt" compensator, as it provides a feedback gain as a function of frequency which is tilted or shaped with respect to the gain/frequency of a conventional or positional compensation unit. The entire compensator is herein referred to as a Tilt-Integral-Derivative (TID) controller. For the Tilt controller, n is a *nonzero real number*, preferably between 2 and 3. Thus, unlike the conventional PID controller, wherein exponent coefficients of the transfer functions of the elements of the compensator are either 0, -1, or $+1$, the TID scheme exploits an exponent coefficient of $-1/n$. By replacing the conventional proportional component with the tilt component of the invention, an overall response is achieved which is closer to the theoretical optimal response determined by Bode as illustrated in Figure 1.2(b).

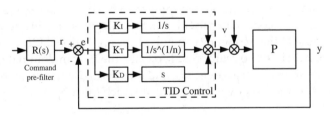

Figure 1.3   Block diagram of TID control scheme

In Figure 1.3, $R(s)$ is a pre-filter provided for a proper command signal pre-filtering which is commonly seen in practice. A preferred transfer function for the pre-filter is

$$R(s) = \frac{s^2 + 2\omega_c s + \omega_c^2}{s^2 + 5.25\omega_c s + \omega_c^2}$$

Since the T-term eliminates static error, the coefficient of the I-term can be set to zero for many problems, thus simplifying controller tuning. A suggested tuning procedure for the TID controller is:

(1) set $K_I = 0$, $K_D = 0$, and set the coefficient K_T for the loop gain to be 0 dB at a desired crossover frequency f_c;
(2) set K_D such that the phase stability margin at the crossover frequency is about *5 degrees larger than desired*; and
(3) set $K_I = 0.25 K_T f_c^{(1-1/n)}$.

Taking $n = 1/3$ as an example, the transfer function $1/s^{1/3}$ can be approximated by a transfer function having alternating real poles and zeros in a complex plane representation. Three poles and three zeros per decade generally suffice to achieve the phase error of less than 1 degree and the amplitude error of less than 0.1 db which is given by

$$T_{6/6}(s) = \frac{.442s^6 + 2.23s^5 + 1.86s^4 + 0.428s^3 + .0295s^2 + .000568s + 2.18 \times 10^{-6}}{s^6 + 2.42s^5 + 1.304s^4 + .201s^3 + .0092s^2 + .0001098s + 1.98 \times 10^{-7}}.$$

Enter the coefficients for the above approximated transfer function $T_{6/6}(s)$ for $1/s^{1/3}$ into `CtrlLAB`© [237], three mouse clicks give the Bode plot, the Nichols chart and the root locus as shown in Figure 1.4.

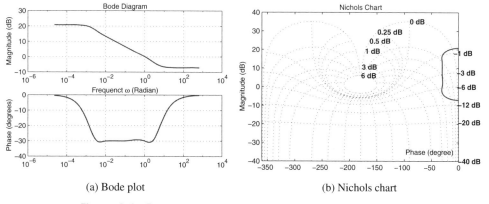

(a) Bode plot   (b) Nichols chart

Figure 1.4 Frequency responses of transfer function $T_{6/6}(s)$

From the Bode plot, we can see that $T_{6/6}(s)$ is a good approximation for $1/s^{1/3}$ in both magnitude and in particular the phase (constant phase angle). The vertical line in the Nichols chart Figure 1.4(b), is a desired robustness property for controller design.

In TID patent [144], an analog circuit using op-amps plus capacitors and resistors is introduced with a detailed component list which is useful in some cases where the computing power to implement $1/s^{1/n}$ digitally is not possible. An example is given in [144] to illustrate the benefits of TID over conventional PID in both time and frequency domains.

1.2.3.2 CRONE Controller

The CRONE control was proposed by Oustaloup in pursuing *fractal robustness* [181], [183]. CRONE is a French abbreviation for "*Contrôle Robuste d'Ordre Non Entier*" (which means non-integer order robust control). In this section, we shall follow the basic concept of *fractal robustness*, which motivated the CRONE control, and then mainly focus on the second generation CRONE control scheme and its synthesis based on the desired frequency template which leads to fractional transmittance [174], [180].

A. **Fractal Robustness**

In [177], "fractal robustness" is used to describe the following two characteristics: the iso-damping and the vertical sliding form of the frequency template in the Nichols chart. This desired robustness motivated the use of fractional order controller in classical control systems to enhance their performance.

(1) **Iso-damping lines**. Consider the characteristic equation

$$1 + (\tau s)^{\alpha} = 0, \tag{1.21}$$

where τ is a constant. The two poles are given by

$$s = \frac{1}{\tau} e^{\pm j\pi/\alpha}, \tag{1.22}$$

with $1 < \alpha < 2$. The poles are complex and conjugated, and form a center angle 2Θ with $\Theta = (\pi - \pi/\alpha)$ as shown in Figure 1.5(a). Clearly, the poles move at a constant angle (fixed by the order α) when τ varies. The robustness in s plane is then illustrated by two half-straight lines which form the same angle Θ in relation to the real axis and are called *iso-damping half-straight lines*.

The *natural frequency* and the *damping ratio* are directly deducible from the poles, through their modulus $1/\tau$ and the half-center angle Θ as follows:

$$\omega_p = \frac{1}{\tau} \sin \Theta = \frac{1}{\tau} \sin \left(\pi - \frac{\pi}{\alpha} \right) = \frac{1}{\tau} \sin \left(\frac{\pi}{\alpha} \right), \tag{1.23}$$

and

$$\zeta\left(\alpha\right) = \cos\Theta = \cos\left(\pi - \frac{\pi}{\alpha}\right) = -\cos\left(\frac{\pi}{\alpha}\right). \tag{1.24}$$

It can be clearly seen that *the damping ratio ζ is exclusively a function of the fractionality order α*, thus allowing the introduction of the notion of *robust oscillatory mode*.

(2) **Frequency template.** With a unit negative feedback, the forward path transfer function, or open-loop transmittance, for the characteristic equation (1.21) is

$$\beta\left(s\right) = \left(\frac{1}{\tau s}\right)^{\alpha} = \left(\frac{\omega_u}{s}\right)^{\alpha}, \tag{1.25}$$

which is the transmittance of a *non-integer integrator* in which $\omega_u = 1/\tau$ denotes the unit gain (or transitional) frequency.

As $\angle\beta\left(jw\right) = -\alpha\pi/2$ with $1 < \alpha < 2$, the Nichols chart of $\beta\left(jw\right)$ is a *vertical straight line* between $-\pi/2$ and $-\pi$. This is illustrated in Figure 1.5(b). When τ, the system parameter, changes, the vertical straight line shown in Figure 1.5(b) slides. Such a vertical displacement ensures a constant phase margin Φ_m, and thus correspondingly a constant damping ratio in the time domain.

In controller design, the objective is to achieve such a similar frequency behavior, in a medium frequency range around ω_u, knowing that the closed-loop dynamic behavior is exclusively linked to the open-loop behavior around ω_u. Therefore, the ideal controller design comprises:

(1) An open-loop Nichols locus which forms a vertical straight line segment around ω_u for the nominal parametric state of the plant, called the open-loop frequency template (or more simply the template) (Figure 1.5(b));
(2) A sliding of the template on itself when there are parameter changes in the plant (assume that the parameter change will lead to gaining variations around ω_u).

Synthesizing such a template defines the non-integer approach that the second generation CRONE control uses.

B. The Second Generation CRONE Control: Basic Concept

For a typical disturbed feedback control system as shown in Figure 1.2(a), its control performance is fully characterized by the sensitivity function $\mathbb{S}\left(s\right)$, also known as *the transmittance in regulation*, or the complementary sensitivity function $\mathbb{T}\left(s\right)$, also known as *the transmittances in tracking*, and we know that $\mathbb{S}\left(s\right) + \mathbb{T}\left(s\right) = 1$. It is practically true that given the open-loop behavior around the unit gain frequency, one can determine the dynamic behavior in closed loop. Therefore, we use the transmittance frequency

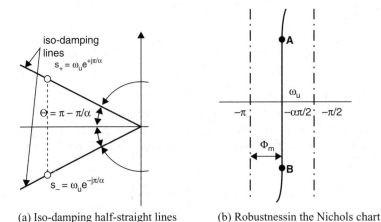

(a) Iso-damping half-straight lines   (b) Robustnessin the Nichols chart

Figure 1.5   Illustrations of fractal robustness

template , $\beta(s)$, as shown in Figure 1.5, to define the desired behavior of $\mathbb{T}(s)$ or $\mathbb{S}(s)$. Let's choose a template such that

$$\beta(s) = \beta(j\omega), \qquad \forall \omega \in [\omega_A, \omega_B], \tag{1.26}$$

where

$$\beta(s) = \left(\frac{\omega_u}{s}\right)^\alpha, \qquad \alpha \in [1, 2]. \tag{1.27}$$

Referring to Figure 1.2(a), the desired or ideal $\mathbb{T}(s)$ and $\mathbb{S}(s)$ are set as follows:

$$\mathbb{T}(s) = \left[\frac{Y(s)}{Y_r(s)}\right]_{V(s)=0} = \frac{\beta(s)}{1+\beta(s)} = \frac{1}{1+(s/\omega_u)^\alpha}, \tag{1.28}$$

and

$$\mathbb{S}(s) = \left[\frac{Y(s)}{V(s)}\right]_{Y_r(s)=0} = \frac{1}{1+\beta(s)} = \frac{(s/\omega_u)^\alpha}{1+(s/\omega_u)^\alpha}. \tag{1.29}$$

In tracking, the gain reaches a maximum for resonance frequency

$$\omega_t = \left(-\cos\alpha\frac{\pi}{2}\right)^{1/\alpha}\omega_u, \tag{1.30}$$

and in regulation, the gain reaches a maximum for resonance frequency

$$\omega_r = \left(-\cos\alpha\frac{\pi}{2}\right)^{-1/\alpha}\omega_u. \tag{1.31}$$

This result reveals the existence of a resonance when $\cos(\alpha\pi/2) < 0$, namely for $1 < \alpha < 3$ and therefore for the CRONE control since $1 < \alpha < 2$. The *resonance ratio in tracking* is

$$Q_t(\alpha) = \frac{|\mathbb{T}(j\omega_t)|}{|\mathbb{T}(j0)|} = \frac{1}{\sin\alpha\dfrac{\pi}{2}}, \tag{1.32}$$

while the *resonance ratio in regulation* is

$$Q_r(\alpha) = \frac{|\mathbb{S}(j\omega_r)|}{|\mathbb{S}(j\infty)|} = \frac{1}{\sin\alpha\dfrac{\pi}{2}}. \tag{1.33}$$

These results show that *the resonance ratio depends exclusively on control order α*, thus allowing the introduction of the notion of *robust resonance*. By observation, it can be found that

$$(\omega_t\omega_r)^{1/2} = \omega_u, \tag{1.34}$$

and

$$Q_t(\alpha) = Q_r(\alpha). \tag{1.35}$$

From (1.34)–(1.35), the resonance frequencies in tracking and in regulation are symmetrically distributed with regard to the open-loop unit gain frequency while the resonance ratios in tracking and in regulation are identical.

C. The Second Generation CRONE Control–Design Steps

Usually, descriptive specifications of the open-loop behavior (for the nominal plant) will be given, such as:

(1) *The accuracy specifications at low frequencies.*
(2) *The vertical template around unit gain frequency ω_u.*
(3) *The input sensitivity specifications at high frequencies.*

For a stable minimum phase plant, it turns out that the behavior thus defined can be described by a *transmittance based on the frequency-limited real non-integer differentiator*, that is,

$$\beta(s) = \left[K_b\left(\frac{\omega_b}{1} + 1\right)\right]^{n_b} \left(\sqrt{\frac{1 + (\omega_u/\omega_b)^2}{1 + (\omega_u/\omega_h)^2}}\,\frac{1 + s/\omega_h}{1 + s/\omega_b}\right)^{\alpha} \left(\frac{K_h}{1 + s/\omega_h}\right)^{n_h}, \tag{1.36}$$

with

$$K_b = \left(1 + (\omega_b/\omega_u)^2\right)^{-1/2} \text{ and } K_h = \left(1 + (\omega_u/\omega_h)^2\right)^{1/2}, \tag{1.37}$$

where ω_b and ω_h are transitional frequencies with $\omega_b < \omega_h$. In the particular case where transitional frequencies ω_b and ω_h are sufficiently distant from frequency ω_u, around this frequency (that is $\omega_b \ll \omega \ll \omega_h$), $\beta(s)$ can be reduced to transmittance

$$\beta(s) = (\omega_u/s)^\alpha, \tag{1.38}$$

which is the same as that described by the template (relation (1.27)).

The order α transmittance of relation (1.36) describes the frequency truncation of the template defined by the transitional frequencies ω_b and ω_h. This transmittance results from the substitution of the part raised at power α for the transmittance ω_b/p which is used in the description of the template between frequencies ω_A and ω_B, as shown in Figure 1.5(b).

Finally, referring to Figure 1.2(a), the controller $C(s)$ in cascade with the plant is synthesized from its frequency response according to

$$C(j\omega) = \frac{\beta(j\omega)}{G_0(j\omega)}, \tag{1.39}$$

where $G_0(j\omega)$ denotes the frequency response of the nominal plant.

There are a number of real-life applications of CRONE controller such as the car suspension control [176], [183], a flexible transmission [181], a hydraulic actuator [115] etc. CRONE control has been developed into a powerful non-conventional control design tool with a dedicate MATLAB toolbox for it [182]. For an extensive overview, refer to [179] and the references therein.

1.2.3.3 $PI^\lambda D^\mu$ Controller

$PI^\lambda D^\mu$ controller, also known as $PI^\lambda D^\delta$ controller, was studied in the time domain in [189] and in the frequency domain in [185].

A. **Basic Formulae**

In general form, referring to Figure 1.2(a), the transfer function of $PI^\lambda D^\delta$ is given by

$$C(s) = \frac{U(s)}{E(s)} = K_p + T_i s^{-\lambda} + T_d s^\delta, \tag{1.40}$$

where λ and δ are positive real numbers, K_p is the proportional gain, T_i the integration constant and T_d the differentiation constant. Clearly, taking $\lambda = 1$ and $\delta = 1$, we obtain a classical PID controller. If $\lambda = 0$ $(T_i = 0)$, we obtain a PD^δ controller, etc. All these types of controllers are particular cases of the $PI^\lambda D^\delta$ controller. The time domain formula is that

$$u(t) = K_p e(t) + T_{it}^{-\lambda} e(t) + T_{dt}^\delta e(t); \qquad (t^{(*)} \equiv_0 t^{(*)}). \qquad (1.41)$$

It can be expected that $PI^\lambda D^\delta$ controller (1.41) may enhance the systems control performance due to the introduction of more tuning knobs. Actually, in theory, $PI^\lambda D^\delta$ itself is an infinite dimensional linear filter due to the fractional order in differentiator or integrator.

B. Simple Controller Synthesis Scheme

Unlike a conventional PID controller, there is no systematic and yet rigorous design or tuning method existing for a $PI^\lambda D^\delta$ controller. Here, a simple scheme based on the dominant root principle to design a $PI^\lambda D^\delta$ controller is briefly introduced. The pole distribution of the characteristic equation of the controlled system in the complex plane should be located at the desired dominant roots which are designed based on the control performance requirement. Assume that the desired dominant roots are a pair of complex conjugate roots as follows:

$$p_{1,2} = -r \pm j\omega. \qquad (1.42)$$

It is clear that the above dominant roots define the stability measure S_t and damping measure T_l. In this simplified situation, the parameters design of the $PI^\lambda D^\delta$ controller [185], [188] can be divided into two steps, that is,

(1) **The design of K_p.** Proportional gain K_p is related to the static error E_t [%], settling time T_r [section], and overshoot P_r [%]. In general, the larger the K, the smaller the control time T_r [section] as well as the static error E_t [%]. Therefore, K_p can be simply set via

$$K_p \geqslant (100/E_t).$$

(2) **The design of T_d, δ, T_i, λ.** From the complex conjugate roots (1.42), the (required) stability measure $S_t = r$ and damping measure $T_l = r/\omega$ can be computed. Given S_t and T_l, using the classical root locus method, we can numerically solve T_d, δ, T_i, λ from the characteristic equation with fractional order controller $C(s)$ which is given by

$$C(s)P_0(s) + 1 = 0 \qquad (1.43)$$

where $P_0(s)$ is a nominal model of $P(s)$ as shown in Figure 1.2(a). More specifically, for simple plant models, this can now be done by solving

$$\min_{T_d, \delta, T_i, \lambda} |C(s)P_0(s) + 1|_{s=-r \pm j\omega}.$$

1.2.3.4 Fractional Lead-Lag Compensator

In the previous subsections, fractional order controllers are directly related to the use of fractional order differentiator or integrator. It is possible to extend the classical lead-lag compensator to the fractional order case which was studied in [192]. The fractional lead-lag compensator is given by

$$C_r(s) = C_0 \left(\frac{1 + s/\omega_b}{1 + s/\omega_h} \right)^r, \tag{1.44}$$

where $0 < \omega_b < \omega_h$, $C_0 > 0$ and $r \in (0, 1)$.

Consider the feedback control loop in Figure 1.2(a). A robust control problem of interest is to find C guaranteeing the robust Q-factor (amplitude magnification factor at the resonance frequency) for the transfer functions from y_y to y and from v to y for all plants in the form $P = kP_0$, with P_0 the nominal plant model and $k \in [k_m, k_M]$.

An ideal solution to this problem is to make the nominal loop transfer function $L_0(s) = C(s)P_0(s) = (\omega_0/s)^n$, where $1 < n < 2$. For $L(s) = C(s)P(s)$, its Nyquist plot is similar to $L_0(s)$. Therefore, the phase margin and Q-factors for the closed-loop transfer functions from y_r to y and from v to y will be independent of k and uniquely determined by n. Furthermore, since the transfer function from y_r to y can be approximated by a second-order stable system, the maximum closed-loop step response overshoot is almost independent of k. Thus, a change of k will result in a slower or faster but equally damped response, which is a desirable robustness property in some applications, such as the car suspension design problem [192].

A. **Design Concept**

The ideal loop transfer function can be approximated by shaping $L_0(j\omega)$ close to $(\omega_0/j\omega)^n$ around the crossover frequency ω_0 at which $|L_0(j\omega_0)| = 1$. To achieve this, when the plant $P(s) = Ms^{-m}$, it is required that $\angle(C(j\omega))$ is close to $m - 2 + \phi/\pi$, where ϕ is the desired phase margin, in a given frequency range around ω_0. One could obviously seek to achieve this by using the classical rational lead-lag filter corresponding to (1.44) with $r = 1$ due to the fact that $\angle[C_1(j\omega)]$ is close to $\angle[C_1(j\omega_0)]$ around the frequency $\omega_0 = \sqrt{\omega_b\omega_h}$. Clearly, when $r = 1$, the width of the frequency range over which the condition $\angle[C_1(j\omega)] \approx \angle[C_1(j\omega_0)] = m - 2 + \phi/\pi$ holds is entirely determined by the choice of ϕ, and thus cannot be adjusted to meet robustness requirements.

When a non-integer value of r in (1.44) is used, it is possible to guarantee the robust phase margin, closed-loop resonance and overshoot for any given range of variation of k.

To apply this control strategy to more general linear systems, the control $C(s)$ in Figure 1.2(a) should take the form $C(s) = C_r(s)G(s)$, where $G(s)$ is to be chosen such that

$$G(s) \approx Ms^{-m}/P_0(s). \tag{1.45}$$

B. Realization of Fractional Lead-Lag Compensator

A state-space representation of the fractional lead-lag compensator is proposed in [192] together with an error bound estimate. However, the stable minimum-phase frequency-domain fitting is an easier and more effective method.

In the suspension controller design example of [192], which is cited from [183], the plant is $P(s) = 1/(Ms^2)$. The range of change in M is from 100 kg to 900 kg. The parameters [192] of the designed fractional lead-lag compensator are: $\omega_b = 0.5$, $\omega_h = 200$, $r = 0.65$. Nominal frequency $\omega_0 = 10$ rad/s at the nominal mass $M = M_0 = 300$kg. λ is set to 20 so that $\omega_b = \omega_0/\lambda$ and $\omega_h = \lambda\omega_0$. C_0 is determined from $|P(j\omega)C_r(j\omega)| = 1$ which gives $C_0 = M_0\omega_0^2\lambda^{-r}$ such that ω_0 is the gain cross-over frequency for the nominal case when $M = M_0$.

Here we give the fitting result for $C_{0.65}(s)$ using the stable frequency fitting method introduced in [239]. The 4/4 fitting result is that

$$C_{0.65}(s) = 4280.1 \left(\frac{1 + 2s}{1 + 0.005s} \right)^{0.65}$$

$$\approx \frac{9.457 \times 10^{-11}s^4 + 1.218 \times 10^{-8}s^3 + 3.07 \times 10^{-7}s^2 + 1.476 \times 10^{-6}s + 9.794 \times 10^{-7}}{4.5 \times 10^{-16}s^4 + 1.161 \times 10^{-13}s^3 + 6.99 \times 10^{-12}s^2 + 9.516 \times 10^{-11}s + 2.14 \times 10^{-10}}$$

with its Bode plot and Nichols chart drawn in CtrlLAB in Figure 1.6. We can see that Figure 1.6 is quite similar to the characteristic of a frequency-band fractional differentiator.

1.3 Fractional Order Motion Controls

Motion control is a sub-field of automation, in which the velocity and position of machines are controlled using some type of device such as a hydraulic pump, a linear

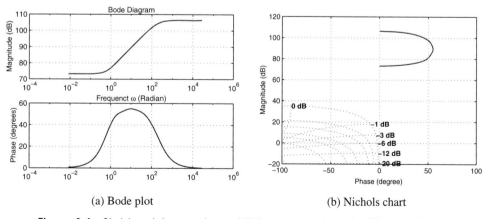

(a) Bode plot   (b) Nichols chart

Figure 1.6 Stable minimum phase 4/4 frequency domain fitting of $C_{0.65}(s)$

actuator, or an electric motor, generally a servo. Motion control is an important part of robotics and Computer Numerical Control (CNC) machine tools, and is widely used in the packaging, printing, semiconductor manufacturing, assembly industries, and so on. In motion control systems, control strategies are implemented by the mechanical setups, and motion is the effect of the implementation. The implementation taches of the control strategies should be stable, fast and precise. In real-time environment, the movement of every tache in the system is influenced by various disturbances. But the high performance motion control systems require stable and robust operations, which are not influenced or just minimally influenced by the disturbances. Therefore, the motion control research normally is explored with the aim of obtaining ascendant command tracking performance and disturbance rejection performance. In this monograph, the fractional calculus and fractional order control strategies are applied to the linear or nonlinear motion systems to improve the command tracking and disturbance compensation performances.

As presented in Section 1.2, fractional order control means the controlled systems and/or controllers are described by fractional order differential equations. In the past few decades, with the rapid development of computer technology and the better understanding of the potential of fractional calculus, the realization of fractional order control system has become much easier than previously, and the fractional calculus has become more useful in various sciences and engineering areas [18], [29], [47], [114], [130], [150], [181], [183], [195], [212]. In the motion control branch, some examples of using fractional calculus can be found in [76], [117], [125], [146], [160], [223], [238]. The application of the fractional calculus in adaptive motion control is just beginning with more research efforts on this subject [114], [116], [196], [219] demonstrates that the response of the system under control is significantly better for the fractional order integration exploited in the parameter adaptation stage than that for the classical integer order integration. [138] devises a fractional order adaptive compensation method for cogging effect minimization for permanent magnetic synchronous motors position and velocity servo system. A fractional order periodic adaptive learning compensation method is proposed for the general state-dependent periodic disturbance minimization on the position and velocity servo platform in [142]. Meanwhile, for the robust fractional order motion controls, the stability issues are discussed in [127], [128]. A fractional order robust control of a 2 degrees of the freedom classical mechanical system, a ball-beam system is presented in [193]. Fractional calculus has also been applied to many intelligent controls for motion systems. In [74], a novel parameter adjustment scheme to improve the robustness of the fuzzy sliding-mode control achieved by the use of an adaptive neuro-fuzzy inference system architecture is presented, where the proposed scheme utilizes fractional order integration in the parameter tuning stage. A chattering analysis of the continuous power-fractional sliding-mode algorithm is carried out in the frequency domain by the describing function method in [32]. An approach is proposed for vibration suppression in a two-inertia system using an integration of a fractional order disturbance observer and a single neuron-based PI fuzzy controller in [121]. A fractional order iterative learning control algorithm is proposed in [122], and a design scheme of fractional order iterative learning control on frequency domain is presented in [120].

For PID control on motion systems, the fractional order PID controller was proposed in [189] as a generalization of the traditional PID controller, where the expanding derivatives and integrals to fractional orders can adjust the frequency response of the control system directly and continuously. This great flexibility makes it possible to demonstrate better performance and more robust control in comparison with the traditional PID controller. Reference [118] presented an optimal fractional PID controller design based on specified gain and phase margin by minimizing the integral squared error (ISE) criterion. In [22], [23] and [24], the integer order PID controller was tuned by minimizing a penalty function that forces the behavior of the integer order PID becoming close to that of a desired fractional order transfer function [160] demonstrated a practical scheme for tuning and auto-tuning of fractional order controllers for industrial applications. More research results on fractional order PID controller tuning and applications can be found in the literatures [89], [137], [157], [159], [161], [162], [197], [219], [222], [240], [249]. However, it is not straightforward to set the parameters of the fractional order controllers in a systematic way because of the complexity [146], [160], [238], [247], [249]. Moreover, designing the fractional order PID controller properly and comparing it with the traditional integer order PID controller fairly to illustrate the advantages and potential of the fractional calculus are even more complicated.

1.4 Contributions

This book summarized the authors' research results on fractional order control theory and applications on motion controls. The major contributions of this monograph include:

(1) In the velocity and position motion systems, fractional calculus is applied to the fractional order modeling and fractional order controllers design in systematic ways to achieve better performances than the designed integer order controllers using traditional optimization methods. Stability and feasibility issues are analyzed and discussed for the fractional order controllers design.

(2) Fractional order disturbance observer, fractional order adaptive feed-forward scheme, fractional order adaptive control and fractional order periodic adaptive learning control for external disturbance compensation are presented.

(3) This book documents the very first optimization approach of the proposed fractional order conditional integrator for nonlinear system controls. Meanwhile, some efforts of the fractional order control on nonlinearities, for example, friction and backlash, are presented.

(4) Simulation illustrations and/or experimental validations are demonstrated for all the proposed control schemes and approaches in this book. The real applications for fractional order motion controls on unmanned aerial vehicles and hard-disk-drives are also presented in this book.

1.5 Organization

This book is organized as follows: The research motivations, fundamentals of fractional calculus, and monograph contributions are introduced in Part I, Part II is dedicated to fractional order velocity controls, which include Chapters 2–5; Part III focuses on fractional order position controls, including Chapters 6–10; the feasible regions of the specifications for the integer order and fractional order controller designs according to the stability analysis are studied in Part IV, containing Chapters 11 and 12; Part V explains how to design a fractional order disturbance observer, a fractional order adaptive feed-forward controller, a fractional order adaptive controller, and a fractional order periodic adaptive learning controller to compensate the external disturbances in motion control systems, and this part contains Chapters 13–16; Part VI is devoted to the fractional order controls on nonlinear systems as shown in Chapters 17–19; the real applications of fractional order motion controls on unmanned aerial vehicles and hard-disk-drives are presented in Part VII, including Chapters 20 and 21.

Part II
Fractional Order Velocity Controls

2

Fractional Order PI Controller Designs for Velocity Systems

In motion control systems, the velocity and/or position of machines are controlled using some type of device, for example, hydraulic pump, linear actuator, or an electric motor, generally a servo. Velocity control is a typical task in motion systems. In this chapter, the velocity system with pure time delay, namely, first order plus time delay (FOPTD) system is focused on. This FOPTD system is the generalization of the simple first order velocity system. For simplification, the pure velocity dynamics can also be obtained with time delay set as zero in FOPTD plant.

In order to improve the control performance of FOPTD systems using traditional proportional integral controller, the fractional order proportional integral controllers are proposed and designed in a systematic way for FOPTD systems in this chapter.

2.1 Introduction

A large number of industrial plants can be modeled by the FOPTD system approximately. In reality, the system parameters of FOPTD are bound to vary due to changes in operating conditions. In the past few decades, various methods for tuning the traditional proportional-integral-derivative (PID) controllers have been proposed based on the FOPTD model [10], [14], [59], [89], [92], [249]. Among the well-known formulas are the Ziegler-Nichols rule, the Cohen-Coon method, the integral-absolute-error (IAE) or integral-time-absolute-error (ITAE) optimum the integral-squared-error (ISE) or integral-time-absolute-error (ITSE) optimum, and the internal model control for the traditional integer order PID control research and industry applications. These formulas are surveyed in the paper [92].

Fractional Order Motion Controls, First Edition. Ying Luo and YangQuan Chen.
© 2013 John Wiley & Sons, Ltd. Published 2013 by John Wiley & Sons, Ltd.

In recent years, fractional calculus has been increasingly used in the control area by more and more researchers [21], [22], [23], [105], [121], [144], [158], [160], [182], [183], [212], [218], [247]. In control practice, it is useful to consider the fractional order controller design for an integer order plant [238]. This is due to the fact that the plant model may have already been obtained as an integer order model in the classical sense. In most cases, our objective of using fractional calculus is to apply the fractional order control (FOC) to enhance the system control performance. The generalized PID controller $PI^\lambda D^\delta$ which can be either the integer order or the fractional order controller was proposed in [189], where a better control performance was demonstrated compared to the classical PID controller. It can be expected that the $PI^\lambda D^\delta$ controller may enhance the systems control performance and increase the system robustness for motion controls due to more tuning knobs being introduced [160], [174], [238], [247]. Actually, in theory, $PI^\lambda D^\delta$ itself is an infinite dimensional linear filter due to the fractional order in differentiator or integrator. Reference [145] gave a fractional order PID controller by minimizing the integral squared error. Some numerical examples of the fractional order controllers were presented in [247]. However, it is not straightforward to set the parameters of the fractional order controllers in a systematic way because of the complexity [159], [160], [238], [247], [249]. It is complicated to illustrate the potential advantages of the fractional calculus by designing the fractional order PID controller properly and comparing it with the traditional integer order PID controller fairly. But we may be able to find a practical parameter tuning scheme of a fractional order controller for a certain type of motion systems.

In this chapter, systemic design schemes of fractional order proportional integral (PI) controller and fractional order [proportional integral] ([PI]) controller for FOPTD systems are presented, respectively. For comparison between the fractional order and the integer order controllers, the integer order PID controller is also designed following the proposed tuning specifications to achieve the robustness requirement. Unfortunately, it is found that the obtained integer order PID controller is not stabilized. The designed FOPI and FO[PI] controllers following the proposed tuning schemes not only make the system stable, but also improve the performance and robustness for the FOPTD system. Simulation results are presented to illustrate the effectiveness of the proposed tuning schemes [226].

2.2 The FOPTD System and Three Controllers Considered

The FOPTD system, namely, the velocity system with pure time delay discussed in this chapter has the following form of transfer function,

$$P(s) = \frac{1}{Ts+1}e^{-Ls}, \qquad (2.1)$$

where T is the time constant, and L is the time delay. Setting $L = 0$, the plant (2.1) is a simple pure velocity servo system. This FOPTD system (2.1) can approximately model a large number of industrially plants. Note that, the plant gain is normalized to 1 since the plant proportional factor can be incorporated in the parameters of the controller.

The integer order PID controller has the following form,

$$C_1(s) = K_p + \frac{K_i}{s} + K_d s, \tag{2.2}$$

where, K_p, K_i and K_d are the parameters for proportional, integral and derivative items, respectively.

The fractional order PI and [PI] controllers have the following forms of transfer function, respectively.

$$C_2(s) = K_p \left(1 + \frac{K_i}{s^r} \right), \tag{2.3}$$

where, K_p and K_i are the parameters for proportional and integral items, respectively, and $r \in (0, 2)$;

$$C_3(s) = \left(K_p + \frac{K_i}{s} \right)^\lambda, \tag{2.4}$$

where, K_p and K_i are proportional and integral coefficients, respectively, $\lambda \in (0, 2)$.

In this study the Caputo definition as shown in (1.6) is adopted for fractional derivative, which allows utilization of initial values of classical integer order derivatives with known physical interpretations [188].

2.3 Design Specifications

In this section, three design specifications are proposed for controller designs, where a "flat phase" tuning rule is important and meaningful for the fractional order controller designs.

Assume that the gain crossover frequency is given by ω_c and the phase margin is specified as ϕ_m for the open-loop control system. For the system stability and robustness, three specifications concerned with the phase and gain of the open-loop transfer function are proposed as follows [119],

(i) Specification on phase margin:

$$\angle[G(j\omega_c)] = \angle[C(j\omega_c)P(j\omega_c)] = -\pi + \phi_m;$$

(ii) Specification on gain crossover frequency:

$$|G(j\omega_c)| = |C(j\omega_c)P(j\omega_c)| = 1;$$

(iii) Specification on robustness to loop gain variations with "flat phase":
   This specification demands that the open-loop phase derivative with respect to the frequency is zero,

$$\frac{\mathrm{d}(\angle(G(j\omega)))}{\mathrm{d}\omega}\Big|_{\omega=\omega_c} = 0,$$

that is, the Bode plot of the open-loop phase is tuned locally flat around the gain crossover frequency. We can expect that, if the loop gain increases or decreases a certain percentage, the phase margin will remain unchanged. Therefore, the step responses under various loop gains changing around the nominal gain will exhibit an iso-damping property. It means that the system is more robust to loop gain variations and the overshoots of the responses are almost the same. Clearly, since loop gain variations are unavoidable in the real world due to possible sensor distortion, environment change and so on, the iso-damping property from "flat phase" is a desirable property which ensures that no harmful excessive overshoot results due to loop gain variations.

2.4 Fractional Order PI and (PI) Controller Designs

In this section, the fractional order PI as (2.3) and fractional order [PI] as (2.4) controllers are designed following a systematic way to achieve some given specifications and the robustness to loop gain variations for FOPTD systems. In order to obtain a fair comparison, the integer order PID controller is also designed following the same scheme first.

2.4.1 Integer Order PID Controller Design

The open-loop transfer function $G_1(s)$ with the integer order PID controller $c_1(s)$ for the FOPTD system $P(s)$ is,

$$G_1(s) = C_1(s)P(s). \tag{2.5}$$

According to the integer order PID controller transfer function (2.2), we can find its frequency response as follows,

$$C_1(j\omega) = K_p + j\left(K_d\omega - \frac{K_i}{\omega}\right). \tag{2.6}$$

The phase and gain are,

$$\angle[C_1(j\omega)] = \arctan((K_d\omega^2 - K_i)/(\omega K_p)),$$
$$|C_1(j\omega)| = \sqrt{K_p^2 + (K_d\omega - (K_i/\omega))^2}. \tag{2.7}$$

According to the FOPTD system transfer function (2.1), we can obtain its frequency response,

$$P(j\omega) = \frac{1}{jT\omega + 1}e^{-jL\omega} = \frac{1}{\sqrt{1 + (T\omega)^2}}e^{-j(\arctan(\omega T) + L\omega)}.$$

The phase and gain of the plant are,

$$\angle[P(j\omega)] = -\arctan(\omega T) - L\omega,$$
$$|P(j\omega)| = \frac{1}{\sqrt{1 + (\omega T)^2}}. \tag{2.8}$$

The open-loop frequency response $G_1(j\omega)$ is,

$$G_1(j\omega) = C_1(j\omega)P(j\omega). \tag{2.9}$$

The phase and gain of the open-loop frequency response are,

$$\angle[G_1(j\omega)] = \arctan[(K_d\omega^2 - K_i)/(\omega K_p)] - \arctan(\omega T) - L\omega, \tag{2.10}$$

$$|G_1(j\omega)| = |C_1(j\omega)||P(j\omega)| = \frac{\sqrt{K_p^2 + (K_d\omega - (K_i/\omega))^2}}{\sqrt{1 + (T\omega)^2}}. \tag{2.11}$$

According to specification (i) in Section 2.3, the phase of $G_1(j\omega)$ at ω_c is,

$$\angle[G_1(j\omega_c)] = \arctan((K_d\omega_c^2 - K_i)/(\omega_c K_p)) - \arctan(\omega_c T) - L\omega_c$$
$$= -\pi + \phi_m, \tag{2.12}$$

we have,

$$\frac{K_d\omega_c^2 - K_i}{K_p\omega_c} = A, \tag{2.13}$$

where $A = \tan[\arctan(\omega_c T) + L\omega_c + \phi_m]$.

According to specification (ii) in Section 2.3, we can establish an equation about K_p,

$$|G_1(j\omega_c)| = |C_1(j\omega_c)||P(j\omega_c)|$$
$$= \frac{\sqrt{K_p^2 + (K_d\omega_c - K_i/\omega_c)^2}}{\sqrt{1 + (T\omega_c)^2}} = 1. \tag{2.14}$$

According to specification (iii) in Section 2.3 on the robustness to loop gain variations,

$$\frac{d(\angle(G_1(j\omega)))}{d\omega}\Big|_{\omega=\omega_c}$$

$$= [\arctan\left(\frac{K_d\omega^2 - K_i}{\omega K_p}\right) - \arctan(\omega T) - L\omega]'|_{\omega=\omega_c}$$

$$= 0, \tag{2.15}$$

then, we have,

$$\frac{K_p(K_d\omega_c^2 + K_i)}{(K_p\omega_c)^2 + (K_d\omega_c^2 - K_i)^2} = \frac{T}{B} + L, \tag{2.16}$$

where $B = 1 + (\omega_c)^2$.

From (2.13), (2.14) and (2.16), we can obtain,

$$K_p = \sqrt{\frac{B}{1 + A^2}},$$

$$K_i = \frac{1}{2}\left[\sqrt{\frac{1 + A^2}{B}}(T\omega_c^2 + BL\omega_c^2) - A\omega_c\sqrt{\frac{B}{1 + A^2}}\right], \tag{2.17}$$

$$K_d = \frac{1}{2}\left[\sqrt{\frac{1 + A^2}{B}}(T + BL) + A\omega_c^{-1}\sqrt{\frac{B}{1 + A^2}}\right]. \tag{2.18}$$

If the parameters are set as follows,

$$\omega_c = 10 \ rad/s, \quad T = 0.4 \ s, \quad L = 0.01 \ s, \quad \phi_m = 50°,$$

the parameters K_d, K_i and K_p can be calculated,

$$K_d = -0.0500, \quad K_i = 25.7861, \quad K_p = 2.7425.$$

As $K_d = -0.0500 < 0$, the system using this designed integer order PID controller is unstable.

2.4.2 Fractional Order PI Controller Design

The open-loop transfer function $G_2(s)$ of the fractional order PI controller for the FOPTD system is,

$$G_2(s) = C_2(s)P(s). \tag{2.19}$$

From the fractional order PI controller transfer function (2.3), we can find its frequency response as follows,

$$C_2(j\omega) = K_p(1 + K_i(j\omega)^{-\lambda})$$
$$= K_p\left(1 + K_i\omega^{-\lambda}\cos\left(\lambda\frac{\pi}{2}\right) - jK_i\omega^{-\lambda}\sin\left(\lambda\frac{\pi}{2}\right)\right). \qquad (2.20)$$

The phase and gain are,

$$\angle[C_2(j\omega)] = -\arctan\frac{K_i\omega^{-\lambda}\sin(\lambda\pi/2)}{1 + K_i\omega^{-\lambda}\cos(\lambda\pi/2)},$$
$$|C_2(j\omega)| = K_p\sqrt{\left(1 + K_i\omega^{-\lambda}\cos\left(\lambda\frac{\pi}{2}\right)\right)^2 + \left(K_i\omega^{-\lambda}\sin\left(\lambda\frac{\pi}{2}\right)\right)^2}. \qquad (2.21)$$

From the phase and gain of the FOPTD in (2.8), the open-loop frequency response $G_2(j\omega)$ is,

$$G_2(j\omega) = C_2(j\omega)P(j\omega). \qquad (2.22)$$

The phase and gain of the open-loop frequency response are,

$$\angle[G_2(j\omega)] = \angle[C_2(j\omega)] + \angle[P(j\omega)]$$
$$= -\arctan\frac{K_i\omega^{-\lambda}\sin(\lambda\pi/2)}{1 + K_i\omega^{-\lambda}\cos(\lambda\pi/2)} - \arctan(\omega T) - L\omega, \qquad (2.23)$$

$$|G_2(j\omega)| = |C_2(j\omega)||P(j\omega)|$$
$$= \frac{K_p\sqrt{[1 + K_i\omega^{-\lambda}\cos(\lambda\pi/2)]^2 + [K_i\omega^{-\lambda}\sin(\lambda\pi/2)]^2}}{\sqrt{1 + (\omega T)^2}}. \qquad (2.24)$$

According to specification (i) in Section 2.3, the phase of $G_2(j\omega)$ at ω_c is,

$$\angle[G_2(j\omega_c)] = -\arctan\frac{K_i\omega_c^{-\lambda}\sin(\lambda\pi/2)}{1 + K_i\omega_c^{-\lambda}\cos(\lambda\pi/2)} - \arctan(\omega_c T) - L\omega_c$$
$$= -\pi + \phi_m. \qquad (2.25)$$

Then, a relationship between K_i and λ can be established as follows,

$$K_i = \frac{-\tan[\arctan(\omega_c T) + \phi_m + L\omega_c]}{M}, \qquad (2.26)$$

where

$$M = \omega_c^{-\lambda}\sin(\lambda\pi/2) + \omega_c^{-\lambda}\cos(\lambda\pi/2)\tan[\arctan(\omega_c T) + \phi_m + L\omega_c].$$

According to specification (ii) in Section 2.3, we can establish an equation for K_p,

$$|G_2(j\omega_c)| = |C_2(j\omega_c)P(j\omega_c)|,$$

$$\frac{K_p\sqrt{[1 + K_i\omega_c^{-\lambda}\cos(\lambda\pi/2)]^2 + \left[K_i\omega_c^{-\lambda}\sin(\lambda\pi/2)\right]^2}}{\sqrt{1 + (\omega_c T)^2}} = 1,$$

then, we have

$$K_p = \sqrt{\frac{1 + (\omega_c T)^2}{\left[1 + K_i\omega_c^{-\lambda}\cos(\lambda\pi/2)\right]^2 + \left[K_i\omega_c^{-\lambda}\sin(\lambda\pi/2)\right]^2}}. \tag{2.27}$$

According to specification (iii) in Section 2.3 for the robustness to loop gain variations,

$$\frac{d(\angle(G_2(j\omega)))}{d\omega}\bigg|_{\omega=\omega_c}$$

$$= [-\arctan\frac{K_i\omega^{-\lambda}\sin(\lambda\pi/2)}{1 + K_i\omega^{-\lambda}\cos(\lambda\pi/2)} - \arctan(\omega T) - L\omega]'|_{\omega=\omega_c}$$

$$= 0. \tag{2.28}$$

Then, we can establish another equation for K_i and λ,

$$AK_i^2 + \left[2\omega_c^\lambda \cos\left(\lambda\frac{\pi}{2}\right)A - \lambda\omega_c^{\lambda-1}\sin\left(\lambda\frac{\pi}{2}\right)\right]K_i + A\omega_c^{2\lambda} = 0, \tag{2.29}$$

simplifying equation (2.29) yields,

$$A\omega_c^{-2\lambda}K_i^2 + BK_i + A = 0, \tag{2.30}$$

where,

$$A = \frac{T}{1 + (\omega_c T)^2} + L,$$
$$B = 2A\omega_c^{-\lambda}\cos\left(\lambda\frac{\pi}{2}\right) - \lambda\omega_c^{-\lambda-1}\sin\left(\lambda\frac{\pi}{2}\right),$$

then, K_i can be expressed as

$$K_i = \frac{-B \pm \sqrt{B^2 - 4A^2\omega_c^{-2\lambda}}}{2A\omega_c^{-2\lambda}}. \tag{2.31}$$

Theoretically, we can solve equations (2.26), (2.31) and (2.27) to find λ, K_i and K_p. But it is difficult to obtain the analytical solutions of K_i and λ as the equations

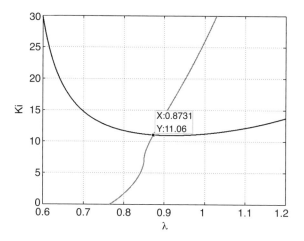

Figure 2.1 K_i versus λ

(2.26) and (2.31) are complicated. Fortunately, the graphical method can be used as a practical way to find K_i and λ.

The design procedures of the fractional order PI controller are summarized with an example as below:

(1) Given the time constant of the plant, $T = 0.4$ s.
(2) Given the pure time delay, $L = 0.01$ s.
(3) Given the gain crossover frequency, $\omega_c = 10$ rad/s.
(4) Given the desired phase margin, $\phi_m = 50°$.
(5) Plot curve 1, K_i with respect to λ according to (2.26), and plot curve 2, K_i with respect to λ according to (2.31), the two curves are shown in Figure 2.1.
(6) Obtain the values of λ and K_i from the intersection point of the two curves in Figure 2.1, $\lambda = 0.8731$, and $K_i = 11.06$.
(7) Calculate the K_p from (2.27), $K_p = 2.1204$.
(8) The open-loop Bode diagram with the designed fractional order PI controller can be plotted as shown in Figure 2.2. It can be seen that the open-loop phase is flat around the gain crossover frequency, which means the system is robust to loop gain variations. All three specifications in Section 2.3 are satisfied with this designed fractional order PI controller.

In order to reveal the relationships between the fractional order λ of the FOPI controller and the gain crossover frequency ω_c, and the parameters T, L of the FOPTD system, Figures 2.3, 2.4 and 2.5 are plotted to show the curves ω_c with respect to λ, T with respect to λ and L with respect to λ, respectively. These quantitative relationships can be served as a ready reference for the FOPI controller design of FOPTD systems.

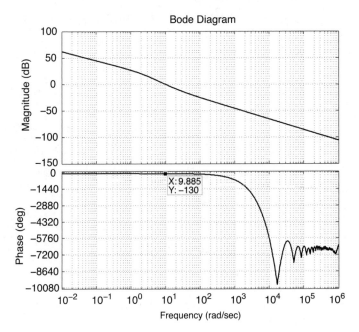

Figure 2.2   Bode plot with the fractional order PI controller

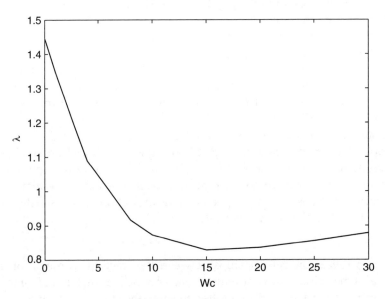

Figure 2.3 ω_c versus λ

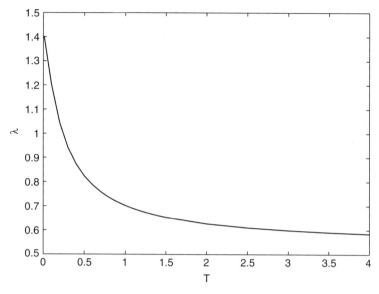

Figure 2.4 *T* versus λ

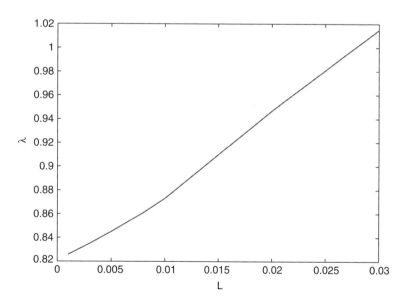

Figure 2.5 *L* versus λ

2.4.3 Fractional Order (PI) Controller Design

The open-loop transfer function $G_3(s)$ of the fractional order [PI] controller for the FOPTD system is,

$$G_3(s) = C_3(s)P(s). \tag{2.32}$$

From the fractional order [PI] controller transfer function (2.4), we can find its frequency response as follows,

$$C_3(j\omega) = \left(K_p + \frac{K_i}{j\omega} \right)^{\lambda}$$

$$= \left[K_p^2 + \left(\frac{K_i}{\omega} \right)^2 \right]^{\lambda/2} e^{-j\lambda \arctan[K_i/(K_p\omega)]}.$$

The phase and gain are,

$$\angle[C_3(j\omega)] = -\lambda \arctan \frac{K_i}{K_p\omega},$$

$$|C_3(j\omega)| = \left[K_p^2 + (K_i/\omega)^2 \right]^{\lambda/2}.$$

From the phase and gain of the FOPTD system in (2.8), the open-loop frequency response $G_3(j\omega)$ is,

$$G_3(j\omega) = C_3(j\omega)P(j\omega).$$

The phase and gain of the open-loop are,

$$\angle[G_3(j\omega)] = \angle[C_3(j\omega)] + \angle[P(j\omega)]$$

$$= -\lambda \arctan \frac{K_i}{K_p\omega} - \arctan(\omega T) - L\omega, \tag{2.33}$$

$$|G_3(j\omega)| = |C_3(j\omega)||P(j\omega)| = \frac{\left[K_p^2 + (K_i/\omega)^2 \right]^{\lambda/2}}{\sqrt{1 + (\omega T)^2}}. \tag{2.34}$$

According to specification (i) in Section 2.3, the phase of $G_3(j\omega)$ at ω_c is,

$$\angle[G_3(j\omega_c)] = -\pi + \phi_m,$$

that is,

$$-\lambda \arctan \frac{K_i}{K_p\omega_c} - \arctan(\omega_c T) - L\omega_c = -\pi + \phi_m. \tag{2.35}$$

According to specification (ii) in Section 2.3, we can establish an equation for K_p,

$$|G_3(j\omega_c)| = |C_3(j\omega_c)||P(j\omega_c)| = \frac{\left[K_p^2 + (K_i/\omega_c)^2\right]^{\lambda/2}}{\sqrt{1 + (\omega_c T)^2}} = 1,$$

we have,

$$\frac{(K_p\omega_c)^2 + K_i^2}{\omega_c^2} = [1 + (T\omega_c)^2]^{1/\lambda} = A^{1/\lambda}. \tag{2.36}$$

According to specification (iii) in Section 2.3 for the robustness to loop gain variations,

$$\frac{d(\angle(G_3(j\omega)))}{d\omega}\Big|_{\omega=\omega_c} = 0,$$

$$[-\lambda \arctan \frac{K_i}{K_p\omega} - \arctan(\omega T) - L\omega]'\Big|_{\omega=\omega_c} = 0,$$

then, we can obtain,

$$\frac{\lambda K_i K_p}{(K_p\omega_c)^2 + K_i} = \frac{T}{1 + (T\omega_c)^2} + L = \frac{T}{A} + L, \tag{2.37}$$

where $A = 1 + (\omega T)^2$.

From (2.35), (2.36) and (2.37),

$$K_i = \omega_c \sqrt{\frac{(T + LA)\omega_c}{\lambda} A^{(1/\lambda - 1)} B}, \tag{2.38}$$

$$K_i = \sqrt{\omega_c^2 B^2 A(1/\lambda)/(1 + B^2)}, \tag{2.39}$$

$$K_p = \frac{K_i}{\omega_c B}. \tag{2.40}$$

Clearly, we can solve equations (2.38), (2.39) and (2.40) to obtain λ, K_i and K_p.

The design procedures of the fractional order [PI] controller are also summarized with an example as below:

(1) Given the time constant of the plant, $T = 0.4$ s.
(2) Given the delay time, $L = 0.01$ s.
(3) Given the gain crossover frequency, $\omega_c = 10$ *rad/s*.
(4) Given the desired phase margin, $\phi_m = 50°$.
(5) Plot curve 1, K_i with respect to λ according to (2.38), and plot curve 2, K_i with respect to λ according to (2.39), the two curves are shown in Figure 2.6.
(6) Obtain the λ and K_i from the intersection point of the two plots in Figure 2.6, $\lambda = 0.7914$, and $K_i = 52.4$.

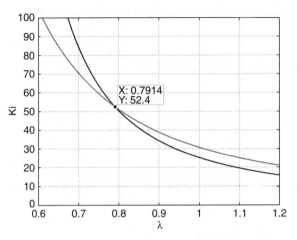

Figure 2.6 K_i versus λ

(7) Calculate the K_p from (2.40), $K_i = 2.8999$.

(8) The open-loop frequency response with the designed fractional order [PI] controller can also be fixed; the open-loop Bode diagram is plotted in Figure 2.7. As can be seen that the phase is flat around the gain crossover frequency which means the system is robust to loop gain changes. All three specifications in Section 2.3 are satisfied with the designed fractional order PI controller.

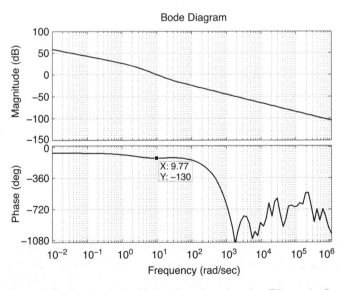

Figure 2.7 Bode plot with the fractional order (PI) controller

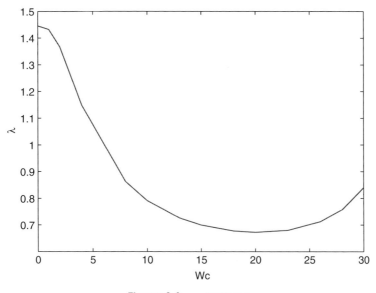

Figure 2.8 ω_c versus λ

In order to reveal the relationships between the fractional order λ of the FO[PI] controller and the gain crossover frequency ω_c, and the parameters T, L of the FOPTD system, so that a ready reference for the FO[PI] controller design of FOPTD systems is presented, Figures 2.8, 2.9 and 2.10 are also plotted to show the curves ω_c with respect to λ, T, with respect to λ, and L with respect to λ respectively.

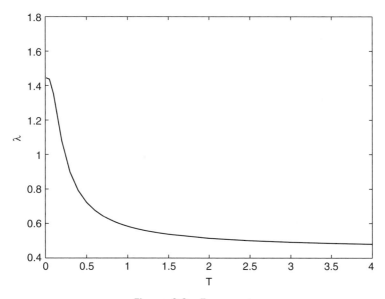

Figure 2.9 T versus λ

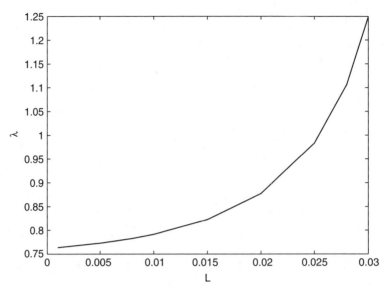

Figure 2.10 L versus λ

2.5 Simulation

In this section, the tuning methods presented in Section 2.4 for fractional order PI and [PI] controllers are illustrated via numerical simulation.

The fractional order operators s^λ for the fractional order PI controller and $(1 + 1/s)^\lambda$ for the fractional order [PI] controller are implemented by the impulse response invariant discretization (IRID) method in the time domain [51], [52].

With the same setting of $T = 0.4s$, $L = 0.01s$, $\omega_c = 10 \, rad/s$, and $\phi_m = 50°$, all the parameters of the fractional order PI and [PI] controllers are calculated in Section 2.4.

As shown in Figure 2.11, using the fractional order PI controller, the unit step responses are plotted with open-loop gain changing from 1.9084 to 2.3324($\pm 10\%$ variations from the desired value 2.1204). In Figure 2.12, applying fractional order [PI] controller, the unit step responses are plotted with open-loop gain changing from 2.6099 to 3.1899($\pm 10\%$ variations from the desired value 2.8999).

It can be seen from Figure 2.11 and Figure 2.12 that the fractional order PI and [PI] controllers designed by the proposed methods in this chapter are effective. The overshoots of the step responses remain almost constant under gain variations, which means the system is robust to loop gain changes. From Figure 2.13, we can see that the overshoot of the unit step response using the designed FOPI controller is even shorter than that using the designed FO[PI] controller. So, following the proposed design algorithms in this chapter, the FOPI controller outperforms the FO[PI] controller.

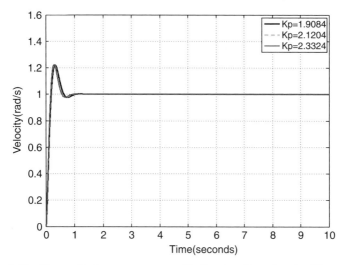

Figure 2.11 The unit step responses with the fractional order PI controller

2.6 Chapter Summary

In this chapter, we presented the systematic design schemes of FOPI and FO[PI] controllers for the FOPTD system, respectively. These design schemes ensure that the desired gain crossover frequency and the phase margin can be achieved, and the open-loop phase is flat around the gain crossover frequency. So that, the closed-loop system is robust to loop gain variations and the step response exhibits an

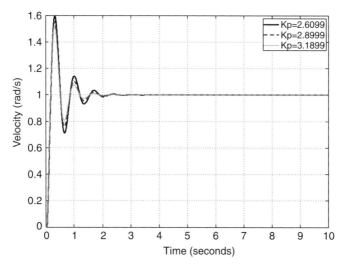

Figure 2.12 The unit step responses with the fractional order (PI) controller (For a color version of this figure, see Plate 1)

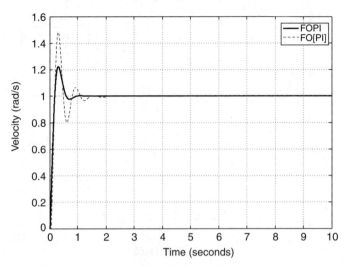

Figure 2.13   The unit step responses with the fractional order PI and (PI) controllers (For a color version of this figure, see Plate 2)

iso-damping property. For the comparison between the fractional order and the integer order controllers, the integer order PID controller is also designed following the proposed tuning thinking, but the obtained parameters of the integer order PID controller cannot stabilize the system. On the other hand, the FOPI and FO[PI] controllers designed by the proposed tuning methods not only make the system stable, but also improve the performance and robustness for the FOPTD system. Simulation results are presented to illustrate the effectiveness of the proposed tuning schemes in this chapter.

3

Tuning Fractional Order PI Controllers for Fractional Order Velocity Systems with Experimental Validation

3.1 Introduction

Along with the application of fractional calculus becoming a hot topic in the control area [119], [144], [183], [189], the research directions include not only the fractional order controller design as in Chapter 2 for the traditional integer order systems [160], [174], [238], but also the controller design for the fractional order systems [155], [189], [247]. Fractional calculus can serve as a powerful tool to characterize the memory and hereditary effects in various materials in real systems [17, 166]. Based on the traditional velocity system $k/(\tau s + 1)$, the fractional order system $k/(\tau s^{\alpha} + 1)$ can be extended with the fractional order $\alpha \in (0, 2)$, which is called the fraction order velocity system (FOVS) in this book. This FOVS can accurately model many real systems in bioengineering [147].

In this chapter, two fractional order proportional integral controllers are tuned to control the FOVS $k/(\tau s^{\alpha} + 1)$. For fair comparison between the proposed two fractional order controllers and the traditional integer order PID (IOPID) controller, these three controllers are all designed following the same set of the imposed tuning constraints. These proposed design schemes are practical and systematic way of the controllers tuning for the considered fractional order plants [134]. From not only the simulation illustration but also the experimental validation, it can be seen that both the designed fractional order proportional integral (FOPI) and the fractional order [proportional integral] (FO[PI]) controllers work effectively, and the performances

Fractional Order Motion Controls, First Edition. Ying Luo and YangQuan Chen.
© 2013 John Wiley & Sons, Ltd. Published 2013 by John Wiley & Sons, Ltd.

are better than that using the designed stabilizing IOPID controller by observation. It also can be seen that under the imposed tuning constraints, a stabilizing IOPID may not exist.

3.2 Three Controllers to be Designed and Tuning Specifications

The fractional order velocity systems (FOVS) to be controlled has the following form of transfer function,

$$P(s) = \frac{K}{Ts^\alpha + 1}. \tag{3.1}$$

Note that, the plant gain K in (3.1) can be normalized to 1 without loss of generality since the proportional factor in the transfer function (3.1) can be incorporated in the proportional coefficient of the controller. α in (3.1) is the fractional order, a known positive real number and $\alpha \in (0, 2)$; s is the usual Laplace transform variable; T is a known parameter.

The motivation of considering this FOVS (3.1) can be found from the Cole-Cole model [83], [147], which provides a fractional order parameter r to improve the fitting between dielectric theory and dispersion data, and interpret the underlying physical mechanism of dielectric relaxation in terms of fractional calculus and fractional order dynamics [147]. Many current works are interpreting the dielectric susceptibilities in terms of the fractional physics at the molecular and bulk macroscopic levels [91], [231].

The work in this chapter is to design controllers to achieve the robustness to loop gain variations for the discussed fractional order system. In this study, we consider three controllers: the IOPID as (2.2), the FOPI as (2.3), and the FO[PI] as (2.4).

Assume that the gain crossover frequency is given by ω_c and the phase margin is specified by ϕ_m. For controllers design, three tuning specifications in Section 2.3 are also used in this chapter.

3.3 Tuning Three Controllers for FOVS

The systematic way to design FO[PI] controller for the FOVS (3.1) according to three tuning specifications introduced in Section 3.2 is presented in this section. The design details for FOPI and IOPID controllers can be obtained following the similar deduction of FO[PI] controller design.

The open-loop transfer function $G_3(s)$ of the FO[PI] controller for the fraction order system (3.1) is,

$$G_3(s) = C_3(s)P(s).$$

The phase and gain of the open-loop are as follows:

$$\angle[G_3(j\omega)] = \angle[C_3(j\omega)] + \angle[P(j\omega)]$$

$$= -\lambda \tan^{-1}\left(\frac{K_i}{K_p\omega}\right) - \tan^{-1}\left(\frac{B_0}{A_0}\right), \tag{3.2}$$

$$|G_3(j\omega)| = |C_3(j\omega)||P(j\omega)|$$

$$= \frac{[K_p^2 + (K_i/\omega)^2]^{\lambda/2}}{\sqrt{A_0^2 + B_0^2}}. \tag{3.3}$$

According to specification (i) in Section 2.3, the phase of $G_3(j\omega)$ at ω_c is,

$$\angle[G_3(j\omega_c)] = -\lambda \tan^{-1}(K_i/(K_p\omega_c)) - \tan^{-1}(B/A),$$

$$= -\pi + \phi_m. \tag{3.4}$$

According to specification (ii) in Section 2.3, we can establish an equation for K_p,

$$|G_3(j\omega_c)| = |C_3(j\omega_c)||P(j\omega_c)| = \frac{[K_p^2 + (K_i/\omega_c)^2]^{\lambda/2}}{\sqrt{A^2 + B^2}} = 1, \tag{3.5}$$

that is,

$$\frac{(K_p\omega_c)^2 + K_i^2}{\omega_c^2} = (A^2 + B^2)^{1/\lambda}. \tag{3.6}$$

According to specification (iii) in Section 2.3 for the robustness to loop gain changes,

$$\left.\frac{d(\angle(G_3(j\omega)))}{d\omega}\right|_{\omega=\omega_c} = \frac{\lambda K_i K_p}{(K_p\omega_c)^2 + K_i^2} - E_3 = 0, \tag{3.7}$$

where

$$E_3 = \frac{\alpha T \omega_c^{\alpha-1}[A \sin(\alpha\pi/2) - B \cos(\alpha\pi/2)]}{A^2 + B^2}.$$

From (3.4), (3.6) and (3.7), we can obtain,

$$K_i = \sqrt{\frac{E_3}{\lambda}\omega_c^3 D_3(A^2 + B^2)^{1/\lambda}}, \tag{3.8}$$

$$K_i = \omega_c\sqrt{(A^2 + B^2)^{1/\lambda}[1 - E_3\omega_c/(\lambda D_3)]}, \tag{3.9}$$

$$K_p = \sqrt{[E_3\omega_c(A^2 + B^2)^{1/\lambda}]/(\lambda D_3)}, \tag{3.10}$$

where $D_3 = \tan[(\pi - \phi_m - \tan^{-1}(B/A))/\lambda]$.

Clearly, we can solve equations (3.8), (3.9) and (3.10) to find λ, K_i and K_p.

3.4 Illustrative Examples and Design Procedure Summaries

In this section, we present illustrative examples to verify the proposed controller tuning schemes for the FOVS in this chapter. The design procedures of the three controllers are briefly summarized.

3.4.1 Fractional Order (PI) Controller Design Procedures

The tuning procedures of the FO[PI] are illustrated through an example below:

(1) Given $K = 1$, s, $\omega_c = 10$ rad/s, $\phi_m = 50°$, and $\alpha = 0.5$.
(2) Plot curve 1, K_i with respect to λ according to (3.8), and plot curve 2, K_i with respect to λ according to (3.9).
(3) λ and K_i from the intersection point of the above two plots read $\lambda = 1.229$, $K_i = 18.19$.
(4) From (3.10), find $K_p = 0.1521$.
(5) The FO[PI] controller is obtained; the open-loop Bode plot using the designed FO[PI] controller shown in Figure 3.1 validates that the phase is flat around the gain crossover frequency and all three specifications are satisfied precisely.

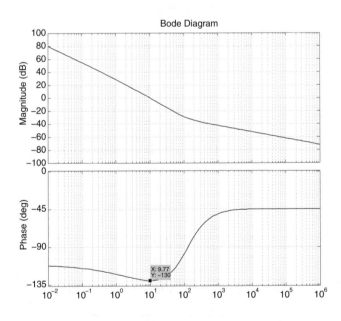

Figure 3.1 Bode plot with the designed fractional order (PI) controller

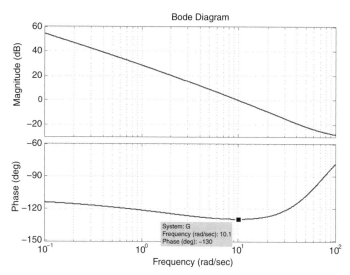

Figure 3.2   Bode plot with the designed fractional order PI controller

3.4.2 Fractional Order PI Controller Design Procedures

The tuning procedures of the FOPI controller are summarized as follows with an example:

(1) Given $K = 1$, $T = 0.4$ s, $\omega_c = 10$ rad/s, $\phi_m = 50°$, and $\alpha = 0.5$.
(2) Plot curve 1, K_i with respect to λ, and plot curve 2, K_i with respect to λ.
(3) Obtain λ and K_i from the intersection point of the above two curves, which reads $\lambda = 1.216$, $K_i = 194.4$.
(4) Calculating K_p from λ, K_i and the relationship among them, $K_p = 0.1817$.
(5) The designed FOPI controller is obtained; the Bode plot of the open-loop system with this designed FOPI is shown in Figure 3.2, where we can see that the phase is flat around the gain crossover frequency and all three specifications are satisfied.

3.4.3 Integer Order PID Controller Design Procedures

With an example, the design procedures of the IOPID controller are also summarized as below:

(1) Given $K = 1$, $T = 0.4$ s, $\omega_c = 10$ rad/s, $\phi_m = 50°$, and $\alpha = 0.5$.
(2) Plot curve 1, K_i with respect to K_d, and plot curve 2, K_i with respect to K_d.
(3) Obtain K_i and K_d from the intersection point of the above two curves, $K_i = 5.9319$, $K_d = -0.1433$.
(4) Calculate K_p according to K_i, K_d and the relationship among them, $K_p = -0.5325$.
(5) The designed IOPID controller can be obtained with K_p, K_i and K_d.

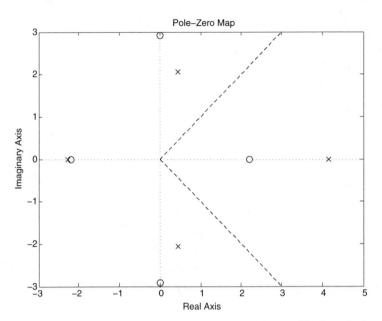

Figure 3.3 Poles and zeros plot of the closed-loop system with the designed IOPID controller

The pole and zero figure of the closed-loop system with the designed IOPID is plotted in Figure 3.3. The two dashed lines stand for the stability boundary of pole locations of this fractional order system with the angle $\alpha\pi/2$ and $-\alpha\pi/2$, respectively. According to the stability criterion in [152], there is a pole located on the positive real axis which is on the right side of these two dashed lines, so this IOPID controller is not stable for the discussed fractional order plant (3.1).

3.5 Simulation Illustration

In this section, the tuning methods presented in Section 3.4 are illustrated via numerical simulation. Two simulation cases are presented below.

- Case-1s: Simulation tests for the designed FOPI and FO[PI] controllers with $\omega_c = 10\ rad/s$ and $\phi_m = 50°$;
- Case-2s: Simulation tests for the designed IOPID and FOPI and FO[PI] controllers with $\omega_c = 15\ rad/s$ and $\phi_m = 65°$;

The fractional order operators s^λ for FOPI controller and $(K_p + K_i/s)^\lambda$ for FO[PI] controller are implemented by the impulse response invariant discretization (IRID) method in time domain [51], [52]. For the implementation of the fractional order system (3.1), the IRID method [51] is also used for the fractional order operator s^α. s^λ.

3.5.1 Case-1s Simulation Tests for the Designed FOPI and FO(PI) Controllers with $\omega_c = 10$ rad/s and $\phi_m = 50°$

As mentioned in Section 3.3 and Section 3.4.3, with $\omega_c = 10$ *rad/s* and $\phi_m = 50°$, the designed IOPID controller $C_1(s) = -0.5325 + 5.9319/s - 0.1433s$ is not stable for the fractional order system (3.1). So, just the designed FOPI and FO[PI] are tested and compared in this simulation case.

Under the same setting of $K = 1$ and $T = 0.4s$ in FOVS (3.1), $\omega_c = 10$ *rad/s* and $\phi_m = 50°$, all the parameters of the FOPI and FO[PI] controllers are calculated in Section 3.4 as shown below for FOPI and FO[PI], respectively,

$$C_2(s) = 0.1817 \left(1 + 194.4 \frac{1}{s^{1.216}} \right); \tag{3.11}$$

$$C_3(s) = \left(0.1521 + 18.19 \frac{1}{s} \right)^{1.229}. \tag{3.12}$$

As shown in Figure 3.4, using the designed FOPI controller, the unit step responses and output disturbance (amplitudes is −0.5) rejections are plotted with open-loop gain K changing from 0.9 to 1.1 (±10% variations from the nominal value 1). In

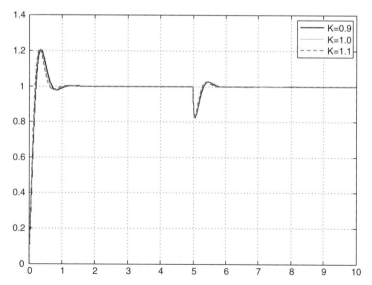

Figure 3.4 Simulation. Step responses and disturbance rejections using FOPI with loop gain variations ($\omega_c = 10$ *rad/s*, $\phi_m = 50°$)

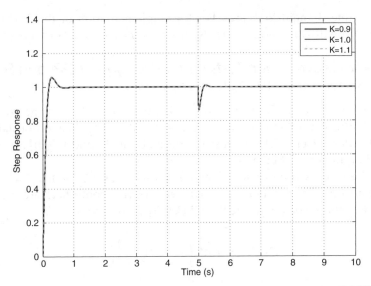

Figure 3.5 Simulation. Step responses and disturbance rejections using FO(PI) with loop gain variations ($\omega_c = 10$ rad/s, $\phi_m = 50°$)

Figure 3.5, applying the designed FO[PI] controller, the unit step responses and output disturbance (amplitudes is -0.5) rejections are also plotted with open-loop gain K changing from 0.9 to 1.1 ($\pm 10\%$ variations from the original value 1).

It can be seen from Figure 3.4 and Figure 3.5 that the FOPI and FO[PI] controllers designed by the proposed method in this chapter work effectively for the fractional order systems considered. The overshoots of the step responses and disturbance rejections remain almost constant under loop gain variations, which means the system is robust to loop gain changes. From Figure 3.6, the overshoots of the unit step response and disturbance rejection using the designed FO[PI] controller are much shorter than that using the designed FOPI controller. So, following our proposed tuning algorithms, the designed FO[PI] controller outperforms the designed FOPI controller for the fractional order systems considered by observation. The input signals to the control plant are plotted in Figure 3.7, the control effort using the FO[PI] is smaller than that using the FOPI.

3.5.2 Case-2s Simulation Tests for the Designed IOPID and FOPI and FO(PI) Controllers with $\omega_c = 15$ rad/s and $\phi_m = 65°$

In order to make fair comparisons between the proposed fractional order controllers and the traditional IOPID controller following the similar design scheme in this chapter, this simulation case is presented. Given $\omega_c = 15$ rad/s, $\phi_m = 65°$ and with the same plant setting of $K = 1$ and $T = 0.4s$ in (3.1), all three controllers IOPID, FOPI

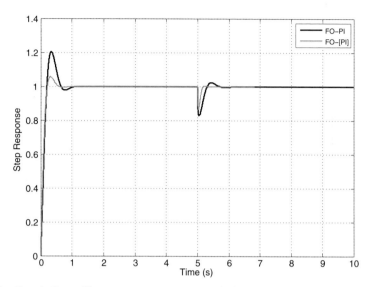

Figure 3.6 Simulation. Step response and disturbance rejection comparison ($\omega_c = 10\ rad/s$, $\phi_m = 50°$)

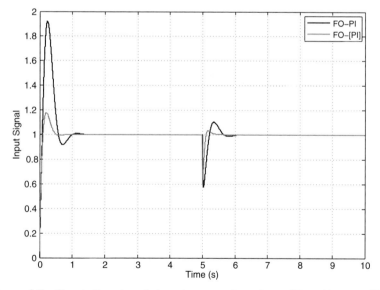

Figure 3.7 Simulation. Input signal comparison ($\omega_c = 10\ rad/s$, $\phi_m = 50°$)

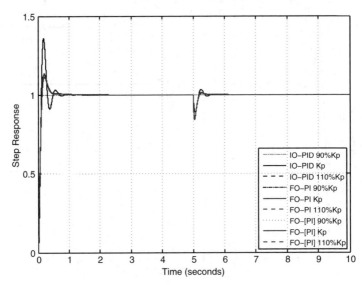

Figure 3.8 Simulation. Step response and disturbance rejection comparison with loop gain variations ($\omega_c = 15$ rad/s, $\phi_m = 65°$)

and FO[PI] can be tuned following the procedures introduced in Section 3.3 as shown below, respectively,

$$C_1(s) = 0.1072 + 56.0224\frac{1}{s} + 0.0915s; \tag{3.13}$$

$$C_2(s) = 0.225\left(1 + 38.655\frac{1}{s^{1.0317}}\right); \tag{3.14}$$

$$C_3(s) = \left(0.2189 + 34.3241\frac{1}{s}\right)^{1.0339}. \tag{3.15}$$

From Figure 3.8, three controllers are all robust to the loop gain changes. It is obvious that the overshoots of the step responses and disturbance (amplitude is -0.5) rejections using the designed two fractional order controllers are much smaller than that using the designed IOPID controller. The overshoots with FO[PI] are a little bit smaller than that with FOPI. The input signals are also compared in Figure 3.9. The control efforts with FOPI and FO[PI] are much smaller than that with IOPID.

3.6 Experimental Validation

In this section, the practicality of the proposed fractional order controller design schemes and the performance advantages of the fractional order controllers for the

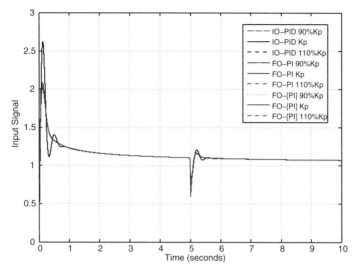

Figure 3.9 Simulation. Input signal comparison with loop gain variations ($\omega_c = 15\ rad/s$, $\phi_m = 65°$)

FOVS presented in the simulation are validated in a hardware-in-the-loop (HIL) experimental test bench.

3.6.1 Experimental Setup

A fractional horsepower dynamometer was developed as a general purpose experimental platform [210]. The architecture of the dynamometer control system is shown in Figure 3.10. The dynamometer includes the DC motor to be tested, a hysteresis

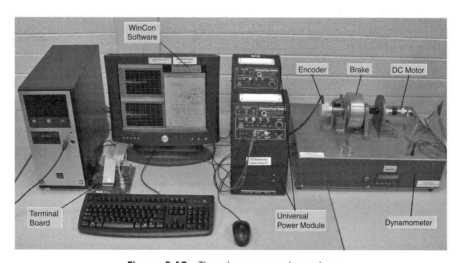

Figure 3.10 The dynamometer setup

brake for applying torque load to the motor, a load cell to provide force feedback, an optical encoder for position feedback and a tachometer for velocity feedback. The dynamometer is connected to a Quanser Q4 terminal board for the system control through the Matlab/Simulink Real-Time Workshop (RTW) based software. This terminal board connects with the Quanser Q4 data acquisition card on the computer motherboard. Then, the Matlab/Simulink environment in the computer, which uses the WinCon application from Quanser, is used to communicate with the data acquisition card. This setup enables rapid prototyping of real-time closed-loop control systems.

3.6.2 HIL Emulation of the FOVS

Through simple system identification process, the dynamometer velocity control system can be approximately modeled by a transfer function $\frac{1.52}{0.4s+1}$. The fractional order system (3.1) can be emulated by modifying the dynamometer as the same model in the simulation with $K = 1$, $T = 0.4$ s and $\alpha = 0.5$. The realization of the fractional order system is shown in Figure 3.11,

$$P(s) = G_m(s)\frac{1}{K_{dy}}\left(\frac{K_0 A(s)}{1 + K_0 A(s)} + \frac{B(s)}{1 + K_0 A(s)}\right), \tag{3.16}$$

where

$$G_m(s) = \frac{K}{Ts + 1}, \; A(s) = s^{-\alpha}, \; B(s) = s^{-\alpha+1}, \; K_0 = \frac{1}{T}, \; K_{dy} = 1.52,$$

and $G_m(s)$ is the model of the dynamometer velocity control system. The impulse response invariant discretization (IRID) method in time domain [51] is also used for the fractional order operator $1/s^{\alpha}$ implementation.

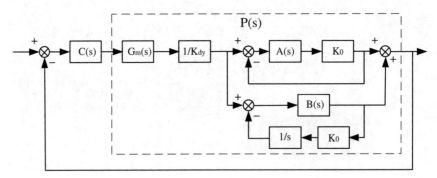

Figure 3.11   The emulated fractional order plant by modifying the dynamometer

3.6.3 Experimental Results

Two cases in the simulation are both validated on the HIL experimental test bench for the emulated FOVS. Corresponding to the simulation, two experimental cases are demonstrated as follows:

- Case-1e: Experimental validation of the designed FOPI and FO[PI] controllers with $\omega_c = 10$ *rad/s* and $\phi_m = 50°$.
- Case-2e: Experimental validation of the designed IOPID, FOPI and FO[PI] controllers with $\omega_c = 15$ *rad/s* and $\phi_m = 65°$.

The fractional order operators s^λ for fractional order PI controller and $(K_p + K_i/s)^\lambda$ for fractional order [PI] controller are implemented by the impulse response invariant discretization (IRID) method in time domain [51], [52]. For all the step response tests in the experiments, the step reference is 1 *rad/s*.

3.6.3.1 Case-1e Experimental Validation of the Designed FOPI and FO(PI) Controllers with $\omega_c = 10$ *rad/s* and $\phi_m = 50°$

As shown in Figures 3.12 and 3.13, applying the FOPI and FO[PI] controllers, the unit step responses and disturbance (amplitude is -0.5 *rad/s*) rejections are plotted with open-loop gain changes from 0.9 to 1.1 ($\pm10\%$ variations from the nominal value $K = 1$), respectively. From Figures 3.12 and 3.13, the fractional order PI and [PI] controllers designed by the proposed schemes work effectively for the fractional

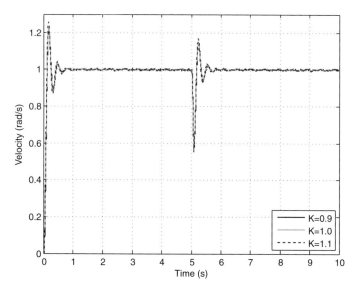

Figure 3.12 Experiment. Step responses and disturbance rejections using FOPI with loop gain variations ($\omega_c = 10$ *rad/s*, $\phi_m = 50°$)

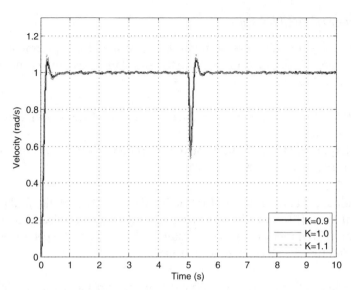

Figure 3.13 Experiment. Step responses and disturbance rejections using FO(PI) with loop gain variations ($\omega_c = 10$ *rad/s*, $\phi_m = 50°$)

order systems considered. The overshoots of the step responses and disturbance rejections remain almost constant under loop gain variations, which means that the closed-loop control systems are robust to loop gain changes. From Figure 3.14, it can be seen clearly that the overshoots of the unit step response and disturbance rejection using the designed FO[PI] are much smaller than that using the designed FOPI. So, by observation, the designed FO[PI] controller outperforms the designed FOPI controller following our proposed tuning algorithms for the fractional order systems considered. The input signals to the plant are also presented in Figure 3.15, the control effort of the designed FO[PI] is also much smaller than that of the designed FOPI.

3.6.3.2 Case-2e Experimental Validation of the Designed IOPID, FOPI and FO(PI) Controllers with $\omega_c = 15$ *rad/s* and $\phi_m = 65°$

For the designed IOPID controller, due to the existence of the derivative item, the noise of the real-time system may be amplified. So, a low pass filter is added in the derivative item when the designed IOPID controller is implemented in the experimental system as shown below,

$$C_1'(s) = 0.1072 + 56.0224\frac{1}{s} + 0.0915\frac{s}{\tau s + 1},$$

where τ is chosen as 0.314s, so the cut-off frequency is 20 *rad/s*, which is chosen according to the gain crossover frequency $\omega_c = 15$ *rad/s* for the designed controllers.

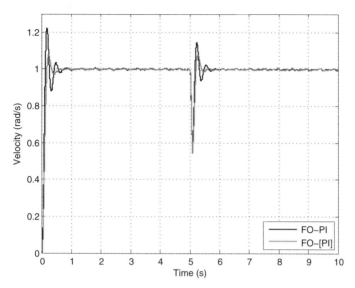

Figure 3.14 Experiment. Step response and disturbance rejection comparison (ω_c = 10 rad/s, ϕ_m = 50°)

From Figures 3.16, 3.17 and 3.18, the IOPID, FOPI and FO[PI] controllers are all robust to the loop gain variations. As shown in Figure 3.19, it is obvious that the overshoots of the step responses and disturbance rejections using the designed two fractional order controllers are much smaller than that using the designed IOPID controller. From the input signals presented in Figure 3.20, the control efforts with

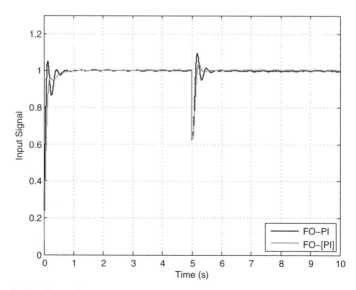

Figure 3.15 Experiment. Input signal comparison (ω_c = 10 rad/s, ϕ_m = 50°)

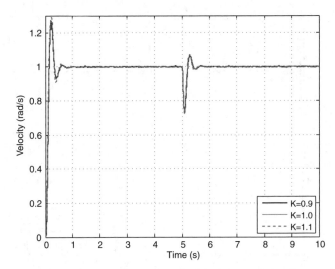

Figure 3.16 Experiment. Step responses and disturbance rejections using IOPID with loop gain variations ($\omega_c = 15$ *rad/s*, $\phi_m = 65°$)

FOPI and FO[PI] are also much smaller than that with IOPID. The advantage of using the fractional order controllers to control the fractional order systems considered is clearly observed by comparing the performances of the designed FOPI and FO[PI] with that of the designed IOPID. Meanwhile, the overshoot and control effort with FO[PI] are a little bit smaller than that with FOPI.

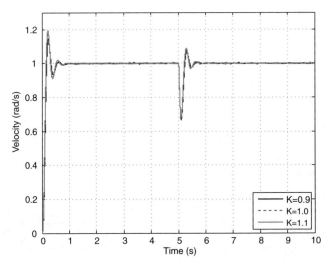

Figure 3.17 Experiment. Step responses and disturbance rejections using FOPI with loop gain variations ($\omega_c = 15$ *rad/s*, $\phi_m = 65°$)

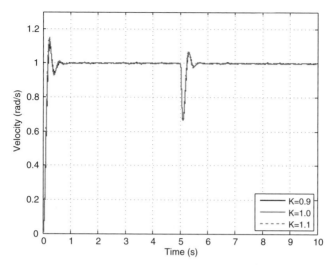

Figure 3.18 Experiment. Step responses and disturbance rejections using FO(PI) with loop gain variations ($\omega_c = 15$ *rad/s*, $\phi_m = 65°$) (For a color version of this figure, see Plate 3)

3.7 Chapter Summary

In this chapter, two fractional order proportional integral controllers are tuned for a class of fractional order velocity systems. For fair comparison, the FOPI and FO[PI] controllers and the traditional IOPID controller are all designed following the same set of imposed tuning specifications, which can guarantee the robustness of the

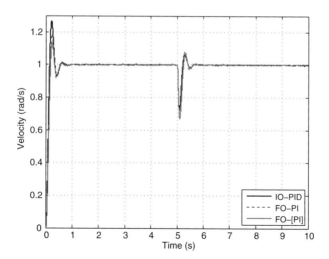

Figure 3.19 Experiment. Step response and disturbance rejection comparison ($\omega_c = 15$ *rad/s*, $\phi_m = 65°$) (For a color version of this figure, see Plate 4)

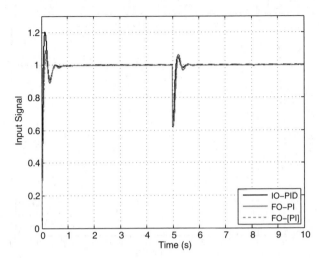

Figure 3.20 Experiment. Input signal comparison ($\omega_c = 15 \ rad/s, \phi_m = 65°$)

designed controllers to loop gain variations. From the results presented in simulation illustration and also in experimental validation, both of the two designed fractional order controllers work efficiently. The designed FOPI and FO[PI] controllers can improve the control performance for the fractional order systems discussed in this chapter compared with the designed stabilizing IOPID controller following the same imposed tuning specifications.

4

Relay Feedback Tuning of Robust PID Controllers

4.1 Introduction

From Chapters 2 and 3, the fractional order PI controllers are shown to offer a lot of potential for the improvement of control performance using traditional integer order PID controllers based on velocity systems. Actually, fractional order PID controllers can also be properly designed with unknown stable plant dynamics by the auto-tuning techniques for better performance over the traditional PID controllers. In order to demonstrate more effectiveness and advantages of the fractional order PID controllers, some ideas for the integer order robust PID auto-tuning algorithm [48] can be presented as a baseline first in this chapter.

According to a survey [240] of the state of process control systems in 1989 conducted by the Japanese Electric Measuring Instrument Manufacturer Association, more than 90 percent of the control loops were of the PID type. It was also indicated [27] that a typical paper mill in Canada has more than 2,000 control loops and that 97 percent use PI control. Therefore, the industrialist had concentrated on PI/PID controllers and had already developed *one-button type* relay auto-tuning techniques for fast, reliable PI/PID control yet with satisfactory performance [65], [85], [117], [207], [244]. Although many different methods have been proposed for tuning PID controllers, till the present, the Ziegler-Nichols method [249] is still extensively used to determine the parameters of PID controllers. The design is based on the measurement of the critical gain and critical frequency of the plant and using simple formulae to compute the controller parameters. In 1984, Åström and Hägglund [11] proposed an automatic tuning method based on a simple relay feedback test which uses describing function analysis to give the critical gain and the critical frequency of the system. This information can be used to compute a PID controller with desired gain and phase margins. In relay feedback tests, it is common practice to use a relay with hysteresis [11] for

Fractional Order Motion Controls, First Edition. Ying Luo and YangQuan Chen.
© 2013 John Wiley & Sons, Ltd. Published 2013 by John Wiley & Sons, Ltd.

noise immunity. Another commonly used technique is to introduce an artificial time delay within the relay closed-loop system, for example, [103], to change the oscillation frequency in relay feedback tests.

After identifying a point on the Nyquist curve of the plant, the so-called modified Ziegler-Nichols method [85], [89] can be used to move this point to another position in the complex plane. Two equations for phase and amplitude assignment can be obtained to retrieve the parameters of a PI controller. For a PID controller, however, an additional equation should be introduced. In the modified Ziegler-Nichols method, α, the ratio between the integral time T_i and the derivative time T_d, is chosen to be constant, that is, $T_i = \alpha T_d$, in order to obtain a unique solution.

The control performance is heavily influenced by the choice of α as observed in [103]. Recently, the role of α has attracted much attention, such as, [113], [184], [223]. For the Ziegler-Nichols PID tuning method, α is generally assigned as a magic number 4 [85]. Wallén, Åström and Hägglund proposed that the tradeoff between the practical implementation and the system performance is the major reason for choosing the ratio between T_i and T_d as 4 [223].

The main work in this chapter is the use of an auto-tuning rule which gives a new relationship between T_i and T_d instead of the equation $T_i = 4T_d$ proposed in the modified Ziegler-Nichols method [85], [89]. The "flat phase" conception is also applied to the controller design in this chapter. We propose to add an extra condition that the phase Bode plot at a specified frequency ω_c at the point where sensitivity circle touches Nyquist curve is locally flat which implies that the system will be robust to gain variations. This additional condition can be expressed as $\frac{d\angle G(j\omega)}{d\omega}|_{\omega=\omega_c} = 0$, which can be equivalently expressed as

$$\angle \frac{dG(j\omega)}{d\omega}|_{\omega=\omega_c} = \angle G(j\omega)|_{\omega=\omega_c} \tag{4.1}$$

where ω_c is the frequency at the point of tangency and $G(s) = C(s)P(s)$ is the transfer function of the open-loop system including the controller $C(s)$ and the plant $P(s)$. The above equivalence in (4.1) is mathematically explained as follows.

Assume that $G(jw) = x(w) + jy(w)$. Then,

$$\angle G(jw) = \tan^{-1}\left(\frac{y(w)}{x(w)}\right). \tag{4.2}$$

The derivative of $G(jw)$ with respect to w is that

$$\frac{d\angle G(jw)}{dw} = \frac{1}{1 + (y/x)^2}\frac{d(y/x)}{dw}$$

$$= \frac{x^2}{x^2 + y^2}\left(\frac{dy/dw}{x} - \frac{ydx/dw}{x^2}\right)$$

$$= \frac{1}{x^2 + y^2}\left(x\frac{dy}{dw} - y\frac{dx}{dw}\right) = 0.$$

Furthermore, one has $x\frac{dy}{dw} - y\frac{dx}{dw} = 0$, which means that $\frac{y}{x} = \frac{dy/dw}{dx/dw}$. Since $\tan\angle(\frac{dG(jw)}{dw}) = \frac{dy/dw}{dx/dw}$, then, $\tan\angle(\frac{dG(jw)}{dw}) = \frac{y}{x}$. So, $\angle(\frac{dG(jw)}{dw}) = \tan^{-1}(\frac{y}{x})$. From (4.2), we can find $\angle(\frac{dG(jw)}{dw}) = \angle G(jw)$.

In this chapter, we consider the PID controller of the following form:

$$C(s) = K_p\left(1 + \frac{1}{T_i s} + T_d s\right). \tag{4.3}$$

This "flat phase" idea as mentioned in Section 2.3 is illustrated in Figure 4.1(a) where the Bode plot of the open-loop system is shown with its phase being tuned locally flat around ω_c. We can expect that, if the gain increases or decreases a certain percentage, the phase margin will remain unchanged. Therefore, in this case, the step responses under various gains changing around the nominal gain will exhibit an iso-damping property, that is, the overshoots of step responses will be almost the same. This can also be explained by Figure 4.1(b) where the sensitivity circle touches the Nyquist curve of the open-loop system at the flat phase point.

Assume that the phase of the open-loop system at ω_c is

$$\angle G(s)|_{s=j\omega_c} = \Phi_m - \pi. \tag{4.4}$$

Then, assume the corresponding gain can be expressed by

$$|G(j\omega_c)| = \cos(\Phi_m). \tag{4.5}$$

With these two conditions (4.4) and (4.5) and the new condition (4.1), all three parameters of PID controller can be calculated.

As in the Ziegler-Nichols method, T_i and T_d are used to tune the phase condition (4.4) and K_p is determined by the gain condition (4.5). However, the condition (4.1) gives a relationship between T_i and T_d instead of $T_i = \alpha T_d$.

Note that in this new tuning method, ω_c is not necessarily the gain crossover frequency although it is close. Precisely, ω_c is the frequency at which the Nyquist curve tangentially touches the sensitivity circle. Similarly, Φ_m, the tangent phase, is not necessarily the phase margin usually used in previous PID tuning methods. According to [85], the phase margin is always selected from 30° to 60°. Due to the flat phase condition (4.1), the derivative of the phase near ω_c will be relatively small. Therefore, if Φ_m is selected to be around 30°, such as 35°, the phase margin will be generally within the desired interval.

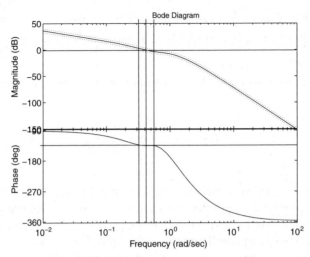

(a) Basic idea: a flat phase curve at gain crossover frequency

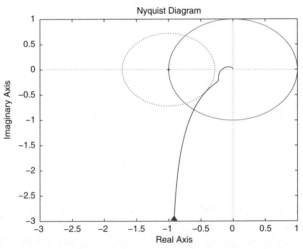

(b) Sensitivity circle tangentially touches Nyquist curve at the flat phase

Figure 4.1 Illustration of the basic idea for iso-damping robust PID tuning

4.2 Slope Adjustment of the Phase Bode Plot

In this section, we will show how T_i and T_d are related under the new condition (4.1). Substitute s by jw so that the closed-loop system can be written as $G(jw) = C(jw)P(jw)$, where

$$C(jw) = K_p \left(1 + \frac{1}{jwT_i} + jwT_d \right) \tag{4.6}$$

is the PID controller obtained from (4.3). The phase of the open-loop system is given by

$$\angle G(jw) = \angle C(jw) + \angle P(jw). \tag{4.7}$$

The derivative of the open-loop system $G(jw)$ with respect to w can be written as follows:

$$\frac{dG(jw)}{dw} = P(jw)\frac{dC(jw)}{dw} + C(jw)\frac{dP(jw)}{dw}. \tag{4.8}$$

From (4.1), the phase of the derivative of the open-loop system cannot obviously be obtained directly from (4.8). So, we need to simplify (4.8).

The derivative of the controller with respect to w is

$$\frac{dC(jw)}{dw} = jK_p\left(T_d + \frac{1}{w^2 T_i}\right). \tag{4.9}$$

To calculate $\frac{dP(jw)}{dw}$, since we have

$$\ln P(jw) = \ln|P(jw)| + j\angle P(jw), \tag{4.10}$$

differentiating (4.10) with respect to w gives

$$\frac{d\ln P(jw)}{dw} = \frac{1}{P(jw)}\frac{dP(jw)}{dw}$$

$$= \frac{d\ln|P(jw)|}{dw} + j\frac{d\angle P(jw)}{dw}. \tag{4.11}$$

Straightforwardly, we arrive at

$$\frac{dP(jw)}{dw} = P(jw)\left[\frac{d\ln|P(jw)|}{dw} + j\frac{d\angle P(jw)}{dw}\right]. \tag{4.12}$$

Substituting (4.6), (4.9) and (4.9) into (4.4) gives

$$\frac{dG(jw)}{dw} = K_p P(jw)\left[j\left(T_d + \frac{1}{w^2 T_i}\right)\right.$$
$$\left. + \left(1 + j\left(T_d w - \frac{1}{w T_i}\right)\right)\left(\frac{d\ln|P(jw)|}{dw} + j\frac{d\angle P(jw)}{dw}\right)\right]. \tag{4.13}$$

Hence, the slope of the Nyquist curve at any specific frequency ω_0 is given by

$$\angle \left. \frac{dG(jw)}{dw} \right|_{\omega_0} = \angle P(j\omega_0)$$

$$+ \tan^{-1} \left[\frac{(T_d T_i \omega_0^2 + 1) + (T_d T_i \omega_0^2 - 1)s_a(\omega_0) + s_p(\omega_0)T_i\omega_0}{s_a(\omega_0)T_i\omega_0 - (T_d T_i \omega_0^2 - 1)s_p(\omega_0)} \right] \tag{4.14}$$

where, following the notations introduced in [105], [106], $s_a(\omega_0)$ and $s_p(\omega_0)$ are used throughout this chapter defined as follows:

$$s_a(\omega_0) = \omega_0 \left. \frac{d\ln|P(jw)|}{dw} \right|_{\omega_0}, \tag{4.15}$$

$$s_p(\omega_0) = \omega_0 \left. \frac{d\angle P(jw)}{dw} \right|_{\omega_0}. \tag{4.16}$$

Here, our task is to adjust the slope of the Nyquist curve to match the condition shown in (4.1). By combining (4.1), (4.7) and (4.14), one obtains

$$\angle C(jw)|_{\omega_0} = \tan^{-1} \left[\frac{(T_d T_i \omega_0^2 + 1) + (T_d T_i \omega_0^2 - 1)s_a(\omega_0) + s_p(\omega_0)T_i\omega_0}{s_a(\omega_0)T_i\omega_0 - (T_d T_i \omega_0^2 - 1)s_p(\omega_0)} \right]. \tag{4.17}$$

After a straightforward calculation, one obtains the relationship between T_i and T_d as follows:

$$T_d = \frac{-T_i\omega_0 + 2s_p(\omega_0) + \sqrt{\Delta}}{2s_p(\omega_0)\omega_0^2 T_i}, \tag{4.18}$$

where $\Delta = T_i^2\omega_0^2 - 8s_p(\omega_0)T_i\omega_0 - 4T_i^2\omega_0^2 s_p^2(\omega_0)$. Note that due to the nature of the quadratic equation, an alternative relationship, is that $T_d = \frac{-T_i\omega_0 + 2s_p(\omega_0) - \sqrt{\Delta}}{2s_p(\omega_0)\omega_0^2 T_i}$. We should discard one to ensure that the T_d gain is a real positive number to avoid the right half plane zeros in $C(s)$. In what follows, equation (4.18) is used in all our examples. Additionally, Δ could be negative if ω_0 is not specified properly.

The approximation of s_p for stable and minimum phase plant can be given as follows [28]:

$$s_p(\omega_0) = \omega_0 \left. \frac{d\angle P(jw)}{dw} \right|_{\omega_0}$$

$$\approx \angle P(j\omega_0) + \frac{2}{\pi}[\ln|K_g| - \ln|P(j\omega_0)|], \tag{4.19}$$

where $|K_g| = P(0)$ is the static gain of the plant, $\angle P(j\omega_0)$ is the phase and $|P(j\omega_0)|$ is the gain of the plant at the specific frequency ω_0.

It is obvious that T_i and T_d are related by s_p alone. For this new tuning method, s_p includes all the information that we need to know about the unknown plant. In what follows, we show that the s_p estimated formula can be extended to plants with integrators and/or time delay.

Consider the plant with m integrators

$$P(s) = \frac{\tilde{P}(s)}{s^m}, \quad m = 1, 2, 3, \cdots. \tag{4.20}$$

Clearly, one cannot find the static gain of such systems to compute s_p directly. But from (4.16),

$$s_p(\omega_0) = \omega_0 \left. \frac{d\angle P(jw)}{dw} \right|_{\omega_0}$$

$$= \omega_0 \left. \frac{d\left(\angle\tilde{P}(jw) - \dfrac{m\pi}{2}\right)}{dw} \right|_{\omega_0} = \omega_0 \left. \frac{d\angle\tilde{P}(jw)}{dw} \right|_{\omega_0}, \tag{4.21}$$

which means that for the systems with integrators, s_p should be estimated according to the systems without any integrator.

For the plant with a time delay τ

$$\tilde{P}(s) = \bar{P}(s)e^{-\tau s}, \tag{4.22}$$

in the same way,

$$s_p(\omega_0) = \omega_0 \left. \frac{d\angle\tilde{P}(jw)}{dw} \right|_{\omega_0} = \omega_0 \left. \frac{d\angle\bar{P}(jw)}{dw} \right|_{\omega_0} - \tau\omega_0. \tag{4.23}$$

Consequently, substituting (4.19), we obtain

$$s_p(\omega_0) \approx \angle\bar{P}(j\omega_0) + \frac{2}{\pi}[\ln|K_g| - \ln|\bar{P}(j\omega_0)|] - \tau\omega_0$$

$$\approx \angle\tilde{P}(j\omega_0) + \frac{2}{\pi}[\ln|K_g| - \ln|\tilde{P}(j\omega_0)|]. \tag{4.24}$$

Obviously, the time delay will not contribute to the estimation of s_p.

So, in general, for the plant with both integrators and a time delay

$$P(s) = \frac{\tilde{P}(s)}{s^m} = \frac{\bar{P}(s)e^{-\tau s}}{s^m}, \quad m = 1, 2, 3, \cdots, \tag{4.25}$$

according to (4.21) and (4.24),

$$s_p(\omega_0) = \omega_0 \frac{d\angle P(jw)}{dw}\bigg|_{\omega_0} = \omega_0 \frac{d\angle \tilde{P}(jw)}{dw}\bigg|_{\omega_0}$$

$$\approx \angle \tilde{P}(j\omega_0) + \frac{2}{\pi}[\ln|K_g| - \ln|\tilde{P}(j\omega_0)|]. \tag{4.26}$$

4.3 The New PID Controller Design Formulae

Suppose that we have known s_p at ω_c. How to experimentally measure $s_p(\omega_c)$ will be discussed in Section 4.4 based on the measurement of $\angle P(j\omega_c)$ and $|P(j\omega_c)|$.

To write down explicitly the formulae for K_p, T_i and T_d, let us summarize what is known at this point. We are given (i) ω_c, the desired tangent frequency; (ii) Φ_m, the desired tangent phase; (iii) measurement of $\angle P(j\omega_c)$ and $|P(j\omega_c)|$; and (iv) the estimation of $s_p(\omega_c)$.

Furthermore, using ((4.4)) and ((4.5)), the PID controller parameters can be set as follows:

$$K_p = \frac{\cos(\Phi_m)}{|P(j\omega_c)\sqrt{1 + \tan^2(\Phi_m - \angle P(j\omega_c))}|}, \tag{4.27}$$

$$T_i = \frac{-2}{\omega_c[s_p(\omega_c) + \hat{\Phi}) + \tan^2(\hat{\Phi})s_p(\omega_c)]}, \tag{4.28}$$

where $\hat{\Phi} = \Phi_m - \angle P(j\omega_c)$. Finally, T_d can be computed from (4.18).

Remark 4.3.1 *The selection of ω_c greatly depends on the system dynamics. For most plants, there exists an interval for the selection of ω_c to achieve flat phase condition. If no better idea about ω_c, exists, desired cutoff frequency can be used as the initial value. For Φ_m, a good choice is within $30°$ to $35°$.*

4.4 Phase and Magnitude Measurement via Relay Feedback Tests

Following the discussion in the above section, the parameters of a PID controller can be calculated straightforwardly if we know $\angle P(j\omega_c)$, $|P(j\omega_c)|$ and $s_p(\omega_c)$.

As indicated in (4.19), $s_p(\omega_c)$ can be obtained from the knowledge of the static gain $|P(0)|$, $\angle P(j\omega_c)$ and $|P(j\omega_c)|$. The static gain $|P(0)|$ or K_g is very easy to measure and it is assumed to be known. The relay feedback test, shown in Figure 4.2, can be used to "measure" $\angle P(j\omega_c)$ and $|P(j\omega_c)|$. In the relay feedback experiments, a relay is connected in closed loop with the unknown plant as shown in Figure 4.2 which is usually used to identify one point on the Nyquist diagram of the plant. To change the

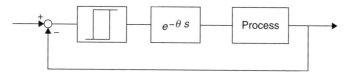

Figure 4.2   Relay plus artificial time delay (θ) feedback system

oscillation frequency due to relay feedback, an artificial time delay is introduced in the loop. The artificial time delay θ is the tuning knob here to change the oscillation frequency. Our problem here is how to find the right value of θ which corresponds to the tangent frequency ω_c. To solve this problem, an iterative method can be used as summarized in the following:

(1) Start with the desired tangent frequency ω_c.
(2) Select two different values (θ_{-1} and θ_0) for the time delay parameter properly and do the relay feedback test twice. Then, two points on the Nyquist curve of the plant can be obtained. The frequencies of these points can be represented as ω_{-1} and ω_0 which correspond to θ_{-1} and θ_0, respectively. The iteration begins with these initial values (θ_{-1}, ω_{-1}) and (θ_0, ω_0).
(3) With the values obtained in the previous iterations, the artificial time delay parameter θ can be updated using a simple interpolation/extrapolation scheme as follows:

$$\theta_n = \frac{\omega_c - \omega_{n-1}}{\omega_{n-1} - \omega_{n-2}} (\theta_{n-1} - \theta_{n-2}) + \theta_{n-1}$$

where n represents the current iteration number. With the new θ_n, after the relay test, the corresponding frequency ω_n can be recorded.
(4) Compare ω_n with ω_c. If $|\omega_n - \omega_c| < \delta$, quit iteration. Otherwise, go to Step 3. Here, δ is a small positive number.

The iterative method proposed above is feasible because in general the relationship between the delay time θ and the oscillation frequency w is one-to-one.

After the iteration, the final oscillation frequency is quite close to the desired one ω_c so that the oscillation frequency is considered as ω_c. Hence, the amplitude and the phase of the plant at the specified frequency can be obtained. Using (4.19), one can calculate the approximation of s_p.

4.5 Illustrative Examples

The new PID design method presented above will be illustrated via some simulation examples. In the simulation, the following classes of plants, studied in [223], will be used.

$$P_n(s) = \frac{1}{(s+1)^{(n+3)}}, \quad n = 1, 2, 3, 4; \tag{4.29}$$

$$P_5(s) = \frac{1}{s(s+1)^3}; \tag{4.30}$$

$$P_6(s) = \frac{1}{(s+1)^3}e^{-s}; \tag{4.31}$$

$$P_7(s) = \frac{1}{s(s+1)^3}e^{-s}. \tag{4.32}$$

4.5.1 High-order Plant $P_2(s)$

Consider plant $P_2(s)$ in (4.29). This plant was also used in [105]. The specifications are set as $\omega_c = 0.4 \ rad/s$ and $\Phi_m = 45°$. The PID controller designed by using the proposed tuning formulae is

$$K_{1p}(s) = 0.921\left(1 + \frac{1}{1.961s} + 1.969s\right). \tag{4.33}$$

The PID controller designed by the modified Ziegler-Nichols method is

$$K_1(s) = 1.131\left(1 + \frac{1}{3.124s} + 0.781s\right). \tag{4.34}$$

The Bode and the Nyquist plots are compared in Figure 4.3. From the Bode plots, it is seen that the phase curve near the frequency $\omega_c = 0.4 \ rad/s$ is flat. The phase margin roughly equals 45°. That means the controller moves the point $P(0.4j)$ of the Nyquist curve to $C(0.4j)P(0.4j)$ with a phase of 135° and at the same time makes the Nyquist curve satisfy (4.1).

However, in Figure 4.3(b), the Nyquist plot of the open-loop system is not tangential to the sensitivity circle at the flat phase but to another point on the Nyquist curve. Define $[\omega_l, \omega_h]$ the frequency interval corresponding to the flat phase. So, the gain crossover frequency ω_c can be moved within $[\omega_l, \omega_h]$ by adjusting K_p by $K_p' = \beta K_p$ where $\beta \in \left[\frac{\omega_l}{\omega_c}, \frac{\omega_h}{\omega_c}\right]$. For this example, if K_p is changed to $K_p' = 0.7K_p = 0.652$, the flat phase segment will tangentially touch the sensitivity circle. The Nyquist plot of the open-loop system with the modified proposed PID controller, that is, $0.7C_{1p}(s)$, is shown in Figure 4.4(a) and the step responses of the closed-loop system are compared in Figure 4.4(b). Comparing the closed-loop system with the modified proposed PID controller to that with the modified Ziegler-Nichols controller, the overshoots of

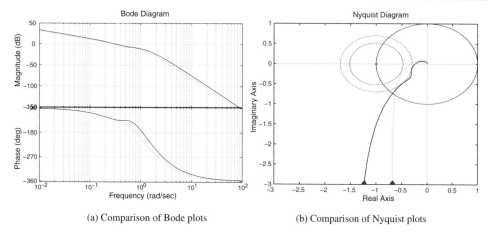

(a) Comparison of Bode plots   (b) Comparison of Nyquist plots

Figure 4.3 Frequency responses of $K_{1p}(s)P_2(s)$ and $K_1(s)P_2(s)$ (Dashed line: The modified Ziegler-Nichols, Solid line: The proposed PID controller)

the step responses from the proposed scheme remain almost invariant under gain variations. However, the overshoots using the modified Ziegler-Nichols controller change remarkably.

4.5.2 Plant with an Integrator $P_5(s)$

For the plant $P_5(s)$, the proposed controller is

$$K_{2p}(s) = 0.33 \left(1 + \frac{1}{6.53s} + 1.89s \right),$$

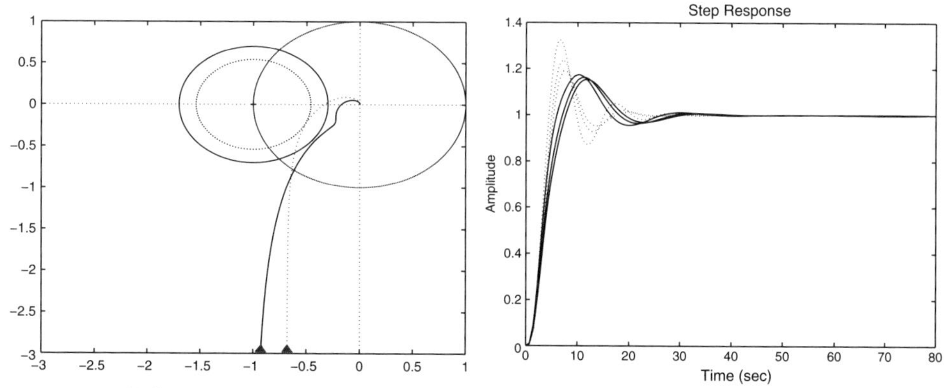

(a) Comparison of Nyquist plots   (b) Comparison of step responses

Figure 4.4 Comparisons of frequency responses and step responses of $0.7K_{1p}(s)P_2(s)$ and $K_1(s)P_2(s)$ (Dashed line: The modified Ziegler-Nichols, Solid line: The proposed PID controller. For both schemes, gain variations 1, 1.1, 1.3 are considered in step responses)

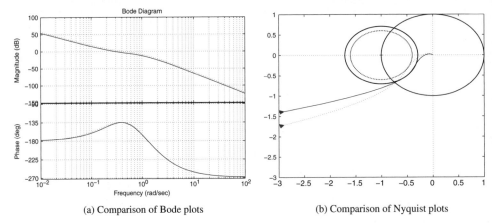

(a) Comparison of Bode plots   (b) Comparison of Nyquist plots

Figure 4.5 Comparisons of frequency responses of $K_{2p}(s)P_5(s)$ and $K_2(s)P_5(s)$ (Dashed line: The modified Ziegler-Nichols, Solid line: The proposed PID controller)

with respect to $\beta = 1$, $\omega_c = 0.4$ *rad/s* and $\Phi_m = 45°$. The controller designed by the modified Ziegler-Nichols method is

$$K_2(s) = 0.528 \left(1 + \frac{1}{7.195s} + 1.799s\right).$$

The Bode plot of this situation, shown in Figure 4.5(a), is quite different from that of plant $P_2(s)$. The flat phase occurs at the peak of the phase Bode plot. The Nyquist diagrams are compared in Figure 4.5(b). The step responses are compared in Figure 4.6 where the proposed controller does not exhibit an obviously better performance than the modified Ziegler-Nichols controller for the iso-damping property because of the effect of the integrator.

4.5.3 Plant with a Time Delay $P_6(s)$

For the plant $P_6(s)$ the proposed controller is

$$K_{3p}(s) = 1.024 \left(1 + \frac{1}{1.241s} + 1.539s\right)$$

with respect to $\beta = 0.7$, $\omega_c = 0.6$ *rad/s* and $\Phi_m = 30°$. The controller designed by the modified Ziegler-Nichols method is

$$K_3(s) = 1.674 \left(1 + \frac{1}{2.57s} + 0.643s\right).$$

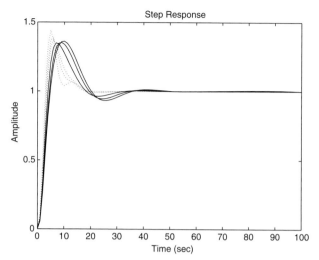

Figure 4.6 Comparison of step responses of $K_{2p}(s)P_5(s)$ and $K_2(s)P_5(s)$ (Solid line: The proposed modified controller with gain variations 1, 0.9, 0.8; Dotted line: The modified Ziegler-Nichols controller with gain variations 1, 0.9, 0.8)

The Bode plots and Nyquist plots are compared in Figure 4.7. The step responses are compared in Figure 4.8 where the iso-damping property can be clearly observed.

4.5.4 Plant with an Integrator and a Time Delay $P_7(s)$

For the plant $P_7(s)$, the proposed controller is

$$K_{4p} = 0.212 \left(1 + \frac{1}{9.52s} + 2.061s\right)$$

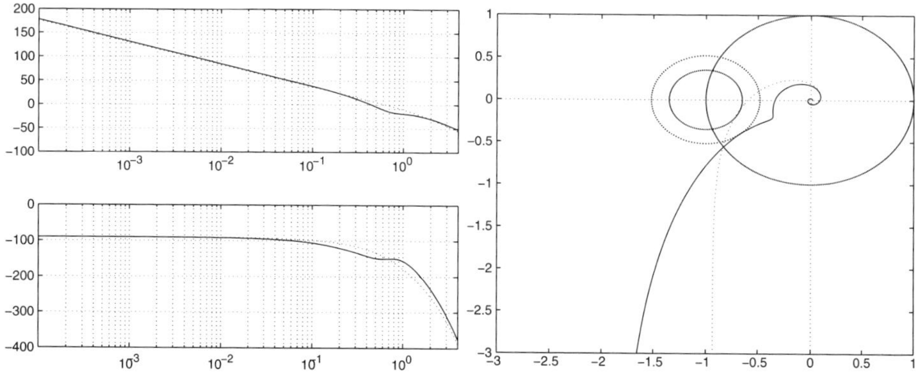

(a) Comparison of Bode plots (b) Comparison of Nyquist plots

Figure 4.7 Comparisons of frequency responses of $K_{3p}(s)P_6(s)$ and $K_3(s)P_6(s)$ (Dashed line: The modified Ziegler-Nichols, Solid line: The proposed PID controller)

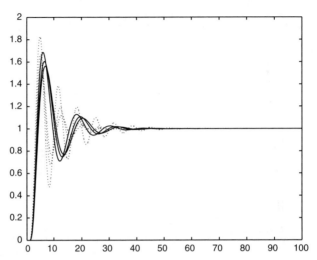

Figure 4.8 Comparison of step responses of $K_{3p}(s)P_6(s)$ and $K_3(s)P_6(s)$ (Solid line: The proposed modified controller with gain variations 1, 1.5, 1.7; Dotted line: The modified Ziegler-Nichols controller with gain variations 1, 1.5, 1.7)

with respect to $\beta = 1$, $\omega_c = 0.25$ *rad/s* and $\Phi_m = 39$ °. The controller designed by the modified Ziegler-Nichols method is

$$K_4 = 0.273 \left(1 + \frac{1}{2.161s} + 8.644s\right).$$

The Bode plots and Nyquist plots are compared in Figure 4.9. The step responses are compared in Figure 4.10 where the iso-damping property can be clearly observed.

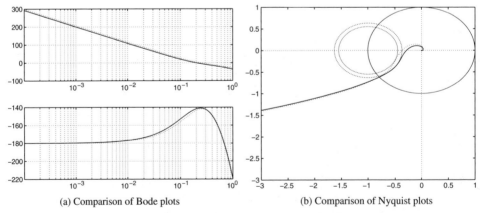

(a) Comparison of Bode plots   (b) Comparison of Nyquist plots

Figure 4.9 Comparisons of frequency responses of $K_{4p}(s)P_7(s)$ and $K_4(s)P_7(s)$ (Dashed line: The modified Ziegler-Nichols, Solid line: The proposed PID controller)

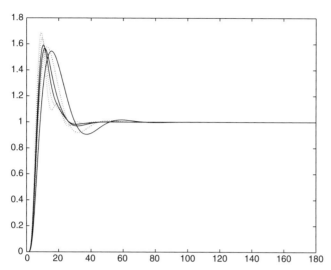

Figure 4.10 Comparison of step responses of $K_{4p}(s)P_7(s)$ and $K_4(s)P_7(s)$ (Solid line: The proposed modified controller with gain variations 1, 1.5, 1.7; Dotted line: The modified Ziegler-Nichols controller with gain variations 1, 1.5, 1.7)

4.6 Chapter Summary

In this chapter, a traditional integer order PID auto-tuning method is proposed for a class of unknown, stable and minimum phase plants. Given the tangent frequency ω_c, the tangent phase Φ_m and with the "flat phase" condition that the phase Bode plot at ω_c is locally flat, we can design the PID controller to ensure that the closed-loop system is robust to gain variations and to ensure that the step responses exhibit an iso-damping property. No plant model is assumed during the PID controller design. Only several relay tests are needed. Simulation examples illustrate the effectiveness and the simplicity of the proposed method for robust PID controller design with an iso-damping property for different types of plants.

5

Auto-Tuning of Fractional Order Controllers with Iso-Damping

In Chapter 4, a auto-tuning rule for integer order PID controller is proposed to give a new relationship between T_i and T_d instead of the equation $T_i = 4T_d$ proposed in the modified Ziegler-Nichols method [85], [89]. It is presented to add an extra condition that the open-loop phase Bode plot at a specified frequency and the point where the sensitivity circle touches the Nyquist curve is locally flat which implies that the system will be robust to gain variations. In this chapter, the fractional order PID controllers are applied to make use of this auto-tuning scheme in Chapter 4 for even better control performance with more flexibility and advanced potential of fractional order PID controllers over traditional integer order PID controllers.

5.1 Introduction

In Chapters 2 and 3 and some other literatures [48], [105], [179], [221], some tuning methods of fractional order controllers for certain class of known and stable plants, for example, velocity systems were studied, and the fractional order proportional integral (FOPI) and fractional order [proportional integral] (FO[PI]) controllers were designed in systematic way [224], [225]. Effectiveness, simplicity and robustness are some of the reasons that have made these tuning methods attractive in the academic and industrial fields. However, for the unknown and stable plants, auto-tuning methods for fractional order controllers design have not been discussed.

In this chapter, two sets of tuning formulae for FOPI and FO[PI] controllers are proposed for a class of unknown and stable plants with iso-damping property [250]. During the design process, we assume that the plant gain and phase at the desired

Fractional Order Motion Controls, First Edition. Ying Luo and YangQuan Chen.
© 2013 John Wiley & Sons, Ltd. Published 2013 by John Wiley & Sons, Ltd.

tangent frequency can be identified by several relay feedback tests in an iterative way. The plant gain and phase are used to estimate the derivatives of amplitude and phase of the plant with respect to frequency at the same tangent frequency point by Bode's integral relationship [48], [106]. Then, these derivatives are applied for FOPI and FO[PI] controller designs. The FOPI and FO[PI] controllers discussed in this chapter have the following transfer functions, respectively:

$$C_1(s) = K_{1p}\left(1 + \frac{K_{1i}}{s^\lambda}\right), \tag{5.1}$$

$$C_2(s) = K_{2p}\left(1 + \frac{K_{2i}}{s}\right)^\gamma. \tag{5.2}$$

where, $\lambda \in (0, 2)$ and $\gamma \in (0, 2)$.

Assuming that the tangent frequency is ω_c, the phase at the frequency ω_c is ϕ_m. Open-loop transfer function is presented as $G(s) = C(s)P(s)$, where $C(s)$ is the controllers which can be $C_1(s)$ for FOPI or $C_2(s)$ for FO[PI], $P(s)$ is the unknown and stable minimum phase plant. Then, for the stability and robustness, three specifications concerned with the phase and gain of the open-loop transfer function $G(s)$ are proposed as follows:

(i) The phase Bode plot is flat at the given tangent frequency ω_c, that is, the phase derivative with respect to the frequency is zero at the tangent frequency,

$$\angle\frac{dG(j\omega)}{d\omega}\bigg|_{\omega=\omega_c} = 0,$$

the above formula can be equivalently expressed as [48],

$$\angle\frac{dG(j\omega)}{d\omega}\bigg|_{\omega=\omega_c} = \angle G(j\omega)|_{\omega=\omega_c}. \tag{5.3}$$

The explanation of this equation can be found in Section 4.1 of Chapter 4.

(ii) The phase specification at the tangent frequency ω_c,

$$\angle G(j\omega_c) = \phi_m - \pi. \tag{5.4}$$

(iii) The gain specification at the tangent frequency ω_c,

$$|G(j\omega_c)| = \cos(\phi_m). \tag{5.5}$$

Following these specifications, we can design the FOPI and FO[PI] controllers to ensure that the closed-loop systems are robust to gain variations and the step responses exhibit iso-damping property for different types of unknown and stable plants.

5.2 FOPI and FO(PI) Controller Design Formulae

In this section, the FOPI and FO[PI] controller designs are presented for unknown and stable minimum phase plants with auto-tuning schemes, respectively.

5.2.1 FOPI Controller Auto-Tuning

From the transfer function (5.1) of FOPI controller, we can obtain its frequency response,

$$
\begin{aligned}
C_1(j\omega) &= K_p(1 + K_i(j\omega)^{-\lambda}) \\
&= K_p\left(1 + K_i\omega^{-\lambda}\cos\left(\lambda\frac{\pi}{2}\right) - jK_i\omega^{-\lambda}\sin\left(\lambda\frac{\pi}{2}\right)\right).
\end{aligned} \tag{5.6}
$$

The phase and gain of $C_1(j\omega)$ are,

$$
|C_1(j\omega)| = K_p\sqrt{(1 + K_i\omega^{-\lambda}\cos(\lambda\pi/2))^2 + (K_i\omega^{-\lambda}\sin(\lambda\pi/2))^2}, \tag{5.7}
$$

$$
\angle C_1(j\omega) = -\arctan\frac{K_i\omega^{-\lambda}\sin(\lambda\pi/2)}{1 + K_i\omega^{-\lambda}\cos(\lambda\pi/2)}. \tag{5.8}
$$

Assume that the transfer function of the plant is $P(s)$, then its frequency response can be expressed by

$$
P(j\omega) = |P(j\omega)|e^{j\angle P(j\omega)}, \tag{5.9}
$$

where, $|P(j\omega)|$ and $\angle P(j\omega)$ are the gain and the phase of the plant, respectively. The open-loop frequency response with FOPI controller is,

$$
G_1(j\omega) = C_1(j\omega)P(j\omega), \tag{5.10}
$$

and the gain and phase of $G_1(j\omega)$ are

$$
|G_1(j\omega)| = |C_1(j\omega)||P(j\omega)|, \tag{5.11}
$$

$$
\angle G_1(j\omega) = \angle C_1(j\omega) + \angle P(j\omega). \tag{5.12}
$$

The derivative of $G_1(j\omega)$ with respect to the frequency is

$$
\frac{dG_1(j\omega)}{d\omega} = P(j\omega)\frac{dC_1(j\omega)}{d\omega} + C_1(j\omega)\frac{dP(j\omega)}{d\omega}. \tag{5.13}
$$

From equation (5.6), we can obtain,

$$
\frac{dC_1(j\omega)}{d\omega} = -\lambda K_p K_i\omega^{(-1-\lambda)}[\cos(\lambda\pi/2) - j\sin(\lambda\pi/2)]. \tag{5.14}
$$

Meanwhile, from (5.9)

$$\ln P(j\omega) = \ln |P(j\omega)| + j\angle P(j\omega), \tag{5.15}$$

$$\frac{d\ln P(j\omega)}{d\omega} = \frac{1}{P(j\omega)} \frac{dP(j\omega)}{d\omega} = \frac{d\ln |P(j\omega)|}{d\omega} + j\frac{d\angle P(j\omega)}{d\omega}, \tag{5.16}$$

$$\frac{dP(j\omega)}{d\omega} = P(j\omega)\left[\frac{d\ln |P(j\omega)|}{d\omega} + j\frac{d\angle P(j\omega)}{d\omega}\right]. \tag{5.17}$$

Combining (5.13), (5.14) and (5.17), we can get,

$$\frac{dG_1(j\omega)}{d\omega} = K_p P(j\omega)\left[-\lambda K_i \omega^{(-1-\lambda)} \cos\left(\frac{\lambda\pi}{2}\right)\right.$$

$$+ \left(1 + \omega^{-\lambda} K_i \cos\left(\frac{\lambda\pi}{2}\right)\right) \frac{d\ln |P(j\omega)|}{d\omega} + \omega^{-\lambda} K_i \sin\left(\frac{\lambda\pi}{2}\right) \frac{d\angle |P(j\omega)|}{d\omega}$$

$$+ j(\lambda K_i \omega^{(-1-\lambda)} \sin\left(\frac{\lambda\pi}{2}\right) - \omega^{-\lambda} K_i \sin\left(\frac{\lambda\pi}{2}\right) \frac{d\ln |P(j\omega)|}{d\omega}$$

$$\left. + \left(1 + \omega^{-\lambda} K_i \cos\left(\frac{\lambda\pi}{2}\right)\right) \frac{d\angle P(j\omega)}{d\omega}\right]. \tag{5.18}$$

According to specification (i) in Section 5.1 and equation (5.81), the slope of the Nyquist curve at tangent frequency ω_c is

$$\angle\frac{dG_1(j\omega)}{d\omega}\bigg|_{\omega_c} = \angle G_1(j\omega)|_{\omega_c} = \angle P(j\omega_c) + \angle C_1(j\omega_c)$$

$$= \angle P(j\omega_c) + \arctan\left(\frac{A}{B}\right). \tag{5.19}$$

Then,

$$\angle C_1(j\omega_c) = \arctan\left(\frac{A}{B}\right) = -\arctan\frac{K_i\omega_c^{-\lambda}\sin(\frac{\lambda\pi}{2})}{1 + K_i\omega_c^{-\lambda}\cos(\frac{\lambda\pi}{2})}, \tag{5.20}$$

where,

$$A = \lambda K_i \omega_c^{(-1-\lambda)} \sin\left(\frac{\lambda\pi}{2}\right) - K_i \omega_c^{(-1-\lambda)} \sin\left(\frac{\lambda\pi}{2}\right) s_a(\omega_c)$$

$$+ \omega_c^{-1} s_p(\omega_c) + K_i \omega_c^{(-1-\lambda)} \cos\left(\frac{\lambda\pi}{2}\right) s_p(\omega_c),$$

$$B = -K_i \omega_c^{(-1-\lambda)} \left(\lambda \cos\left(\frac{\lambda\pi}{2}\right) - \sin\left(\frac{\lambda\pi}{2}\right) s_p(\omega_c)\right)$$

$$+ \left(\omega_c^{-1} + \omega_c^{(-1-\lambda)} K_i \cos\left(\frac{\lambda\pi}{2}\right)\right) s_a(\omega_c),$$

with

$$s_a(\omega_c) = \omega_c \left. \frac{d \ln |P(j\omega)|}{d\omega} \right|_{\omega=\omega_c},$$

$$s_p(\omega_c) = \omega_c \left. \frac{d \angle P(j\omega)}{d\omega} \right|_{\omega=\omega_c}.$$

From (5.20), we have

$$\omega_c^{-2\lambda} s_p(\omega_c) K_i^2 + C K_i + s_p(\omega_c) = 0,$$

where,

$$C = \lambda \omega_c^{-\lambda} \sin\left(\frac{\lambda\pi}{2}\right) + 2\omega_c^{-\lambda} \cos\left(\frac{\lambda\pi}{2}\right) s_p(\omega_c),$$

$$K_i = \frac{-(\lambda \sin(\frac{\lambda\pi}{2}) + 2\cos(\frac{\lambda\pi}{2}) s_p(\omega_c)) + \sqrt{\Delta}}{2\omega_c^{-\lambda} s_p(\omega_c)}, \tag{5.21}$$

with

$$\Delta = \lambda^2 \sin^2\left(\frac{\lambda\pi}{2}\right) + 2\lambda \sin(\lambda\pi) s_p(\omega_c) - 4\sin^2\left(\frac{\lambda\pi}{2}\right) s_p^2(\omega_c). \tag{5.22}$$

The approximation of $s_p(\omega_c)$ for an unknown and stable minimum phase plant can be given as follows [105], [106],

$$s_p(\omega_c) \approx \angle P(j\omega_c) + \frac{2}{\pi} [\ln |K_g| - \ln |P(j\omega_c)|], \tag{5.23}$$

where $|K_g| = P(0)$, which is the static gain of the plant, $\angle P(j\omega_c)$ and $|P(j\omega_c)|$ are the phase and the gain of the plant at the specific frequency ω_c, respectively.

According to specification (ii) in Section 5.1,

$$\angle G_1(j\omega_c) = \angle C_1(j\omega_c) + \angle P(j\omega_c) = \phi_m - \pi,$$
$$\angle C_1(j\omega_c) = \phi_m - \pi - \angle P(j\omega_c).$$

From (5.20)

$$-\arctan \frac{K_i \omega_c^{-\lambda} \sin(\lambda\pi/2)}{1 + K_i \omega_c^{-\lambda} \cos(\lambda\pi/2)} = \phi_m - \pi - \angle P(j\omega_c), \tag{5.24}$$

$$K_i = \frac{-\tan(\phi)}{\omega_c^{-\lambda} (\sin(\lambda\pi/2) + \cos(\lambda\pi/2) \tan(\phi))}, \tag{5.25}$$

with

$$\phi = \phi_m - \pi - \angle P(j\omega_c).$$

According to specification (iii) in Section 5.1,

$$|G_1(j\omega_c)| = |C_1(j\omega_c)||P(j\omega_c)| = \cos(\phi_m)$$
$$= |P(j\omega_c)|K_p\sqrt{1 + 2K_i\omega_c^\lambda \cos(\lambda\pi/2) + \left(K_i\omega_c^{-\lambda}\right)^2}. \tag{5.26}$$

We can get,

$$K_p = \frac{\cos(\phi_m)}{|p(j\omega_c)|\sqrt{1 + 2K_i\omega_c^{-\lambda} \cos(\lambda\pi/2) + \left(K_i\omega_c^{-\lambda}\right)^2}}. \tag{5.27}$$

Therefore, if $s_p(\omega_c)$ is known, K_i, K_p and λ can be calculated from (5.21), (5.27) and (5.25).

5.2.2 FO(PI) Controller Auto-Tuning

The FO[PI] controller $C_2(j\omega)$ has the following phase and gain expressions,

$$\angle[C_2(j\omega)] = -\lambda \arctan\left(\frac{K_i}{\omega}\right), \tag{5.28}$$

$$|C_2(j\omega)| = K_p\left(\sqrt{1 + \frac{K_i^2}{\omega^2}}\right)^\lambda. \tag{5.29}$$

The open-loop frequency response with the FO[PI] controller is,

$$G_2(j\omega) = C_2(j\omega)P(j\omega),$$

the phase and gain of $G_2(j\omega)$ are,

$$|G_2(j\omega_c)| = |C_2(j\omega_c)||P(j\omega_c)|, \tag{5.30}$$
$$\angle G_2(j\omega) = \angle C_2(j\omega) + \angle P(j\omega). \tag{5.31}$$

The derivative of $G_2(j\omega)$ with respect to the frequency is

$$\frac{dG_2(j\omega)}{d\omega} = P(j\omega)\frac{dC_2(j\omega)}{d\omega} + C_2(j\omega)\frac{dP(j\omega)}{d\omega}. \tag{5.32}$$

From equation (5.29), we can get,

$$\frac{dC_2(j\omega)}{d\omega} = j\frac{\lambda K_p K_i}{\omega^2}\left(1 + \frac{K_i}{j\omega}\right)^{\lambda-1}. \tag{5.33}$$

Repeating equation (5.17) in Section 5.2.1

$$\frac{dP(j\omega)}{d\omega} = P(j\omega)\left[\frac{d\ln|P(j\omega)|}{d\omega} + j\frac{d\angle P(j\omega)}{d\omega}\right]. \tag{5.34}$$

Substituting (5.33) and (5.34) to (5.32) gives,

$$\frac{dG_2(j\omega)}{d\omega} = K_p P(j\omega)\left(1 + \frac{K_i}{j\omega}\right)^{\lambda-1}\left[\left(\frac{d\ln|P(j\omega)|}{d\omega} + \frac{K_i}{\omega}\frac{dP(j\omega)}{d\omega}\right)\right.$$

$$\left. + j\left(\frac{K_i\lambda}{\omega^2} + \frac{dP(j\omega)}{d\omega} - \frac{K_i}{\omega}\frac{d\ln|P(j\omega)|}{d\omega}\right)\right]. \tag{5.35}$$

According to specification (i) in Section 5.1 and equation (5.35), the slope of the Nyquist curve at tangent frequency ω_c is

$$\angle\frac{dG_2(j\omega)}{d\omega}\bigg|_{\omega=\omega_c} = \angle G_2(j\omega)|_{\omega=\omega_c} = \angle P(j\omega_c) + \angle C_2(j\omega_c)$$

$$= \angle P(j\omega_c) + (1-\lambda)\arctan\left(\frac{K_i}{\omega_c}\right) + \arctan\left(\frac{A'}{B'}\right)$$

$$= \angle P(j\omega_c) - \lambda\arctan\left(\frac{K_i}{\omega_c}\right), \tag{5.36}$$

where,

$$A' = \frac{\lambda K_i}{\omega_c^2} + \frac{d\angle P(j\omega)}{d\omega}\bigg|_{\omega=\omega_c} - \frac{K_i}{\omega_c}\frac{d\ln|P(j\omega)|}{d\omega}\bigg|_{\omega=\omega_c},$$

$$B' = \frac{d\ln|P(j\omega)|}{d\omega}\bigg|_{\omega=\omega_c} + \frac{K_i}{\omega_c}\frac{\angle P(j\omega)}{d\omega}\bigg|_{\omega=\omega_c}.$$

So we have

$$\arctan\left(\frac{K_i}{\omega_c}\right) = -\arctan\left(\frac{A'}{B'}\right). \tag{5.37}$$

From (5.37), we can get,

$$s_p(\omega_c)K_i^2 + \lambda\omega_c K_i + \omega_c^2 s_p(\omega_c) = 0, \tag{5.38}$$

$$K_i = \frac{-\lambda \omega_c \pm \omega_c \sqrt{\lambda^2 - 4s_p^2(\omega_c)}}{2s_p(\omega_c)}, \tag{5.39}$$

where,

$$s_p(\omega_c) = \omega_c \left. \frac{d\angle P(j\omega)}{d\omega} \right|_{\omega=\omega_c}.$$

According to specification (ii) in Section 5.1,

$$\angle G_2(j\omega_c) = \angle C_2(j\omega_c) + \angle P(j\omega_c) = \phi_m - \pi, \tag{5.40}$$

so,

$$\angle C_2(j\omega_c) = \phi_m - \pi - \angle P(j\omega_c) = -\lambda \arctan\left(\frac{K_i}{\omega_c}\right), \tag{5.41}$$

yields

$$K_i = \omega_c \tan(\varphi), \tag{5.42}$$

where,

$$\varphi = (\pi + \angle P(j\omega_c) - \phi_m)/\lambda.$$

According to specification (iii) in Section 5.1,

$$|G_2(j\omega_c)| = |P(j\omega_c)|K_p \left(\sqrt{1 + \frac{K_i^2}{\omega_c^2}}\right)^{\lambda} = \cos(\phi_m),$$

then, we can get,

$$K_p = \frac{\cos(\phi_m)}{|P(j\omega_c)|\left(\sqrt{1 + K_i^2/\omega_c^2}\right)^{\lambda}}. \tag{5.43}$$

Assuming $s_p(\omega_c)$ is known, K_i, K_p and λ can be solved from (5.39), (5.43) and (5.42).

5.3 Measurements for Auto-Tuning

In this section, the method of measuring $s_p(\omega_c)$ is summarized. Clearly, from (5.23), the static gain $|P(0)|$ or K_g is easy to be measured, and we assume it is known. If we also know $\angle P(j\omega_c)$ and $|P(j\omega_c)|$, $s_p(\omega_c)$ can be calculated. Then, the parameters of the FOPI and FO[PI] controllers can be obtained.

As introduced in Section 4.4 of Chapter 4, the relay feedback test can be used to "measure" $\angle P(j\omega_c)$ and $|P(j\omega_c)|$ [49]. In the relay feedback experiments, a relay is connected in closed-loop with the unknown plant as shown in Figure 4.2. This method is usually used to identify one point on the Nyquist diagram of the plant. In order to change the oscillation frequency of the relay feedback, an artificial time delay can be introduced in the loop. The artificial time delay θ is the tuning knob here for changing the oscillation frequency.

Our problem here becomes how to find the right value of θ which corresponds to the given tangent frequency ω_c. To solve this problem, an iterative method can be used as the summary in Section 4.4.

After a few iterations, the final oscillation frequency becomes quite close to the desired frequency ω_c. Hence, the amplitude and phase of the plant at the specified frequency ω_c can be obtained. Using (5.23), we can calculate the approximation of $s_p(\omega_c)$. Furthermore, the parameters of the FOPI and FO[PI] controllers can be retrieved straightforwardly.

5.4 Simulation Illustration

The auto-tuning methods for FOPI and FO[PI] controllers presented in this chapter will be illustrated via some simulation examples in this section. In the simulation, the following classes of plants, studied as in Section 4.5 of Chapter 4 [48], [223], will be used.

$$P_n(s) = \frac{1}{(s+1)^{(n+3)}}, \quad n = 1, 2, 3, 4, \tag{5.44}$$

$$P_5(s) = \frac{1}{s(s+1)^3}, \tag{5.45}$$

$$P_6(s) = \frac{1}{(s+1)^3}e^{-s}. \tag{5.46}$$

5.4.1 High-Order Plant $P_2(s)$

Consider plant $P_2(s)$ in (5.44), the parameters are given as $\omega_c = 0.2 \ rad/s$ and $\phi_m = 45°$. The FOPI controller transfer function designed using the proposed tuning method is

$$C_{1p_2}(s) = 0.5616(1 + 0.1869/s^{1.3106}). \tag{5.47}$$

The designed FO[PI] controller transfer function is

$$C_{2p_2}(s) = 0.4464(1 + 0.1815/s)^{1.8579}. \tag{5.48}$$

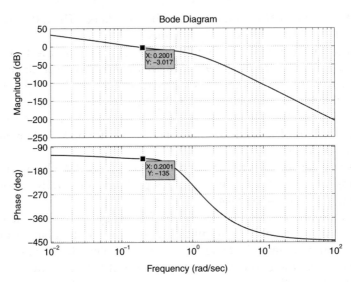

Figure 5.1 Bode plot with FOPI controller for $P_2(s)$

The fractional order operators for FOPI and FO[PI] controllers are also implemented by the impulse response invariant discretization (IRID) method in time domain as introduced in Chapters 2 and 3 [52], [53].

The open-loop Bode plots are shown in Figures 5.1 and 5.2 with the designed two controllers. It can be seen that the phase curve around the frequency $\omega_c = 0.2\,rad/s$ is flat, and the phase margin roughly equals $45°$.

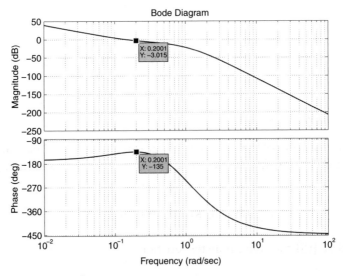

Figure 5.2 Bode plot with FO(PI) controller for $P_2(s)$

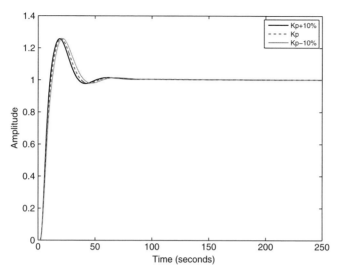

Figure 5.3 Step responses with FOPI controller for $P_2(s)$ under gain variations

The step responses of the closed-loop system with FOPI and FO[PI] controllers for the plant $P_2(s)$ are shown in Figures 5.3 and 5.4. We can see that the FOPI and FO[PI] controllers designed by the proposed method in this chapter are effective. The overshoots of the step responses remain almost constant under gain variations, which means that the system is robust to gain changes.

Figure 5.5 is the comparison of the step responses of two closed-loop systems with FOPI and FO[PI] controllers for the plant $P_2(s)$. It can be seen that the overshoot of

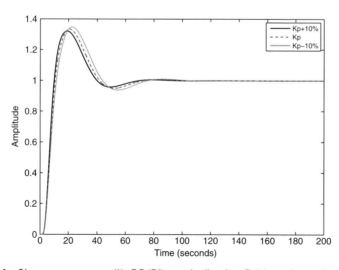

Figure 5.4 Step responses with FO(PI) controller for $P_2(s)$ under gain variations

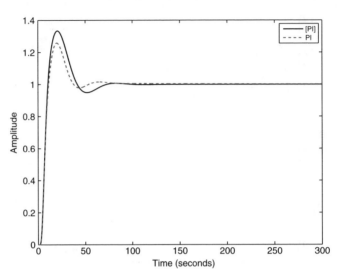

Figure 5.5 Comparison of the step responses with FOPI and FO(PI) controllers for $P_2(s)$

the unit step response using the designed FOPI controller is shorter than that using the designed FO[PI] controller.

5.4.2 Plant with an Integrator $P_5(s)$

For the plant $P_5(s)$, the design parameters are given as $\omega_c = 0.15 \; rad/s$ and $\phi_m = 30°$. The FOPI controller designed by the proposed tuning method is

$$C_{1p_5}(s) = 0.1057(1 + 0.1172/s^{0.9569}).\tag{5.49}$$

The designed FO[PI] controller is

$$C_{2p_5}(s) = 0.1070(1 + 0.1248/s)^{0.8652}.\tag{5.50}$$

The open-loop Bode plots are shown in Figures 5.6 and 5.7 with the designed two controllers. It can be seen that the phase curve around the frequency $\omega_c = 0.15 \; rad/s$ is flat. The phase margin roughly equals 30°.

The step responses of the closed-loop system with FOPI and FO[PI] controllers for the plant $P_5(s)$ are shown in Figures 5.8 and 5.9. The overshoots of the step responses using the designed FOPI and FO[PI] controllers remain almost constant under gain variation, which means the system is robust to gain changes.

Figure 5.10 is the comparison of the step responses of two closed-loop systems with FOPI and FO[PI] controllers for the plant $P_5(s)$. The overshoots of the unit step response using the FOPI and FO[PI] controllers are nearly the same with the designed parameters.

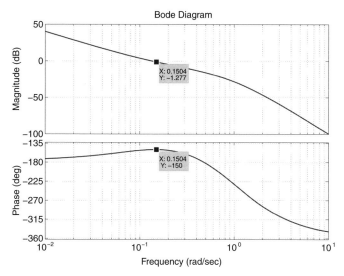

Figure 5.6   Bode plot with FOPI controller for $P_5(s)$

5.4.3 Plant with a Time Delay $P_6(s)$

For the plant $P_6(s)$, the design parameters are set as $\omega_c = 0.3\ rad/s$ and $\phi_m = 30°$. Using the proposed tuning method in this chapter, the FOPI controller can be designed as

$$C_{1p_6}(s) = 0.7870(1 + 0.2852/s^{1.3789}).\tag{5.51}$$

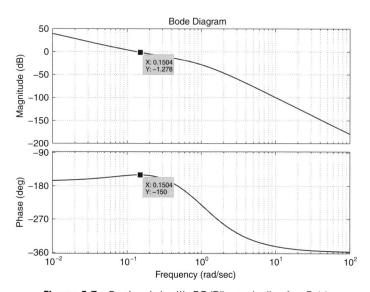

Figure 5.7   Bode plot with FO(PI) controller for $P_5(s)$

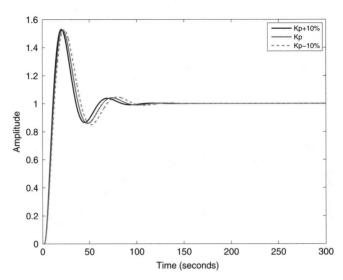

Figure 5.8 Step responses with FOPI controller for $P_5(s)$ under gain variations

The FO[PI] controller can be designed as

$$C_{2p_6}(s) = 0.4298(1 + 0.4242/s)^{1.5113}. \tag{5.52}$$

The open-loop Bode plots are shown in Figures 5.11 and 5.12 with two designed controllers. The phase curve near the frequency $\omega_c = 0.3 \ rad/s$ is flat, and the phase margin roughly equals $30°$.

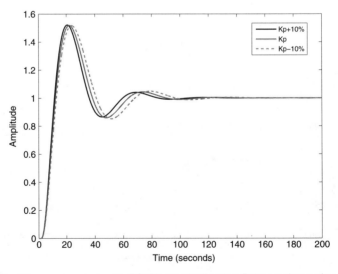

Figure 5.9 Step responses with FO(PI) controller for $P_5(s)$ under gain variations

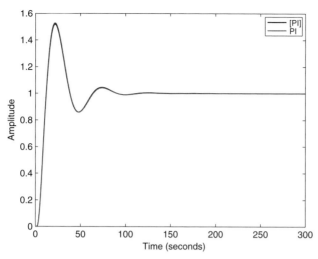

Figure 5.10 Comparison of the step responses with FOPI and FO(PI) controllers for $P_5(s)$

The step responses of the closed-loop system with FOPI and FO[PI] controllers for $P_6(s)$ are shown in Figures 5.13 and 5.14. The overshoots of the step responses remain almost constant under gain variations, which means that the system is also robust to gain changes.

Figure 5.15 is the comparison of the step responses of two closed-loop systems with FOPI and FO[PI] controllers for the plant $P_6(s)$. The overshoot of the step response using the designed FOPI controller is smaller than that using the designed FO[PI] controller.

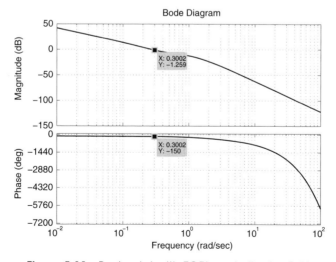

Figure 5.11 Bode plot with FOPI controller for $P_6(s)$

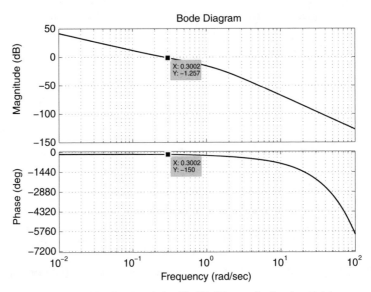

Figure 5.12   Bode plot with FO(PI) controller for $P_6(s)$

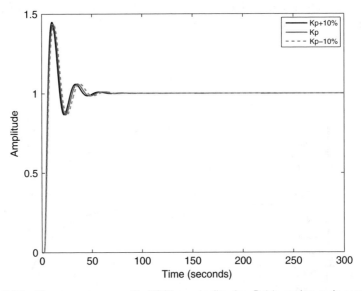

Figure 5.13   Step responses with FOPI controller for $P_6(s)$ under gain variations

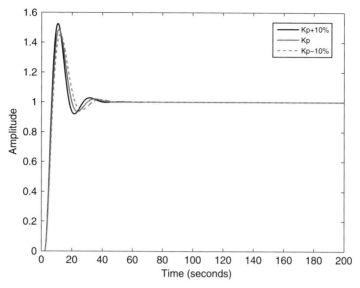

Figure 5.14 Step responses with FO(PI) controller for $P_6(s)$ under gain variations

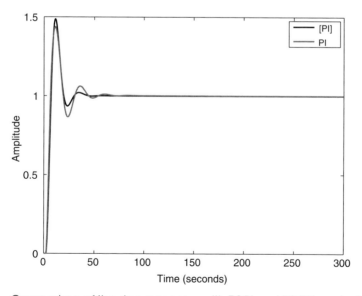

Figure 5.15 Comparison of the step responses with FOPI and FO(PI) controllers for $P_6(s)$

5.5 Chapter Summary

Two sets of auto-tuning formulae for FOPI and FO[PI] controllers are derived for unknown and stable minimum phase plants with iso-damping property in this chapter. The FOPI and FO[PI] controllers designed by the proposed methods ensure that the phase Bode plot is flat around a given frequency called the "tangent frequency," so that the closed-loop system is robust to gain variations and the unit step responses exhibit iso-damping property. In the design process, no plant models are assumed; and several relay feedback tests can be used to identify the plant gain and phase at the tangent frequency in an iterative way. Simulation results are presented to validate the proposed auto-tuning methods for robust FOPI and FO[PI] controllers with iso-damping property for different types of plants.

Part III
Fractional Order Position Controls

Part III

6

Fractional Order PD Controller Tuning for Position Systems

6.1 Introduction

From Chapter 6 to Chapter 9, fractional order position controls will be discussed, and the advantages of fractional order controllers will be presented. In this chapter, let us focus on a fractional order proportional derivative (FOPD) controller for the typical second order position control systems. A practical and simple tuning scheme for the FOPD controller is proposed. Simulation and experimental results show that the closed-loop system can achieve favorable dynamic performance and robustness [119].

The fractional order proportional derivative (PD) controller has the following form of transfer function:

$$C(s) = K_p(1 + K_d s^\mu),$$ (6.1)

where $\mu \in (0, 2)$. Clearly, this is a specific form of the most common $PI^\lambda D^\mu$ controller which involves an integrator of order λ ($\lambda = 0$, in this chapter) and a differentiator of order μ. The typical second-order plant discussed here is:

$$P(s) = \frac{1}{s(Ts + 1)},$$ (6.2)

which can approximately model a DC motor position servo system. Note that the plant gain is normalized to 1 without loss of generality since the proportional factor in the transfer function (6.2) can be incorporated in the gain of the controller.

Fractional Order Motion Controls, First Edition. Ying Luo and YangQuan Chen.
© 2013 John Wiley & Sons, Ltd. Published 2013 by John Wiley & Sons, Ltd.

6.2 Fractional Order PD Controller Design for Position Systems

Let us restrict our attention to position control systems $P(s)$ described by (6.2). The transfer function of FOPD controller has the form of (6.1). The phase and gain of the plant in frequency domain can be given from (6.2) by,

$$\angle[P(j\omega)] = -\tan^{-1}(\omega T) - \frac{\pi}{2}, \tag{6.3}$$

$$|P(j\omega)| = \frac{1}{\omega\sqrt{1 + (\omega T)^2}}. \tag{6.4}$$

The FOPD controller described by (6.1) can be written as,

$$C(j\omega) = K_p(1 + K_d(j\omega)^\mu)$$

$$= K_p\left[\left(1 + K_d\omega^\mu \cos\frac{\mu\pi}{2}\right) + jK_d\omega^\mu \sin\frac{\mu\pi}{2}\right]. \tag{6.5}$$

The phase and gain are,

$$\angle[C(j\omega)] = \tan^{-1}\frac{\sin\frac{(1-\mu)\pi}{2} + K_d\omega^\mu}{\cos\frac{(1-\mu)\pi}{2}} - \frac{(1-\mu)\pi}{2}, \tag{6.6}$$

$$|C(j\omega)| = K_p\sqrt{\left(1 + K_d\omega^\mu \cos\frac{\mu\pi}{2}\right)^2 + \left(K_d\omega^\mu \sin\frac{\mu\pi}{2}\right)^2}. \tag{6.7}$$

The open-loop transfer function $G(s)$ is,

$$G(s) = C(s)P(s). \tag{6.8}$$

From (6.3) and (6.6), the phase of $G(s)$ is,

$$\angle[G(j\omega)] = \tan^{-1}\frac{\sin\frac{(1-\mu)\pi}{2} + K_d\omega^\mu}{\cos\frac{(1-\mu)\pi}{2}} + \frac{\mu\pi}{2} - \pi - \tan^{-1}(\omega T). \tag{6.9}$$

Here, the three specifications in Section 2.3 are applied to design the FOPD controller in this chapter.

6.2.1 Integer Order PD Controller Design

The design of the general PD controller with $\mu = 1$ in (6.1) for position control systems is analyzed in this section.

From (6.9) and according to specification (ii) in Section 2.3,

$$\frac{d[\angle(G(j\omega))]}{d\omega}\Bigg|_{\omega=\omega_c} = \frac{K_d}{1 + (K_d\omega_c)^2} - \frac{T}{1 + (T\omega_c)^2} = 0,$$

we arrive at

$$K_d = \frac{1}{T\omega_c^2},$$

and

$$\angle[G(j\omega)]|_{\omega=\omega_c} = \tan^{-1}\left(\frac{1}{T\omega_c}\right) - \frac{\pi}{2} - \tan^{-1}(\omega_c T).$$

Given a gain crossover frequency ω_c, the phase margin is fixed, which means the specifications (i) and (ii) in Section 2.3 cannot be satisfied simultaneously except for only one given phase margin for traditional integer order PD controller.

6.2.2 Fractional Order PD Controller Design

According to specification (i) in Section 2.3, the phase of $G(s)$ can be expressed as,

$$\angle[G(j\omega_c)] = \tan^{-1}\frac{\sin\frac{(1-\mu)\pi}{2} + K_d\omega_c^\mu}{\cos\frac{(1-\mu)\pi}{2}} + \frac{\mu\pi}{2} - \pi - \tan^{-1}(\omega_c T)$$

$$= -\pi + \phi_m$$

$$= \phi. \tag{6.10}$$

From (6.10), the relationship between K_d and μ can be established as follows:

$$K_d = \frac{1}{\omega_c^\mu}\tan\left[\phi + \tan^{-1}(\omega_c T) - \frac{\mu\pi}{2} + \pi\right]\cos\frac{(1-\mu)\pi}{2}$$

$$- \frac{1}{\omega_c^\mu}\sin\frac{(1-\mu)\pi}{2}. \tag{6.11}$$

According to specification (iii) in Section 2.3 about the robustness to loop gain variations in the plant,

$$\frac{d(\angle[G(j\omega)])}{d\omega}\Bigg|_{\omega=\omega_c}$$

$$= \frac{\mu K_d\omega_c^{\mu-1}\cos\frac{(1-\mu)\pi}{2}}{\cos^2\frac{(1-\mu)\pi}{2} + \left(\sin\frac{(1-\mu)\pi}{2} + K_d\omega_c^\mu\right)^2} - \frac{T}{1 + (T\omega_c)^2}$$

$$= 0. \tag{6.12}$$

From (6.12), we can establish an equation for K_d,

$$A\omega_c^{2\mu} K_d^2 + B K_d + A = 0, \tag{6.13}$$

that is

$$K_d = \frac{-B \pm \sqrt{B^2 - 4A^2\omega_c^{2\mu}}}{2A\omega_c^{2\mu}}, \tag{6.14}$$

where

$$A = \frac{T}{1 + (\omega_c T)^2},$$

$$B = 2A\omega_c^{\mu} \sin \frac{(1 - \mu)\pi}{2} - \mu\omega_c^{\mu-1} \cos \frac{(1 - \mu)\pi}{2}.$$

According to specification (ii) in Section 2.3, we can establish an equation for K_p:

$$|G(j\omega_c)|$$

$$= |C(j\omega_c)||P(j\omega_c)|$$

$$= \frac{K_p\sqrt{\left(1 + K_d\omega_c^{\mu} \cos \frac{\mu\pi}{2}\right)^2 + \left(K_d\omega_c^{\mu} \sin \frac{\mu\pi}{2}\right)^2}}{\omega_c\sqrt{1 + (\omega_c T)^2}}$$

$$= 1. \tag{6.15}$$

Clearly, we can solve equations (6.11), (6.14) and (6.15) to obtain μ, K_d and K_p.

6.3 Design Procedures

It can be observed from (6.11) and (6.14) that μ, K_d can be obtained jointly. The graphical method can be used as a practical and simple way to find μ and K_d because of the plain forms for (6.11) and (6.14). The procedures to tune the fractional order PD controller are as follows:

(1) Given ω_c, the gain crossover frequency.
(2) Given Φ_m, the desired phase margin.
(3) Plot the curve 1, K_d with respect to μ, according to (6.11).
(4) Plot the curve 2, K_d with respect to μ, according to (6.14).
(5) Obtain the μ and K_d from the intersection point on the above two curves.
(6) Calculate the K_p from (6.15).

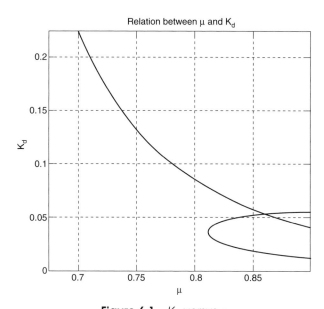

Figure 6.1 K_d versus μ

Remark 6.3.1 *Design specifications should not be chosen that are too aggressive because of the constraint in (6.14) and the existence of the intersection point needed on the two curves above.*

6.4 Simulation Illustration

The fractional order PD controller design method presented in Section 6.3 is illustrated via a numerical simulation. In the simulation, the plant parameter T in (6.2) is 0.05s. The specifications of interest are set as $\omega_c = 60\ rad/s$, $\Phi_m = 70°$, and the robustness to loop gain variations in the plant is required.

According to (6.11) and (6.14), two curves are plotted easily in Figure 6.1. One can obtain the μ and K_d obviously from the intersection point on the two curves, that is $\mu = 0.86$ and $K_d = 0.053$. Then K_p can be calculated from (6.15) easily, that is $K_p = 84.89$.

Actually, the fractional order PD controller itself is an infinite dimensional linear filter due to the fractional order differentiator s^μ. A band-limit implementation is important in practice. Finite dimensional approximation of the fractional order operator s^μ should be utilized in a proper range of frequency of practical interest. The approximation method used here is the Oustaloup Recursive Algorithm [174]. Assuming the frequency range to fit is selected as (ω_b, ω_h). The approximate transfer function of a continuous filter for s^μ with the Oustaloup Algorithm is as follows:

$$G_f(s) = K \prod_{k=-N}^{N} \frac{s + \omega_k'}{s + \omega_k},$$

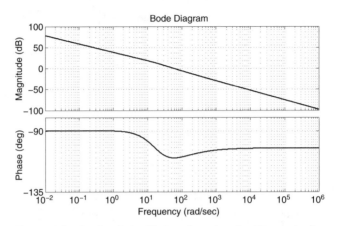

Figure 6.2 Bode plot with fractional order PD controller

where the zeros, poles and the gain can be evaluated from

$$\omega_k' = \omega_b \left(\frac{\omega_h}{\omega_b} \right)^{\frac{k+N+\frac{1}{2}(1-\mu)}{2N+1}},$$

$$\omega_k = \omega_b \left(\frac{\omega_h}{\omega_b} \right)^{\frac{k+N+\frac{1}{2}(1+\mu)}{2N+1}},$$

$$K = \left(\frac{\omega_h}{\omega_b} \right)^{-\frac{\mu}{2}} \prod_{k=-N}^{N} \frac{\omega_k}{\omega_k'}.$$

In our simulation, for the approximation of the fractional order differentiator s^μ, the frequency range of practical interest is set to be from 0.0001Hz to 10000Hz. The Bode plots of system designed are shown in Figure 6.2. As can be seen, the gain crossover frequency specification, $\omega_c = 60 \, rad/s$, and phase margin specification, $\Phi_m = 70°$, are fulfilled. The phase is forced to be flat at ω_c.

In order to compare the integer order controller and our designed fractional order controller fairly, two simulation cases are presented below.

6.4.1 Step Response Comparison

It is well known that a proportional controller is commonly adopted for the typical second-order plant discussed in this chapter, and the ITAE optimum proportional controller parameter is $K = 1/(2T)$ [68]. Therefore, the proportional parameter is set to 10 in this example if the commonly used proportional controller is used.

In Figure 6.3, applying the ITAE optimal P controller, the unit step responses are plotted with the open-loop gain varying from 8 to 12 (±20% variations from the

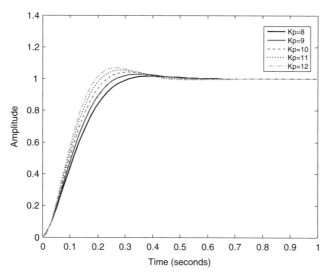

Figure 6.3 Simulation. Step responses with the ITAE optimum proportional controller

desired value 10). In Figure 6.4, applying the fractional order PD controller, the unit step responses are plotted with open-loop gains changing from 67.9 to 101.8 (±20% variations from desired value 84.89).

It can be seen from Figure 6.3 and Figure 6.4 that the fractional order PD controller designed by the proposed method in this chapter is effective. With the designed FOPD controller, faster responses are achieved and, meanwhile, the overshoots of the step responses remain almost constant under gain variations, that is, an iso-damping property is exhibited, that means the system is more robust to gain changes.

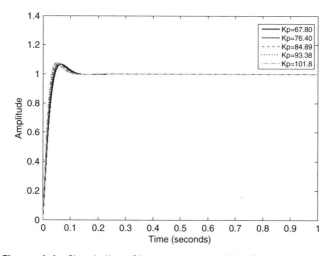

Figure 6.4 Simulation. Step responses with FOPD controller

6.4.2 Ramp Response Comparison

In this simulation case, we use the unit ramp input response to compare the integer order PI controller with the fractional order PD controller, the ITAE optimum parameters of integer order PI controller are designed [68],

$$C_{PI}(s) = K_p(1 + K_i s), \tag{6.16}$$

where $K_p = 21.2245$ and $K_i = 74.6356$.

In Figure 6.5, the ITAE optimum PI controller is applied, the unit ramp responses are plotted with open-loop gain varying from 16.9796 to 25.4694 ($\pm 20\%$ variations from the desired value 21.2245). In Figure 6.6, the fractional order PD controller is applied and the unit ramp responses are plotted with open-loop gains changing from 67.8 to 101.8 ($\pm 20\%$ variations from the desired value 84.89).

From Figure 6.5 and Figure 6.6, it is obvious that, with the designed FOPD controller using the proposed tuning rule, the overshoots are almost constant under gain variations and are much lower than those with the ITAE optimal PI controller.

6.5 Experimental Validation

6.5.1 Introduction of the Experimental Platform

A fractional horsepower dynamometer was developed as a general purpose experiment platform [210]. The architecture of the dynamometer control system is shown in

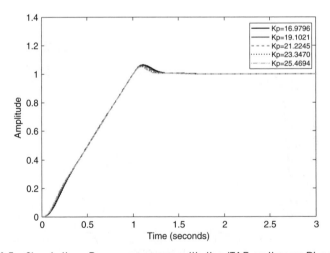

Figure 6.5   Simulation. Ramp response with the ITAE optimum PI controller

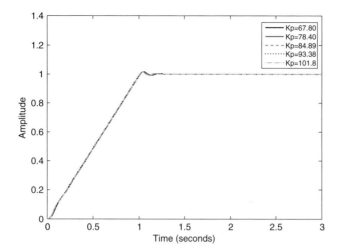

Figure 6.6 Simulation. Ramp responses with FOPD controller

Figure 3.10. The setup of this experimental platform is a hardware-in-the-loop (HIL) control system introduced in Section 3.6.1.

Through simple system identification process, the position control system of the dynamometer using the hardware-in-the-loop setup can be approximately modeled by a transfer function $\frac{1.52}{s(0.4s+1)}$.

6.5.2 Experimental Model Simulation

Since we already have experimentally modeled the position control system with a transfer function $\frac{1.52}{s(0.4s+1)}$, so we can test the simulation effect in Simulink first, then the simulation results can be compared with the real-time experiments on the dynamometer. Thus the verification of our proposed method is more effective.

For the fractional order PD controller, the gain crossover frequency is set as $\omega_c = 10 \ rad/s$. Correspondingly, for an approximation of the fractional order differentiator s^μ, the frequency range of practical interest is set to be from 1Hz to 100Hz and the desired phase margin is set as $\Phi_m = 70°$. Moreover, the robustness to loop gain variations is required. According to the numerical method in Section 6.3, we can find that $\mu = 0.844$, $K_d = 0.368$ and $K_p = 13.860$.

First, the unit step responses are tested to compare the ITAE optimum P controller with the FOPD controller, the ITAE optimum P parameter is $K_p = \frac{1}{2T} = 1.25$ [68]. In Figure 6.7, applying the ITAE optimal P controller, the unit step responses are plotted with open-loop gain varying from 1 to 1.5 ($\pm 20\%$ variations from the desired value 1.25). In Figure 6.8, applying fractional order PD controller, the unit step responses are plotted with open-loop gains varying from 11.088 to 16.632 ($\pm 20\%$ variations from the desired value 13.86).

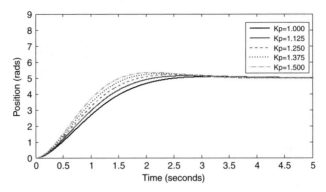

Figure 6.7 Dynamometer simulation model. Step position responses with the ITAE optimal proportional controller

From Figure 6.7 and Figure 6.8, it is obvious that, with the designed FOPD controller, faster unit step responses are achieved, and the overshoots remain almost constant under gain variations and are much lower than that with the optimal P controller. It is demonstrated that the controlled system using FOPD controller is more robust to gain changes in the loop.

Next, the unit ramp responses are tested to compare the ITAE optimum PI controller with the FOPD controller. The ITAE optimum PI parameters are designed as $K_p = 2.6531$ and $K_i = 1.1662$ [68]. In Figure 6.9, applying the ITAE optimal PI controller, the unit ramp responses are plotted with open-loop gain varying from 1.8531 to 3.1837 ($\pm 20\%$ variations from the desired value 2.6531). In Figure 6.8, applying fractional order PD controller, the unit ramp responses are plotted with open-loop gains changing from 11.088 to 16.632 ($\pm 20\%$ variations from the desired value 13.86).

It can be seen from Figure 6.9 and Figure 6.10 that the fractional order PD controller designed by the proposed tuning method is effective.

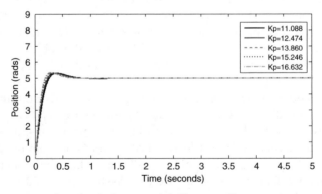

Figure 6.8 Dynamometer simulation model. Step position responses with FOPD controller

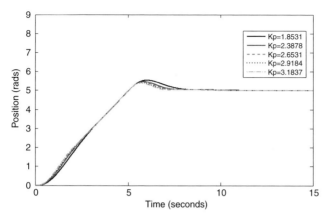

Figure 6.9 Dynamometer simulation model. Ramp position responses with the ITAE optimal PI controller

6.5.3 Experiments on the Dynamometer

Substituting the real dynamometer platform for the DC motor simulation model, the proposed FOPD controller is tested in the hardware-in-the-loop manner.

Figures 6.11 and 6.13 show the ITAE optimal P controller for unit step position responses and the ITAE optimal PI controller for the unit ramp position responses, respectively, Figures 6.12 and 6.14 present the fractional order PD controller designed by the proposed tuning method in this chapter for the unit step and ramp position responses, respectively. It is obvious that, with the designed FOPD controller, faster responses are achieved and the overshoots remain almost constant under gain variations. The overshoots are much lower than those with the ITAE optimal P or PI controllers.

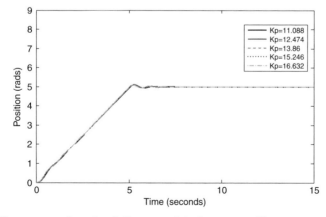

Figure 6.10 Dynamometer simulation model. Ramp position responses with FOPD controller

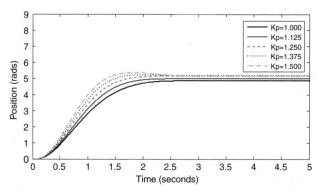

Figure 6.11 Dynamometer real-time experiment. Step position responses with ITAE optimal proportional controller

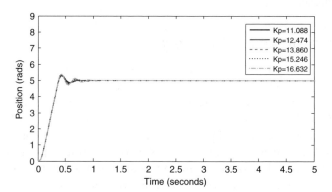

Figure 6.12 Dynamometer real-time experiment. Step position responses with FOPD controller (For a color version of this figure, see Plate 5)

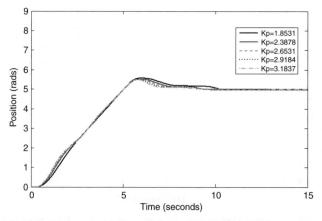

Figure 6.13 Dynamometer real-time experiment. Ramp position responses with ITAE optimal PI controller (For a color version of this figure, see Plate 6)

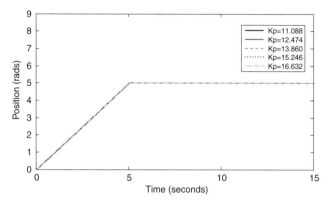

Figure 6.14 Dynamometer real-time experiment. Ramp position response with FOPD controller (For a color version of this figure, see Plate 7)

6.6 Chapter Summary

In this chapter, we have presented a tuning method for fractional order proportional and derivative controller for typical position servo plants. The FOPD controller is tuned to ensure that the given gain crossover frequency and the phase margin are achieved and the open-loop phase Bode plot is locally flat around the gain crossover frequency, so that the closed-loop system is robust to gain variations and the step/ramp response exhibits an iso-damping property. The FOPD controller tuning method proposed, aimed at typical second-order position servo plants, is simple and practical. Simulation and experimental results show that the closed-loop system with the designed fractional order PD controller can achieve favorable dynamic performance and robustness.

7

Fractional Order (PD) Controller Synthesis for Position Systems

7.1 Introduction

As presented in Chapter 6, the controller making use of fractional order derivative and integral can achieve better performances and more robust results over the conventional integer order controllers [119]. In this chapter, a fractional order [proportional derivative] (FO[PD]) controller is proposed for position control systems. We also developed a practical and systematic tuning scheme for the proposed FO[PD] controller. For the first time, other controllers such as the traditional integer order PID (IOPID) controller and the fractional order proportional (FOPD) [119] controller are compared with the proposed FO[PD] controller fairly, under the same number of design parameters and the same specifications for position systems [137]. Side-to-side fair comparisons of three controllers (that is, IOPID, FOPD and FO[PD]) via both simulation and experimental tests have revealed that both the designed FOPD and FO[PD] controllers are always stabilizing to achieve "flat phase" specification, but the designed IOPID controller may not always be stabilizing; both the designed FOPD and FO[PD] controllers outperform the designed IOPID controller; when the time constant of the position servo plant increases, the designed FO[PD] controller outperforms the designed FOPD controller.

Fractional Order Motion Controls, First Edition. Ying Luo and YangQuan Chen.
© 2013 John Wiley & Sons, Ltd. Published 2013 by John Wiley & Sons, Ltd.

7.2 Position Systems and Design Specifications

The position control model discussed in this chapter is shown in (6.2) as follows:

$$P(s) = \frac{1}{s(Ts + 1)}.$$

where T is the time constant. Note that, the plant gain is normalized to 1 without loss of generality since the DC gain of the considered system can be incorporated in the gain of the controller.

The traditional IOPID controller has the following form of transfer function in this chapter,

$$C_1(s) = K_{p1}\left(1 + \frac{K_{i1}}{s} + K_{d1}s\right), \tag{7.1}$$

with three parameters K_{p1}, K_{i1} and K_{d1}.

The FOPD controller has the following form,

$$C_2(s) = K_{p2}\left(1 + K_{d2}s^\lambda\right), \tag{7.2}$$

with three parameters K_{p2}, K_{d2} and λ, where $\lambda \in (0, 2)$. Clearly, this is a specific form of the most common $PI^\gamma D^\lambda$ controller [189] which involves an integrator of order γ ($\gamma = 0$, in this chapter) and a differentiator of order λ.

The proposed FO[PD] controller in this chapter is defined below,

$$C_3(s) = K_{p3}[1 + K_{d3}s]^\mu, \tag{7.3}$$

with three parameters K_{p3}, K_{d3} and μ, where $\mu \in (0, 2)$.

With the position control system model $P(s)$ in (6.2) and the generalized form $C(s)$ of the three controllers in (7.1)–(7.3), the open-loop transfer function $G(s)$ has the form below,

$$G(s) = C(s)P(s). \tag{7.4}$$

Here, three specifications in Section 2.3 are applied to the design of the above three controllers with the iso-damping property [137].

7.3 Fractional Order (PD) Controller Design

In this section, the position system model $P(s)$ described in (6.2) is considered. The proposed FO[PD] controller has the transfer function form in (7.3). The phase and gain of the plant in frequency domain can be given as,

$$\angle[P(j\omega)] = -\tan^{-1}(\omega T) - \frac{\pi}{2}, \tag{7.5}$$

$$|P(j\omega)| = \frac{1}{\omega\sqrt{1+(\omega T)^2}}. \tag{7.6}$$

The FO[PD] controller described in (7.3) can be written as,

$$C_3(j\omega) = K_{p3}[1 + K_{d3}(j\omega)]^\mu, \tag{7.7}$$

the phase and gain of this controller are,

$$\angle[C_3(j\omega)] = \mu\tan^{-1}(\omega K_{d3}), \tag{7.8}$$

$$|C_3(j\omega)| = K_{p3}[1 + (K_{d3}\omega)^2]^{\frac{\mu}{2}}. \tag{7.9}$$

The open-loop transfer function $G_3(s)$ is

$$G_3(s) = C_3(s)P(s). \tag{7.10}$$

From (7.5) and (7.8), we can find the phase of $G_3(s)$,

$$\angle[G_3(j\omega)] = \mu\tan^{-1}(\omega K_{d3}) - \tan^{-1}(\omega T) - \frac{\pi}{2}. \tag{7.11}$$

According to specification (i) in Section 2.3, the phase of $G_3(s)$ can be expressed as,

$$\angle[G_3(j\omega)]|_{\omega=\omega_c} = \mu\tan^{-1}(\omega_c K_{d3}) - \tan^{-1}(\omega_c T) - \frac{\pi}{2}$$

$$= -\pi + \phi_m. \tag{7.12}$$

From (7.12), we can establish the first relationship between K_{d3} and μ,

$$K_{d3} = \frac{1}{\omega_c}\tan\left(\frac{1}{\mu}\left(\Phi_m - \frac{\pi}{2} + \tan^{-1}(T\omega_c)\right)\right). \tag{7.13}$$

According to specification (iii) in Section 2.3 for the robustness to the loop gain variations,

$$\left.\frac{d(\angle(G_3(j\omega)))}{d\omega}\right|_{\omega=\omega_c} = \frac{\mu K_{d3}}{1 + (K_{d3}\omega_c)^2} - \frac{T}{1 + (T\omega_c)^2}$$

$$= 0, \tag{7.14}$$

the second relationship between K_{d3} and μ can be obtained in the following form,

$$A_3\omega_c^2 K_{d3}^2 - \mu K_{d3} + A_3 = 0, \tag{7.15}$$

that is

$$K_{d3} = \frac{\mu \pm \sqrt{\mu^2 - 4A_3\omega_c^2}}{2A_3\omega_c^2}, \tag{7.16}$$

where

$$A_3 = \frac{T}{1 + (T\omega_c)^2}.$$

According to specification (ii) in Section 2.3, we can establish an equation for K_{p3}, K_{d3} and μ,

$$|G_3(j\omega_c)| = |C_3(j\omega_c)||P(j\omega_c)|$$
$$= K_{p3}\frac{(1 + (K_{d3}\omega_c)^2)^{\frac{\mu}{2}}}{\sqrt{(T\omega_c^2)^2 + \omega_c^2}}$$
$$= 1. \tag{7.17}$$

Clearly, we can solve equations (7.13), (7.16) and (7.17) to find K_{p3}, K_{d3} and μ theoretically.

It can be observed from (7.13) and (7.16) that K_{d3} and μ can be obtained jointly in theory. The graphical method can be used as a practical and simple way to find K_{d3} and μ. The procedure to tune the parameters of the FO[PD] controller is as follows:

(1) Given ω_c, the gain crossover frequency.
(2) Given Φ_m, the desired phase margin.
(3) Plot the curve 1, K_{d3} with respect to μ, according to (7.13).
(4) Plot the curve 2, K_{d3} with respect to μ, according to (7.16).
(5) Obtain the K_{d3} and μ from the intersection point on the above two curves.
(6) Calculate the K_{p3} from (7.17).

The theoretical deductions and procedure summaries for IOPID and FOPD controller designs can follow the above process of FO[PD] controller design.

7.4 Controller Design Examples and Bode Plot Validations

7.4.1 FO(PD) Controller Design

The time constant T in (6.2) is chosen as 0.4s, and the specifications of interest are set as $\omega_c = 10\,rad/s$, $\Phi_m = 70°$. According to (7.13) and (7.16), two curves are plotted. The designed values of K_{d3} and μ can be obtained from the intersection point on the two curves, $K_{d3} = 0.2991$ and $\mu = 0.7825$. Then K_{p3} can be calculated from (7.17), that is $K_{p3} = 16.7839$. The Bode plot of the designed open-loop system is shown in Figure 7.1. As can be seen, the gain crossover frequency specification, $\omega_c = 10\,rad/s$, the phase margin specification, $\Phi_m = 70°$, are fulfilled, and the phase is forced to be flat around ω_c.

7.4.2 FOPD Controller Design

The time constant T in (6.2) is also chosen as 0.4s, and the specifications of interest are set as $\omega_c = 10\,rad/s$, $\Phi_m = 70°$. Following the design scheme of FO[PD] controller, the FOPD controller design process can be deduced similarly, or Section 6.2 can be referred. We can obtain the K_{d2} and λ from the intersection point on two curves in terms of K_{d2} and λ from the relationships, that is $K_{d2} = 0.368$ and $\lambda = 0.835$. Then K_{p2} can be calculated as $K_{p2} = 13.8601$. The Bode plot of system designed is shown in Figure 7.2. As can be seen, the gain crossover frequency specification $\omega_c = 10\,rad/s$, phase margin specification $\Phi_m = 70°$, and the flat-phase specification are all fulfilled.

7.4.3 IOPID Controller Design

The position system time constant T in (6.2) is chosen as 0.4s, and the specifications of interest are also set as $\omega_c = 10\,rad/s$, $\Phi_m = 70°$. Following the design scheme of

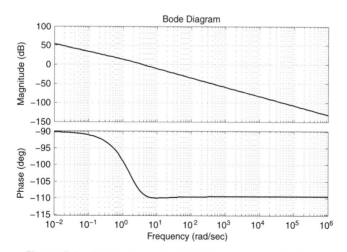

Figure 7.1   Bode plot with FO(PD) controller (T = 0.4s)

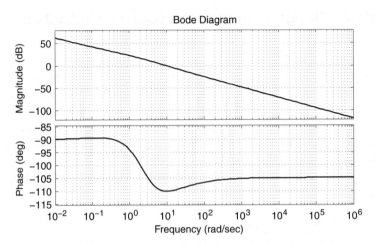

Figure 7.2 Bode plot with FOPD controller (T = 0.4s)

FO[PD] controller, the IOPID controller design process can also be deduced similarly. Two curves in terms of K_{d1} and K_{i1} can be plotted. One can obtain the K_{d1} and K_{i1} from the intersection point on the two curves, $K_{d1} = 0.1018$ and $K_{i1} = -4.625$. Then K_{p1} can be calculated as $K_{p1} = 23.0782$. The Bode plot of the designed open-loop system is shown in Figure 7.3. As can be seen, the three specifications in Section 2.3 are all fulfilled.

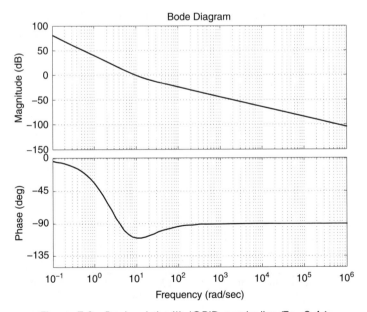

Figure 7.3 Bode plot with IOPID controller (T = 0.4s)

Remark 7.4.1 *The closed-loop transfer function of the system with the designed IOPID controller is,*

$$G_{CL}(s) = \frac{G_1(s)}{1 + G_1(s)} = \frac{K_{p1}K_{d1}s^2 + K_{p1}s + K_{p1}K_{i1}}{Ts^3 + (1 + K_{p1}K_{d1})s^2 + K_{p1}s + K_{p1}K_{i1}},$$

where the parameter $K_{i1} = -4.625$. From the Routh table technique, $K_{p1}K_{i1} < 0$, one can conclude that the system with the designed IOPID controller is unstable.

7.5 Implementation of Two Fractional Order Operators

In this chapter, the fractional order operators s^λ for the FOPD controller and $(1 + \tau s)^\mu$ for the FO[PD] controller are implemented by the impulse response invariant discretization (IRID) method in time domain.

7.5.1 Implementation of s^λ for FOPD

In Chapter 6, the fractional order operator s^λ for the FOPD controller (7.2) is implemented by the frequency domain approximation method with the Oustaloup Recursive Algorithm [174], which is a band-limit finite dimensional approximation, and a proper frequency range of practical interest is needed. In this chapter, s^λ is realized by the impulse response invariant discretization (IRID) method [51] in time domain, where a discrete-time finite dimensional (z) transfer function is computed to approximate the continuous irrational transfer function s^λ, s is the Laplace transform variable, and λ is a real number in the range of $(-1,1)$. s^λ is called a fractional order differentiator if $0 < \lambda < 1$ and a fractional order integrator if $-1 < \lambda < 0$. This approximation keeps the impulse response invariant.

7.5.2 Implementation of $(1 + \tau s)^\mu$ for FO(PD)

The proposed fractional order operator $(1 + \tau s)^\mu$ for the FO[PD] controller (7.3) is implemented by modifying the code of the IRID for fractional order low-pass filter (IRID-FOLPF) [52]. In the code of IRID-FOLPF, a discrete-time finite dimensional (z) transfer function is computed to approximate a continuous-time fractional order low-pass filter (FOLPF) $[1/(\tau s + 1)]^{\mu'}$, s is the Laplace transform variable, and μ' is a real number in the range of $(0, 1)$, τ is the time constant of low-pass filter. This approximation also keeps the impulse response invariant and only supports $\mu' \in (0, 1)$.

When μ' is in $(-1, 0)$, $[1/(\tau s + 1)]^{\mu'}$ is just the fractional order operator for the FO[PD] controller. The implementation of this operator $[1/(\tau s + 1)]^{\mu'}$ is realized as follows: First, FOLPF $[1/(\tau s + 1)]^{-\mu'}$ $(\mu' \in (-1, 0))$ is realized via the

IRID-FOLPF introduced above, we can obtain the discretized transfer function of the FOLPF,

$$G_{FO\text{-}LPF} = \frac{A(z)}{B(z)}.$$

Then, FO[PD] operator $[1/(\tau s + 1)]^{\mu'} = (\tau s + 1)^{\mu}$ ($\mu \in (0, 1)$) can be obtained,

$$G_{FO[PD]} = \frac{B(z)}{A(z)},$$

where $G_{FO[PD]}$ is the discretized transfer function of the fractional order operator for the FO[PD] controller.

7.6 Simulation Illustration

The tuning schemes in Section 7.5 for three controllers are illustrated via numerical simulation in this section.

In order to compare the IOPID, FOPD and the proposed FO[PD] controllers fairly, two simulation cases are presented below.

7.6.1 Case-I: Step Response Comparison with $T = 0.4s$

In this case, the position system time constant T in (6.2) is chosen as 0.4s. The specifications of interest are set as $\omega_c = 10\,rad/s$, $\Phi_m = 70°$, and the robustness to loop gain variations is required.

The parameters of the FO[PD], IOPID and FOPD controllers have been calculated in the examples in Section 7.4. As mentioned in Remark 7.4.1, the system with the designed IOPID controller is unstable. So, in this case, the comparison can only be performed using the designed FOPD and FO[PD] controllers without the designed IOPID controller. In terms of the implementation of two fractional order operators for two designed fractional order controllers, the finite dimensional discretized (z) approximate transfer function of the operator s^{λ} ($\lambda = 0.835$) has the form below following the scheme introduced in Section 7.5.1 with sampling time 0.001s and approximate order 5 of the (z) transfer function,

$$s^{0.835} \approx \frac{N_1}{D_1},$$

where

$$N_1 = z^5 - 3.27z^4 + 4.003z^3 - 2.228z^2 + 0.5329z - 0.03784,$$
$$D_1 = 0.002796z^5 - 0.006369z^4 + 0.004599z^3 - 9.034*10^{-4}z^2$$
$$- 1.464*10^{-4}z + 3.372*10^{-5};$$

and the finite dimensional discretized (z) approximate transfer function of the operator $(1 + K_{d3}s)^{\mu}$ ($K_{d3} = 0.2991$, $\mu = 0.7825$) has the form below following the method presented in Section 7.5.2 with sampling time 0.001s and approximate order 5 of the (z) transfer function,

$$(1 + 0.2991s)^{0.7825} \approx \frac{N_2}{D_2},$$

where

$$N_2 = z^5 - 3.281z^4 + 4.033z^3 - 2.255z^2 + 0.5413z - 0.03851,$$
$$D_2 = 0.01038z^5 - 0.02435z^4 + 0.01831z^3 - 0.003942z^2$$
$$- 5.217 * 10^{-4}z + 1.425 * 10^{-4}.$$

In Figure 7.4, applying the designed FOPD controller, the unit step responses are plotted with the the the open-loop gain varying from 11.0881 to 16.6321 ($\pm20\%$ variations from the desired value 13.8601). In Figure 7.5, applying the designed FO[PD] controller, the unit step responses are plotted with open-loop gain changing from 13.4271 to 20.1407 ($\pm20\%$ variations from the desired value 16.7839).

It can be seen from Figures 7.4 and 7.5 that both the designed FOPD and FO[PD] controllers following the proposed method in this chapter are effective. The overshoots of the step responses remain almost constant under gain variations, that is, the iso-damping property is exhibited, which means the system is robust to gain changes. Furthermore, from Figure 7.6, it can be seen obviously that the overshoot of the dashed line with the proposed FO[PD] controller is smaller than that of the solid line with the FOPD controller. With the saturation setting as ±10, the control input signals of the step responses with the designed FOPD and FO[PD] controllers can also be seen in Figure 7.7.

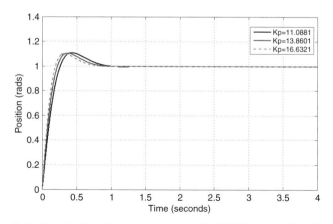

Figure 7.4 Simulation. Step responses with FOPD controller (T = 0.4s)

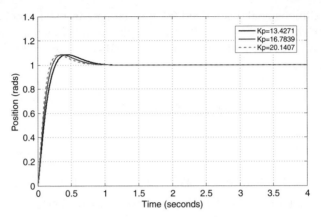

Figure 7.5 Simulation. Step responses with FO(PD) controller (T = 0.4s)

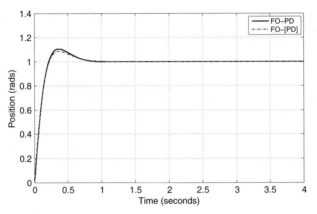

Figure 7.6 Simulation. Step responses comparison with two FO controllers (T = 0.4s)

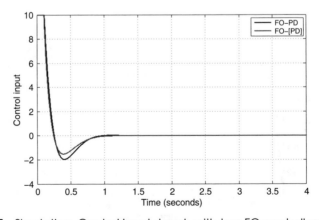

Figure 7.7 Simulation. Control input signals with two FO controllers (T = 0.4s)

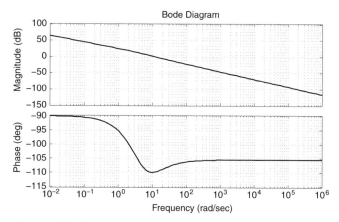

Figure 7.8   Bode plot with FO(PD) controller (T = 0.04s)

7.6.2 Case-II: Step Response Comparison with $T = 0.04s$

In this case, the position system time constant T in (6.2) is chosen as 0.04s. The specifications of interest are also set as $\omega_c = 10\,rad/s$, $\Phi_m = 70°$, and the robustness to gain variations of the system is also required.

7.6.2.1 Parameters Tuning and Bode Plot Validation for FO(PD)

Following the tuning procedure summary in Section 7.3.2, the three parameter values of the designed FO[PD] controller can be obtained, $K_{d3} = 0.0061$, $\mu = 0.5081$ and $K_{p3} = 10.7603$. The Bode plot of the open-loop system with the designed FO[PD] controller is shown in Figure 7.8. We also can see that the gain crossover frequency specification, $\omega_c = 10\,rad/s$, and phase margin specification, $\Phi_m = 70°$, are fulfilled. The phase is forced to be flat at ω_c.

7.6.2.2 Parameters Tuning and Bode Plot Validation for FOPD

According to the parameter tuning examples in Section 7.4, we can obtain the three parameters of the designed FOPD controller, $K_{d2} = 0.0057$, $\lambda = 0.7796$ and $K_{p2} = 10.6417$. The Bode plot of the open-loop system with the designed FOPD controller is shown in Figure 7.9. As can be seen, the three specifications in Section 2.3 are all satisfied.

7.6.2.3 Parameters Tuning and Bode Plot Validation for IOPID

Following the parameter tuning examples in Section 7.4, we can obtain the three parameters of the designed IOPID controller, $K_{d1} = 0.0189$, $K_{i1} = 1.5670$ and $K_{p1} = 10.7649$. The Bode plot of the open-loop system with the designed IOPID controller is shown in Figure 7.10. It can be seen that, the gain crossover frequency specification $\omega_c = 10\,rad/s$, phase margin specification $\Phi_m = 70°$, and the flat-phase specification

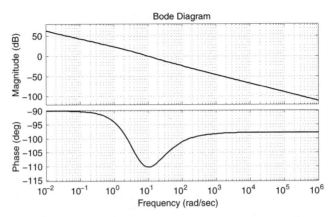

Figure 7.9 Bode plot with FOPD controller (T = 0.04s)

are all fulfilled. In this designed controller, the three parameters K_{d1}, K_{i1} and K_{p1} are all positive, from the Routh table technique, this designed IOPID controller can guarantee the system as stable and satisfy the three specifications in Section 2.3.

7.6.2.4 Implementation of Three Designed Controllers

As mentioned in Section 7.5, the discrete-time finite dimensional (z) transfer functions are computed to approximate the continuous irrational transfer functions of the FOPD and FO[PD] controllers with the impulse response invariant discretization (IRID) method in time domain. In order to compare with two designed fractional order controllers fairly, the discrete-time (z) transfer function discretized from the continuous form (7.1) is used to implement the designed IOPID controller. As calculated and validated by the Bode plot, the IOPID controller is designed as $K_{p1}(1 + K_{i1}\frac{1}{s} + K_{d1}s)$

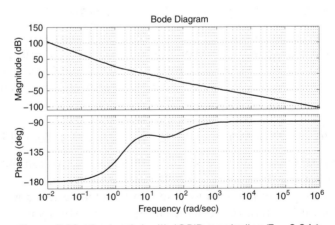

Figure 7.10 Bode plot with IOPID controller (T = 0.04s)

($K_{p1} = 10.7649$, $K_{i1} = 1.567$, $K_{d1} = 0.0189$) for the continuous form, so the (z) transfer function of this controller is shown below,

$$10.7649 \left(1 + 1.567\frac{1}{s} + 0.0189s\right) \approx \frac{417.7z^2 - 813.8z + 396.2}{z^2 - 1}.$$

In terms of the implementation of two designed fractional order controllers, the finite dimensional discretized (z) approximate transfer functions are used following the impulse response invariant discretization (IRID) method in time domain. As the order of the (z) transfer function for the designed IOPID controller implementation above is 2, for the fair comparison, the orders of the (z) approximate transfer functions for two designed fractional order controllers are also set as 2. The finite dimensional discretized (z) approximate transfer function of the designed FOPD controller $K_{p2}(1 + K_{d2}s^\lambda)$ ($K_{p2} = 10.6417$, $K_{d2} = 0.0057$, $\lambda = 0.7796$) has the form below following the scheme introduced in Section 7.5.1 with sampling time 0.001s and order 2,

$$10.6417(1 + 0.0057s^{0.7796}) \approx \frac{0.1046z^2 - 0.1213z + 0.0297}{0.0041z^2 - 0.00255z - 0.000381}.$$

The finite dimensional discretized (z) approximate transfer function of the designed FO[PD] $K_{p3}(1 + K_{d3}s)^\mu$ ($K_{p3} = 10.7603$, $K_{d3} = 0.0061$, $\mu = 0.5081$) has the form below following the method presented in Section 7.5.2 with sampling time 0.001s and order 2,

$$10.7603(1 + 0.0061s)^{0.5081} \approx \frac{10.76z^2 - 11.64z + 2.52}{0.3483z^2 - 0.1826z - 0.0112}.$$

7.6.2.5 Step Response Comparison

In Figure 7.11, applying the designed IOPID controller, the unit step responses are plotted with the open-loop gain varying from 8.6199 to 12.9179 ($\pm$20% variations from the desired value 10.7649). In Figure 7.12, applying the designed FOPD controller, the unit step responses are plotted with the open-loop gain varying from 8.5134 to 12.7701 ($\pm$20% variations from the desired value 10.6417). In Figure 7.13, applying the designed FO[PD] controller, the unit step responses are plotted with open-loop gains changing from 8.6082 to 12.9124 ($\pm$20% variations from desired value 10.7603).

It can be seen from Figure 7.11 that the designed IOPID controller following the examples in Section 7.4 works, and the overshoots just change almost 1.2% as the open-loop gain varying $\pm$20%. From Figure 7.12 and Figure 7.13, we can see that the step responses using the designed FOPD and FO[PD] controllers almost have no overshoot, when the open-loop gain changes $\pm$20%. So, the iso-damping property is exhibited obviously, the robustness of the systems using the designed FOPD and FO[PD] controllers is better than that using the designed IOPID controller. Furthermore, from Figure 7.6, it can be seen clearly that the performances using

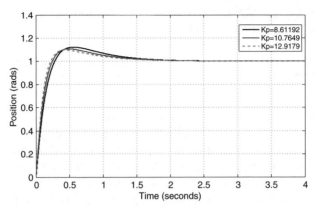

Figure 7.11 Simulation. Step responses with IOPID controller (T = 0.04s)

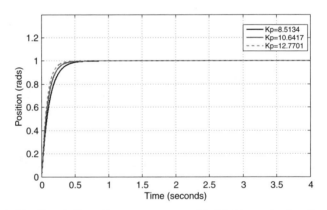

Figure 7.12 Simulation. Step responses with FOPD controller (T = 0.04s)

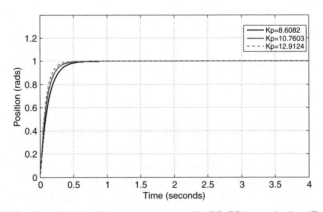

Figure 7.13 Simulation. Step responses with FO(PD) controller (T = 0.04s)

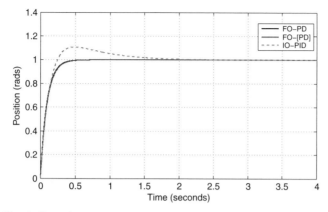

Figure 7.14 Simulation. Step responses comparison with three controllers (T = 0.04s)

both the designed FOPD and FO[PD] controllers are much better than that using the IOPID controller which is designed according to the same specifications. With the saturation setting as ±10, the control input signals of the step responses with the three controllers can also be seen in Figure 7.15.

7.6.3 Step Response Comparison with Time Delay

In order to explore the more advantages of the designed fractional order controllers, pure time delay is added in the considered second order position model (6.2), then, the control system becomes,

$$P'(s) = \frac{1}{s(Ts + 1)}e^{-\tau s}, \tag{7.18}$$

where τ is the delay time.

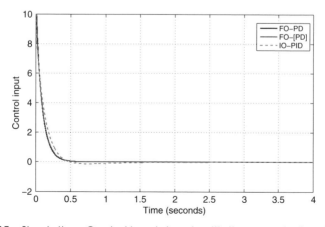

Figure 7.15 Simulation. Control input signals with three controllers (T = 0.04s)

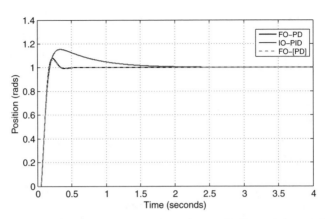

Figure 7.16   Simulation. Step responses comparison with time delay of 0.05s (T = 0.04s)

Figures 7.16 and 7.17 show the comparisons of the step responses using the three designed controllers with the time delay $\tau = 0.05s$ and $\tau = 0.2s$, respectively. It can be seen that both the designed FOPD and FO[PD] controllers have better time delay tolerance than the designed IOPID controller.

7.6.4 Step Response Comparison with Backlash Nonlinearity

We also added the backlash nonlinearity with the deadband 1 into the original second order position system (6.2). The step response comparison using three designed controllers is also tested. From Figure 7.18, we can see that the performances using the designed two fractional order controllers are much better than that using the designed IOPID controller.

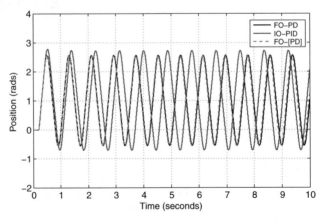

Figure 7.17   Simulation. Step responses comparison with time delay of 0.2s (T = 0.04s)

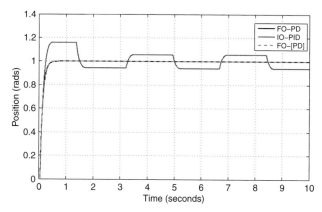

Figure 7.18 Simulation. Step responses comparison with backlash of deadband 1 (T = 0.04s)

7.7 Experimental Validation

7.7.1 Introduction to the Experimental Platform

A fractional horsepower dynamometer was developed as a general purpose experiment platform [210]. The architecture of the hardware-in-the-loop dynamometer control system is shown in Figure 3.10. The setup of this experimental platform is introduced in Section 3.6.1.

Through system identification, the position system on the dynamometer hardware-in-the-loop platform can be approximately modeled as the transfer function $\frac{1.52}{s(0.4s+1)}$. So, the position system of the dynamometer has the same structure with the considered system model (6.2). Thus, the simulation results in Section 7.6.1 with the system time constant $T = 0.4s$ can be validated via the real-time experiments on the dynamometer position system.

7.7.2 Experiments on the Dynamometer Platform

In Figure 7.19, applying the designed FOPD controller, the unit step responses are plotted with the open-loop gain varying from 11.0881 to 16.6321 ($\pm20\%$ variations from the desired value 13.8601). In Figure 7.20, applying the designed FO[PD] controller, the unit step responses are plotted with the open-loop gain changing from 13.4271 to 20.1407 ($\pm20\%$ variations from desired value 16.7839).

It can be seen from Figures 7.19 and 7.20 that, both the designed FOPD and FO[PD] controllers are effective. The overshoots of the step responses remain almost constant under gain variations, the systems are robust to gain changes. Furthermore, from Figure 7.21, it can be seen obviously that the overshoot of the gray line with the proposed FO[PD] controller is smaller than that of the dark line with the FOPD

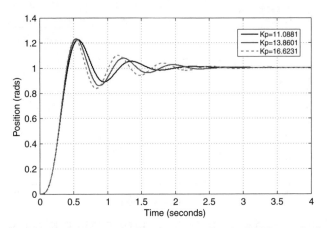

Figure 7.19   Experiment. Step responses with FOPD controller

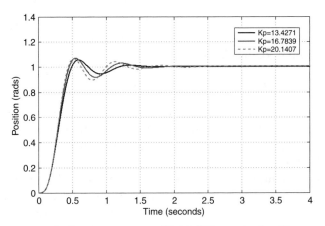

Figure 7.20   Experiment. Step responses with FO(PD) controller (For a color version of this figure, see Plate 8)

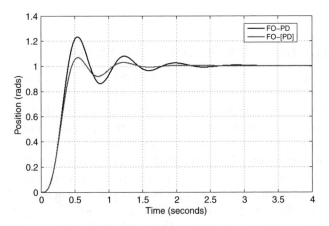

Figure 7.21   Experiment. Step responses comparison with two FO controllers (For a color version of this figure, see Plate 9)

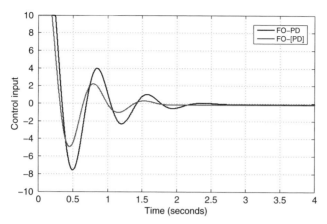

Figure 7.22 Experiment. Control input signals with two FO controllers (For a color version of this figure, see Plate 10)

controller. With the saturation setting as ±10, the control input signals of the step responses with the designed FOPD and FO[PD] controllers can be seen in Figure 7.22. So, we can see that the designed FO[PD] controller outperforms the designed FOPD controller with this system time constant $T = 0.4$.

7.8 Chapter Summary

In this chapter, a fractional order [proportional derivative] controller is proposed for robust position control systems. Focusing on the typical second order position systems, we developed a practical and systematic tuning synthesis for the proposed FO[PD] controller. Comparing the proposed FO[PD] controller with other controllers such as the traditional IOPID controller and the FOPD controller, the fairness issue has been for the first time addressed under the same number of design parameters and the same specifications. It can be seen that the designed IOPID controller may not be stabilizing to achieve "flat phase" specification while both the designed FOPD and FO[PD] controllers are stabilizing. Through the fair comparisons of three designed controllers (that is, IOPID, FOPD and FO[PD]) in both simulation and real-time experimental validation, we can see that both the designed FOPD and FO[PD] controllers outperform the designed IOPID controller; and the designed FO[PD] controller outperforms the designed FOPD controller more when the time constant of the position system increases.

8

Time-Constant Robust Analysis and Design of Fractional Order (PD) Controller

8.1 Introduction

Nowadays, a central issue in the field of control systems is the effect of uncertain parameters. Practically, in a real control systems, the parameters of the plants may not be measured exactly, or in some systems the parameters will change as the environment changes. Faced with this issue, many researchers have published papers about the robust PID controllers to overcome it [94], [242]. In previous chapters of this book, fractional order control systems with robustness on loop gain variations have been studied. Properly designed fractional order controllers show advantages over optimized or similar designed integer order controllers. For some systems which are sensitive to the plant time-constant variations as in [162], this issue was mentioned, but it did not provide a means of showing how to tune the control system for achieving the robustness on time-constant variations.

The phase margin and the gain crossover frequency of the control system are well-known controller tuning specifications [208], [230]. This chapter also uses these specifications, with the fractional order [proportional derivative] (FO[PD]) controller. In addition, our tuning specifications include the robustness requirement with respect to the time-constant variations. Especially, with the increased use of industrial networks, time delay is a severe issue in networked industrial control systems. Therefore, we focus on position systems with pure time delay to design the fractional order controller with the hope that better robustness can be achieved using fractional order

Fractional Order Motion Controls, First Edition. Ying Luo and YangQuan Chen.
© 2013 John Wiley & Sons, Ltd. Published 2013 by John Wiley & Sons, Ltd.

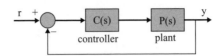

Figure 8.1   A general fractional order control system structure

controllers. To achieve the specified robustness on time-constant variations, our basic idea is to use the gradient of the phase margin and the gain crossover frequency specifications with respect to the time-constant T and the gain crossover frequency itself. However, the equations include a tangent function, so it is difficult to get the analytical solution. In this chapter, we discuss the feasible parameter conditions and focus on how to choose the gain crossover frequency to simplify the computation. The method and computation results are presented. According to the experimental results, the system robustness has been validated. Furthermore, from the system dynamic responses, we can see that the desired system dynamic performance is also guaranteed [251].

8.2 Problem Statement

The fractional order control system to be considered in this chapter is shown in Figure 8.1. The open-loop transfer function is

$$G(s) = C(s)P(s), \tag{8.1}$$

where $C(s)$ is the FO[PD] controller transfer function, $P(s)$ the controlled plant. In this chapter, we choose a DC motor position system with the following approximate model:

$$P(s) = \frac{K}{s(Ts + 1)}e^{-Ls}, \tag{8.2}$$

where T is the uncertain time-constant, L is the known constant dead-time. K is the plant gain which can be normalized to 1 without loss of generality because the DC gain of the system can be incorporated in the proportional gain of the controller. Note that the control performance will deteriorate as the time-constant T changes.

In this chapter, we focus on the FO[PD] controller to illustrate the robust control system tuning rules and robust controller design. The transfer function of the FO[PD] controller is defined as in (7.3),

$$C(s) = (K_p + K_d s)^\alpha, \tag{8.3}$$

where α is the order, which belongs to $(0, 2)$; $(K_p + K_d s)$ is the traditional PD controller with $\alpha = 1$.

The phase and gain of the plant in frequency domain are

$$\angle[P(j\omega)] = -\arctan(T\omega) - \frac{\pi}{2},$$ (8.4)

$$|P(j\omega)| = \frac{1}{\omega\sqrt{(T\omega)^2 + 1}}.$$ (8.5)

Fractional order [PD] controller described in (8.3) is given by

$$C(j\omega) = [K_p + K_d(j\omega)]^\alpha.$$ (8.6)

The phase and gain of the controller are

$$\angle[C(j\omega)] = \alpha \cdot \arctan\left(\frac{K_d\omega}{K_p}\right),$$ (8.7)

$$|C(j\omega)| = \left[K_p^2 + (K_d\omega)^2\right]^{\frac{\alpha}{2}}.$$ (8.8)

8.3 FO(PD) Tuning Specifications and Rules

According to the definitions of gain crossover frequency and phase margin, the following tuning specifications are presented:

(i) Phase margin specification

$$\angle[G(j\omega_c)] = \angle[C(j\omega_c)P(j\omega_c)]$$
$$= \alpha \arctan\left(\frac{K_d\omega_c}{K_p}\right) - \arctan(T\omega_c) - \frac{\pi}{2} - L\omega_c = -\pi + \phi,$$ (8.9)

where ω_c is the specified gain crossover frequency and ϕ is the specified phase margin.

(ii) Gain crossover frequency specification

$$|G(j\omega_c)| = |C(j\omega_c)P(j\omega_c)| = \frac{\left[(K_d\omega_c)^2 + K_p^2\right]^{\alpha/2}}{\omega_c\sqrt{1 + T^2\omega_c^2}} = 1.$$ (8.10)

(iii) Robustness to time-constant variations

Now, specification (iii) is discussed with details in the following subsection.

8.3.1 FO(PD) Robustness to Time-Constant Variations

When we identify the system, we cannot always find the accurate system parameters, which means the obtained parameters are approximate. Actually, in many systems, parameters will vary in an interval. If the controller is sensitive to the system time-constant variations, then the control performance will be different from the desired one when the time-constant changes. From equations (8.9) and (8.10), when the time-constant T varies, both the gain crossover frequency and phase margin will change. Similarly, the system performance will also vary as the time-constant changes. Since the time-constant changes will bring the gain crossover frequency variations. Therefore, according to the gain crossover frequency specification, we should ensure that

$$\frac{\partial |G(j\omega)|}{\partial \omega}\bigg|_{(\omega_c, T_0)} \Delta\omega + \frac{\partial |G(j\omega)|}{\partial T}\bigg|_{(\omega_c, T_0)} \Delta T = 0, \tag{8.11}$$

$$\frac{\Delta\omega}{\Delta T} = -\frac{\dfrac{\partial |G(j\omega)|}{\partial \omega}}{\dfrac{\partial |G(j\omega)|}{\partial T}}\bigg|_{(\omega_c, T_0)}, \tag{8.12}$$

where T_0 is the nominal time-constant. If the time-constant and gain crossover frequency variations do not make the phase margin change, we can establish the following equations according to the phase margin specification

$$\frac{\partial \angle[G(j\omega)]}{\partial \omega}\bigg|_{(\omega_c, T_0)} \Delta\omega + \frac{\partial \angle[G(j\omega)]}{\partial T}\bigg|_{(\omega_c, T_0)} \Delta T = 0, \tag{8.13}$$

$$\frac{\Delta\omega}{\Delta T} = -\frac{\dfrac{\partial \angle[G(j\omega)]}{\partial \omega}}{\dfrac{\partial \angle[G(j\omega)]}{\partial T}}\bigg|_{(\omega_c, T_0)}. \tag{8.14}$$

Observing equations (8.11) and (8.13), as the robust controller should satisfy both equations, we can set

$$-\frac{\dfrac{\partial |G(j\omega)|}{\partial \omega}}{\dfrac{\partial |G(j\omega)|}{\partial T}}\bigg|_{(\omega_c, T_0)} = -\frac{\dfrac{\partial \angle[G(j\omega)]}{\partial \omega}}{\dfrac{\partial \angle[G(j\omega)]}{\partial T}}\bigg|_{(\omega_c, T_0)}. \tag{8.15}$$

If the system satisfies equation (8.15), we are able to ensure that the system is robust to time-constant variations. So, the FO[PD] controller retrieved from the three specifications (8.9), (8.10) and (8.15) will simultaneously satisfy the robustness property and the phase margin criteria.

8.3.2 Numerical Computation

From equation (8.45), $\dfrac{\partial |G(j\omega)|}{\partial \omega}$ and $\dfrac{\partial |G(j\omega)|}{\partial T}$ can be obtained as

$$
\frac{\partial |G(j\omega)|}{\partial \omega}\bigg|_{(\omega_c, T_0)}
$$

$$
= \frac{\left(K_d^2\omega_c^2 + K_p^2\right)^{\frac{\alpha}{2}} \alpha K_d^2}{\left(K_d^2\omega_c^2 + K_p^2\right)\left(1 + T_0^2\omega_c^2\right)^{\frac{1}{2}}} - \frac{\left(K_d^2\omega_c^2 + K_p^2\right)^{\frac{\alpha}{2}}}{\omega_c^2\left(1 + T_0^2\omega_c^2\right)^{\frac{1}{2}}} - \frac{\left(K_d^2\omega_c^2 + K_p^2\right)^{\frac{\alpha}{2}} T_0^2}{\left(1 + T_0^2\omega_c^2\right)^{\frac{3}{2}}},
$$

$$
\frac{\partial |G(j\omega)|}{\partial T}\bigg|_{(\omega_c, T_0)} = -\frac{\left(K_d^2\omega_c^2 + K_p^2\right)^{\frac{\alpha}{2}} T_0\omega_c}{\left(1 + T_0^2\omega_c^2\right)^{\frac{3}{2}}}. \tag{8.16}
$$

Similarly, according to equation (8.9), $\dfrac{\partial A}{\partial \omega}$ and $\dfrac{\partial A}{\partial T}$ are

$$
\frac{\partial A}{\partial \omega}\bigg|_{(\omega_c, T_0)} = \frac{\alpha K_d}{K_p\left(1 + \dfrac{K_d^2\omega_c^2}{K_p^2}\right)} - \frac{T_0}{1 + T_0^2\omega_c^2} - L, \tag{8.17}
$$

$$
\frac{\partial A}{\partial T}\bigg|_{(\omega_c, T_0)} = -\frac{\omega_c}{1 + T_0^2\omega_c^2}. \tag{8.18}
$$

In terms of equation (8.15), the following equation can be derived,

$$
-\alpha \frac{\dfrac{K_d^2\omega_c^2}{K_p^2}}{\left(1 + \dfrac{K_d^2\omega_c^2}{K_p^2}\right)\omega_c T_0} + \frac{1}{T_0\omega_c} = -\alpha \frac{\dfrac{K_d\omega_c}{K_p}}{1 + \dfrac{K_d^2\omega_c^2}{K_p^2}} + L\omega_c. \tag{8.19}
$$

To simplify the expression, $\frac{K_d\omega_c}{k_p}$ is replaced with A. Then the following two simultaneous equations can be used to solve A and α,

$$
\alpha \arctan(A) - \arctan(T_0\omega_c) + \pi/2 - L\omega_c - \phi = 0, \tag{8.20}
$$

$$
\left(T_0 L\omega_c^2 - 1 + \alpha\right) A^2 - \alpha T_0\omega_c A + T_0 L\omega_c^2 - 1 = 0. \tag{8.21}
$$

Because the tangent function appears in the above nonlinear simultaneous equations, we are only able to solve them numerically.

8.4 The Solution Existence Range and An Online Computation Method

According to the property of the equations (8.20) and (8.21), solutions do not always exist, if ω_c is chosen to be too aggressive. So, in this section, we will discuss the solution existence range and an online fast computation method.

8.4.1 The Solution Existence Range

In equation (8.21), assuming A_{01} and A_{02} are solutions of A,

$$A_{01} = \frac{\alpha \omega_c T_0 + \sqrt{\Delta_1}}{2(L\omega_c T_0 \omega_c - 1 + \alpha)}, \tag{8.22}$$

$$A_{02} = \frac{\alpha \omega_c T_0 - \sqrt{\Delta_1}}{2(L\omega_c T_0 \omega_c - 1 + \alpha)}, \tag{8.23}$$

where $\Delta_1 = (\alpha \omega_c T_0)^2 - 4(L\omega_c T_0 \omega_c - 1 + \alpha)(L\omega_c T_0 \omega_c - 1)$. Since $\Delta_1 \geqslant 0$, the order of the controller has to satisfy

$$\alpha \geqslant \frac{4(L\omega_c T_0 \omega_c - 1) + \sqrt{\Delta_2}}{2\omega_c^2 T_0^2}, \tag{8.24}$$

$$\alpha \leqslant \frac{4(L\omega_c T_0 \omega_c - 1) - \sqrt{\Delta_2}}{2\omega_c^2 T_0^2}, \tag{8.25}$$

where $\Delta_2 = 16(L\omega_c T_0 \omega_c - 1)^2 + 16\omega_c^2 T_0^2(L\omega_c T_0 \omega_c - 1)^2$. Because $\alpha \in (0, 2]$, equation (8.25) cannot be held. If ω_c is too large, α will be beyond the range of $(0, 2]$. Therefore, the first conditional expression is

$$2 \geqslant \frac{4(L\omega_c T_0 \omega_c - 1) + \sqrt{\Delta_2}}{2\omega_c^2 T_0^2} \geqslant 0. \tag{8.26}$$

Similarly, assuming A_1 is the solution of (8.20), we have

$$A_1 = \tan\left(\frac{\phi'}{\alpha}\right), \tag{8.27}$$

where $\phi' = \arctan(T_0\omega_c) - \pi/2 + L\omega_c + \phi$. In terms of different values of ω_c, there are three cases that should be discussed.

1. $L\omega_c T_0 \omega_c - 1 > 0$

As we know, equation (8.27) is a decreasing function of ω_c. But equation (8.22) is an increasing function of ω_c, and equation (8.23) is a decreasing function of ω_c. So the vertex function has to be discussed. According to equation (8.24), assuming (A_0, α_0) is the extremum of function (8.21), we have $\Delta_1 = 0$ and

$$\alpha_o = \frac{4(L\omega_c T_0\omega_c - 1) + \sqrt{\Delta_2}}{2\omega_c^2 T_0^2}. \tag{8.28}$$

Substituting α_0 into equation (8.20), A_1' can be obtained as

$$A_1' = \tan\left(\frac{\phi'}{\alpha_0}\right). \tag{8.29}$$

Therefore, comparing A_0 with A_1', there are two conditions that should be discussed.
(1) If $A_1' \geqslant A_0$, the following inequation should be held

$$\tan\left(\frac{\phi'}{2}\right) \leqslant \frac{2\omega_c T_0 + \sqrt{\Delta_3}}{2(L\omega_c T_0\omega_c + 1)}, \tag{8.30}$$

where $\Delta_3 = (2\omega_c T_0)^2 - 4(L\omega_c T_0\omega_c + 1)(L\omega_c T_0\omega_c - 1)$.
(2) If $A_1' \leqslant A_0$, the following inequation should be held too

$$\tan\left(\frac{\phi'}{2}\right) \geqslant \frac{2\omega_c T_0 - \sqrt{\Delta_3}}{2(L\omega_c T_0\omega_c + 1)}. \tag{8.31}$$

2. $L\omega_c T_0\omega_c < 1$
According to equations (8.22) and (8.23), it can be deduced that $L\omega_c T_0\omega_c - 1 + \alpha > 0$ is satisfied. Also because A_0 should be positive, only equation (8.22) can be held. Here, function (8.22) is a decreasing function of ω_c. If $\alpha = 1 - L\omega_c T_0\omega_c$ is held, A_{01} will be positive infinite. It is the same that if $\alpha = \frac{2\phi + 2L\omega_c - \pi + 2\arctan(T_0\omega_c)}{\pi}$, A_1 will be positive infinite. Thus according to the relationship between $1 - L\omega_c T_0\omega_c$ and $\frac{2\phi + 2L\omega_c - \pi + 2\arctan(T_0\omega_c)}{\pi}$, there are two conditions that need to be discussed.
(1) If $1 - L\omega_c T\omega_c \geqslant \frac{2\phi + 2L\omega_c - \pi + 2\arctan(T_0\omega_c)}{\pi}$, the following inequation should be held

$$\tan\left(\frac{\phi'}{2}\right) \leqslant \frac{2\omega_c T_0 + \sqrt{\Delta_3}}{2(L\omega_c T_0\omega_c + 1)}. \tag{8.32}$$

(2) If $1 - L\omega_c T_0\omega_c \leqslant \frac{2\phi + 2L\omega_c - \pi + 2\arctan(T\omega_c)}{\pi}$, the following inequation is obtained

$$\tan\left(\frac{\phi'}{2}\right) \geqslant \frac{2\omega_c T_0 + \sqrt{\Delta_3}}{2(L\omega_c T_0\omega_c + 1)}. \tag{8.33}$$

3. $L\omega_c T_0 \omega_c - 1 = 0$

Physically, ω_c is limited by T and L. The solution of equations (8.20) and (8.21) always exists. Moreover, since α belongs to $(0, 2]$, we have to limit the range of phase margin ϕ. We will discuss the solution in the online computation part.

8.4.2 Numerical Computation Example and Simulation Tests

Assuming $T = 1, L = 0.5$, for validating the results, we choose an ω_c to see if it satisfies the three specifications in Section 8.3. Let $\omega_c = 1.2\ rad/s$ and $\phi = \pi/3\ rad$, then they can be substituted into the three conditions to check whether it meets the specifications. As $L\omega_c T\omega_c < 1$, it belongs to condition 2. We also have

$$1 - L\omega_c T\omega_c = 0.2800, \tag{8.34}$$

$$\frac{2\phi + 2L\omega_c - \pi + 2\arctan(T\omega_c)}{\pi} = 0.6064. \tag{8.35}$$

Meanwhile, A_1 and A_0 at point $\alpha = 2$ can be calculated as

$$A_1 = \tan\left(\frac{\arctan(T_0\omega_c) - \pi/2 + L\omega_c + \phi}{2}\right) = 2.6992, \tag{8.36}$$

$$A_0 = \frac{2\omega_c T_0 + \sqrt{(2\omega_c T_0)^2 - 4(L\omega_c T_0\omega_c + 1)(L\omega_c T_0\omega_c - 1)}}{2(L\omega_c T_0\omega_c + 1)} = 1.5036. \tag{8.37}$$

Obviously, at point $\alpha = 2$, A_1 is larger than A_0. So α in the range of $(0, 2]$ has a solution. In terms of equations (8.20) and (8.21), α and A can be calculated. A and α can be read from the intersection of two lines in Figure 8.2 ($A = 2.023$, $\alpha = 0.8568$). Then K_p and K_d can be calculated from

$$K_p = \sqrt{\frac{(\omega_c^2 (1 + T^2\omega_c^2))^{\frac{1}{\alpha}}}{A^2 + 1}} = 0.9226, \quad K_d = \frac{A * K_p}{\omega_c} = 1.5554. \tag{8.38}$$

The transfer function of the designed robust FO[PD] controller is

$$C(s) = [0.9226 + 1.5554s]^{0.8568}. \tag{8.39}$$

The Bode plot is shown in Figure 8.3. The Matlab/Simulink is used to simulate the FO[PD] control system. In Figure 8.4, the step response using this designed FO[PD] controller (8.39) is shown at $T = 1$. The aim of this FO[PD] controller tuning rule is that the phase margin will not vary with respect to time-constant T variation. To validate it, the time-constant T is changed $\pm 20\%$, and the Bode plots are shown in Figure 8.5 and Figure 8.6. It can be seen that the phase margin is still $\frac{\pi}{3}\ rad$ despite time-constant varying. For further validation, the simulation results are shown in

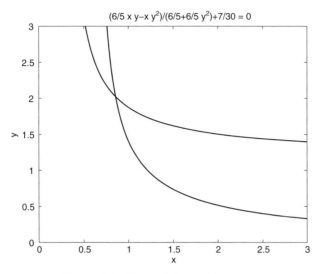

Figure 8.2 The solution of A and α

Figure 8.7. We can see that the overshoots of the three control systems are almost the same when the time-constant varies.

8.4.3 Simulation Comparison

The integer order PID controller tuned by ITAE optimization is employed to compare with the FO[PD] controller. Using the ITAE performance index, the optimum PID parameters are $K_p = 1.3586$; $K_i = 0$ and $K_d = 1.1552$. In Figure 8.8, applying the ITAE

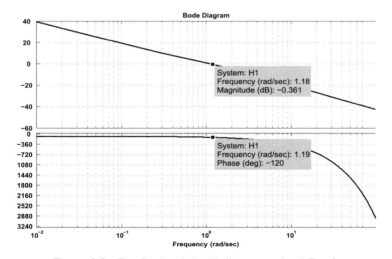

Figure 8.3 The Bode plot with time-constant $T = 1s$

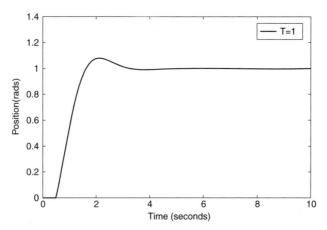

Figure 8.4 The step response using the designed FO(PD) controller (8.39)

optimum PID controller, the step responses are plotted with time-constant varying from 0.8 to 1.2($\pm 20\%$ variation from the desired value 1). Obviously, its overshoot varies with the time-constant changing. Comparing Figure 8.7 with Figure 8.8, we can conclude that the controlled system using the proposed FO[PD] controller is more robust to time-constant changes.

8.4.4 Online Computation

To design the robust controller, equations (8.20) and (8.21) have been derived. We cannot find the analytical solution from (8.20) and (8.21). So, for different systems, we

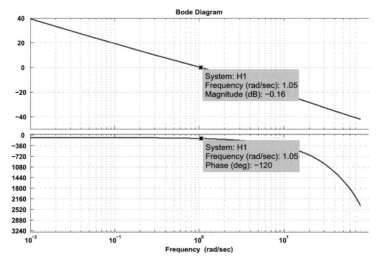

Figure 8.5 The Bode plot with $T = 1.2s$

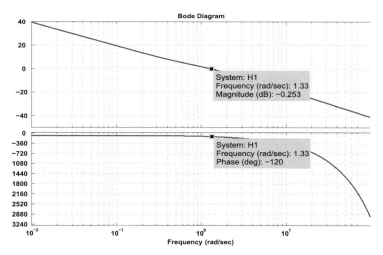

Figure 8.6 The Bode plot with $T = 0.8s$

have to tune the FO[PD] parameters offline. It may prevent the online auto-tuning rules. Therefore, simplifying the computation processing or the online computation is meaningful to FO[PD] controller applications.

As analyzed above, if $L\omega_c T_0\omega_c > 1$ and $L\omega_c T_0\omega_c < 1$, the solution existence requirements have been met. However, one needs to spend much time on parameters calculation.

There is another condition as $L\omega_c T_0\omega_c = 1$ that can be used to implement the controller parameters online tuning.

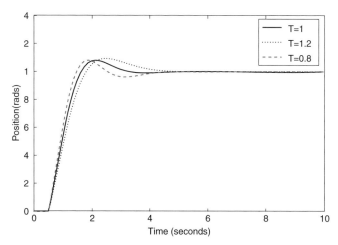

Figure 8.7 The step response comparison with different time-constants using FO(PD) controller

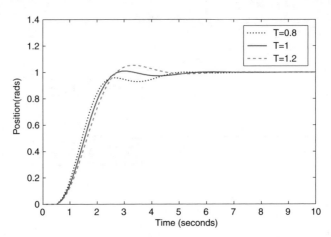

Figure 8.8 The step response comparison with different time-constants using the ITAE optimum PID controller

If $L\omega_c T_0 \omega_c = 1$, ω_c equals $\sqrt{\frac{1}{LT_0}}$. Then, A can be easily computed from (8.21) as $A = T_0 \omega_c = \sqrt{\frac{T_0}{L}}$. Substituting A into (8.47), α is easily calculated as follows:

$$\alpha = \frac{\phi + \sqrt{\frac{L}{T_0}} + \arctan\left(\sqrt{\frac{T_0}{L}}\right) - \frac{\pi}{2}}{\arctan\left(\sqrt{\frac{T_0}{L}}\right)}. \tag{8.40}$$

Meanwhile, K_p and K_d are

$$K_p = \sqrt{\frac{\left[\frac{1}{T_0 L}\left(1 + \frac{T_0}{L}\right)\right]^{\frac{1}{\alpha}}}{\frac{T_0}{L} + 1}}, \tag{8.41}$$

$$K_d = T_0 K_p. \tag{8.42}$$

Since α belongs to $(0, 2]$, we can make the following approximation to estimate the range of $\frac{L}{T_0}$

$$\arctan\sqrt{\frac{L}{T_0}} \approx \begin{cases} \sqrt{\frac{L}{T_0}}, & \text{if } 0 < \frac{L}{T_0} \leqslant 1 \\ \frac{\pi}{2} - \sqrt{\frac{T_0}{L}}, & \text{if } 1 > \frac{L}{T_0} \end{cases} \tag{8.43}$$

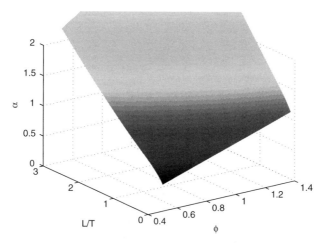

Figure 8.9 The relationship of ϕ, $\frac{L}{T_0}$ and α

The range of $\frac{L}{T_0}$ can be calculated as

$$
0 < \frac{L}{T_0} \leqslant \frac{\left(\frac{\pi}{2} - \phi + \sqrt{(\phi - \frac{\pi}{2})^2 + 4}\right)^2}{4}.
\tag{8.44}
$$

The relationship of ϕ, $\frac{L}{T_0}$ and α is shown in Figure 8.9. On the surface of the plot, we are able to find the α corresponding to different ϕ and $\frac{L}{T_0}$.

8.5 Experimental Validation

For validating the proposed tuning algorithm, we also use the Quanser hardware-in-the-loop system to implement the robust real-time motion control system. The architecture of the Qunser motion control system is shown in Figure 8.10, and the system block diagram is presented in Figure 8.11. To test the robustness of the controller, three DC motors in the lab are employed. Each of them provides a tachometer for velocity feedback. They connect to the Quanser terminal board in order to control the system through Matlab/Simulink Realtime Workshop (RTW) based software. In the computer, there is a data acquisition card, which will convert the analog signal to digital signal that is provided to Matlab/Simulink environment. In the experimental system, the controller algorithm is realized in the Matlab. The motors are driven by UPM power modules, which are equipped with DC power supplies.

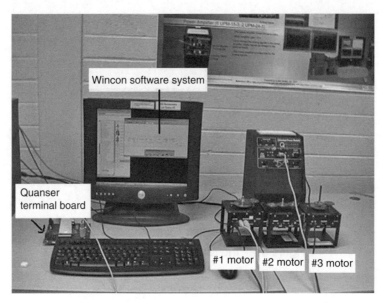

Figure 8.10 The Quanser hardware-in-the-loop system

Through simple system identification to the three motors, we can obtain their respective transfer functions as

$$\sharp 1 : P_1(s) = \frac{1}{s(0.0241s + 1)}e^{-0.01s}, \tag{8.45}$$

$$\sharp 2 : P_2(s) = \frac{1}{s(0.0265s + 1)}e^{-0.01s}, \tag{8.46}$$

$$\sharp 3 : P_3(s) = \frac{1}{s(0.0301s + 1)}e^{-0.01s}. \tag{8.47}$$

The time-constant of motor $\sharp 1$ is less than 9.1% of that of motor $\sharp 2$, and motor $\sharp 2$ is less than 12% of motor $\sharp 3$. For better validation of the proposed algorithm, the plant

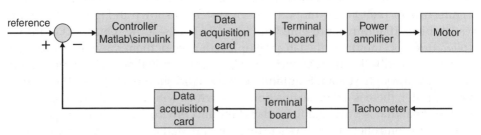

Figure 8.11 The architecture of the experiment block diagram

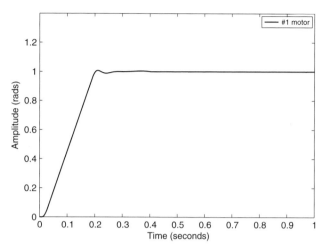

Figure 8.12 Motor #1 step response

gain is normalized to 1 without loss of generality, and the parameters of motor #2 have been chosen to design the controller. We set $\phi = 45^\circ$. In terms of equations (8.41), (8.42) and (8.40), K_p, K_d and α are easily calculated as

$$K_p = 160.5248; K_d = 4.2539; \alpha = 0.8332.$$

So the controller transfer function is

$$C(s) = (160.5248 + 4.2539s)^{0.8332}. \tag{8.48}$$

The FO[PD] controller is implemented applying the impulse response invariant discretization (IRID) method of fractional order low-pass filter following the modification in Section 7.5.2 [51], [52]. A discrete-time finite dimensional (z) transfer function is computed to approximate a continuous-time fractional order low-pass filter. We can obtain the discrete-time transfer function of FO[PD] controller as

$$C(z) = \frac{160.5z^5 - 486z^4 + 547.6z^3 - 278z^2 + 59.83z - 3.732}{0.05676z^5 - 0.1163z^4 + 0.07297z^3 - 0.01014z^2 - 0.00302z + 0.0004798}. \tag{8.49}$$

Using this controller to control the three DC motors, we can obtain the step responses in Figures 8.12, 8.13 and 8.14. These step responses are presented in Figure 8.15 for clear comparison. It can be seen that the overshoots on the three motors are all less than 2%. Therefore, we can conclude that this tuning method guarantees system dynamic performance.

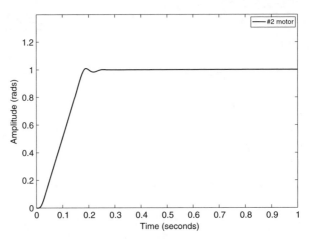

Figure 8.13   Motor ♯2 step response

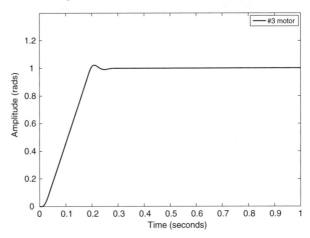

Figure 8.14   Motor ♯3 step response

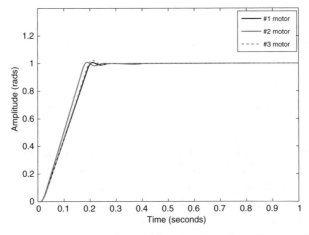

Figure 8.15   Step response comparison of the three motors (For a color version of this figure, see Plate 11)

8.6 Chapter Summary

In this chapter, we presented a tuning method of fractional order [proportional derivative] controller for the position system with pure time delay. The controller is tuned to ensure that the given gain crossover frequency and the phase margin are achieved, meanwhile, the controller meets the robustness requirement with respect to the time-constant changes. The solution range is studied and the online tuning method is discussed. The simulation and experimental results show that this fractional order [PD] controller tuning scheme can achieve desired dynamic performance and robustness.

9

Experimental Study of Fractional Order PD Controller Design for Fractional Order Position Systems

9.1 Introduction

Recently, studies on real systems have revealed inherent fractional order dynamic behavior, and fractional order systems have attracted more and more attention. Mechanical systems are described by the fractional order state equations [17], [18], [148]; in electrical systems, several applications have proposed a concept of fractance, which has intermediate properties between resistance and capacitance [155], [163], [170], [232]; many real systems in bioengineering are modeled or fitted by fractional order systems [56], [147]; some new fractional derivative-based models have been demonstrated in [37], [38], [81], [166], which provide powerful instruments for the description of memory and hereditary effects in various substances.

Fractional order systems can model various real systems more adequately than integer order ones, and thus provide reliable modelling tool in describing many real dynamical processes. It is intuitively true that these fractional order models require the corresponding fractional order controllers to achieve the desired performance. In most cases, however, researchers consider that the fractional order controllers applied to the integer order plants enhance the system control performance [119], [133], [238].

Fractional Order Motion Controls, First Edition. Ying Luo and YangQuan Chen.
© 2013 John Wiley & Sons, Ltd. Published 2013 by John Wiley & Sons, Ltd.

In this chapter, an experimental study is presented to validate the effectiveness of a fractional order proportional derivative (FOPD) controller systematic design scheme for fractional order systems with generalized fractional capacitor membrane model. In the simulation and experiment sections, the performance of the FOPD controller, designed based on the fractional order system (FOS), is compared with the integer order and fractional order controllers which are designed based on the approximate integer order system (IOS) [143].

9.2 Fractional Order Systems and Fractional Order Controller Considered

The fractional order system discussed in this chapter is the fractional calculus model of membrane charging [147] as follows:

$$P(s) = \frac{1}{s(Ts^\alpha + 1)}. \tag{9.1}$$

If choosing $\alpha = 1$, this plant is the typical position system discussed in Chapter 8. Therefore, the fractional order system (9.1) is called a fractional order position system in this book.

Since the proportional factor in the transfer function (9.1) can be incorporated in the controller gain, the plant gain is normalized 1 as we did in previous chapters without loss of generality.

The motivation of considering the fractional order systems (9.1) can be found from the schematic diagram for the circuit of membrane charging [147] shown in Figure 9.1 where the fractional order differential equation can be derived below,

$$C_m^\alpha \frac{d^\alpha V_m(t)}{dt^\alpha} + \frac{V_m(t)}{R_m} = I_{mo} u(t). \tag{9.2}$$

Applying the Laplace transform to (9.2) and the Caputo definition for the fractional derivative [188], we can obtain,

$$V_m(s) = \frac{K}{s(Ts^\alpha + 1)}, \tag{9.3}$$

where $K = I_{mo} R_m$, $T = R_m C_m^\alpha$, with the initial condition $V_m(0^+) = 0$.

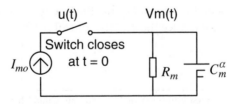

Figure 9.1   Membrane-charging circuit

Using the table of inverse Laplace transforms, we can find a solution in terms of the two parameter Mittag-Leffler function [161],

$$V_m(t) = I_{mo} R_m \frac{t^\alpha}{T} E_{\alpha,\alpha+1}\left(-\frac{t^\alpha}{T}\right). \tag{9.4}$$

This generalized fractional capacitor membrane model plays an important role in describing the dielectric behaviors of membranes, cells, tissues and a variety of biological materials, for example, nerve, muscle, skin and so on [147].

The fractional order proportional and derivative controller considered in this chapter has the form as (6.1). As an additional motivation to consider the fractional order system model used in this chapter, most recently, [99] has considered a fractional order model of the input respiratory impedance from non-invasive measurements of the airflow and air-pressure from both healthy and pathologic subject groups. The authors of [99] postulated that fractional order control might have "possible improvements in the current ventilatory assisting devices."

9.3 FOPD Controller Design Procedure for the Fractional Order Position Systems

In this chapter, we restrict our attention to the fractional order plants $P(s)$ described in (9.1). The transfer function of FOPD considered has the form of (6.1).

9.3.1 Preliminary and Design Specifications

First, recall that

$$j^\alpha = \left(e^{j\frac{\pi}{2}}\right)^\alpha = \cos\frac{\alpha\pi}{2} + j\sin\frac{\alpha\pi}{2}.$$

The phase and gain of the plant in (9.1) can be given by

$$P(jw) = \frac{1}{(jw)(T(jw)^\alpha + 1)} = \frac{1}{-Tw^{1+\alpha}\sin\frac{\alpha\pi}{2} + j\left(w + Tw^{1+\alpha}\cos\frac{\alpha\pi}{2}\right)},$$

$$\angle[P(jw)] = \tan^{-1}\frac{1 + Tw^\alpha\cos\frac{\alpha\pi}{2}}{Tw^\alpha\sin\frac{\alpha\pi}{2}}, \tag{9.5}$$

$$|P(jw)| = \frac{1}{\sqrt{\left(Tw^{1+\alpha}\sin\frac{\alpha\pi}{2}\right)^2 + \left(w + Tw^{1+\alpha}\cos\frac{\alpha\pi}{2}\right)^2}}. \tag{9.6}$$

The phase and gain of the FOPD controller in (6.1) can be written as

$$\angle[C(j\omega)] = \tan^{-1} \frac{\sin \frac{(1-\mu)\pi}{2} + K_d \omega^\mu}{\cos \frac{(1-\mu)\pi}{2}} - \frac{(1-\mu)\pi}{2}, \tag{9.7}$$

$$|C(j\omega)| = K_p \sqrt{\left(1 + K_d \omega^\mu \cos \frac{\mu\pi}{2}\right)^2 + \left(K_d \omega^\mu \sin \frac{\mu\pi}{2}\right)^2}. \tag{9.8}$$

From (9.5) and (9.7), the phase of open-loop transfer function $G(s) = C(s)P(s)$ can be written in frequency domain

$$\angle[G(j\omega)] = \tan^{-1} \frac{\sin \frac{(1-\mu)\pi}{2} + K_d \omega^\mu}{\cos \frac{(1-\mu)\pi}{2}} + \frac{\mu\pi}{2} - \frac{\pi}{2} + \tan^{-1} \frac{1 + T\omega^\alpha \cos \frac{\alpha\pi}{2}}{T\omega^\alpha \sin \frac{\alpha\pi}{2}}. \tag{9.9}$$

Here, the three specifications in Section 2.3 are again applied to design the fractional order PD controller for the fractional order system (9.1) in this chapter.

9.3.2 Numerical Computation Process

9.3.2.1 Integer Order PD Controller Design

From (9.9) and according to specification (ii) in Section 2.3 with $\mu = 1$ in (6.1),

$$\left. \frac{d(\angle(G(j\omega)))}{d\omega} \right|_{\omega=\omega_c} = \frac{K_d}{1 + (K_d \omega_c)^2} - \frac{\alpha T \sin \frac{\alpha\pi}{2} \omega_c^{\alpha-1}}{1 + T^2 \omega_c^{2\alpha} + 2T\omega_c^\alpha \cos \frac{\alpha\pi}{2}} = 0,$$

we arrive at

$$K_d = \frac{1 \pm \sqrt{1 - 4N^2 \omega_c^2}}{2N\omega_c^2}, \tag{9.10}$$

where $N = \frac{\alpha T \sin \frac{\alpha\pi}{2} \omega_c^{\alpha-1}}{1 + T^2 \omega_c^{2\alpha} + 2T\omega_c^\alpha \cos \frac{\alpha\pi}{2}}$.

When the system parameters T and α in (9.1) are fixed, and the crossover frequency ω_c is given, then K_d can be calculated in (9.10), and

$$\angle[G(j\omega_c)] = \tan^{-1}(K_d \omega_c) + \tan^{-1} \left(\frac{1 + T\omega_c^\alpha \cos \frac{\alpha\pi}{2}}{T\omega_c^\alpha \sin \frac{\alpha\pi}{2}} \right).$$

The phase of the open-loop system also can be calculated at the given gain crossover frequency. So, the phase margin of the system cannot be designed according to the requirement, which means that specifications (i) and (ii) in Section 2.3 cannot be satisfied simultaneously for traditional integer order PD controller.

9.3.2.2 Fractional Order PD Controller Design

The fractional order PD controller (6.1) is designed following the three specifications in Section 2.3. The brief equations are presented in this section.

According to specification (i), the phase of $G(s)$ can be expressed as

$$\angle[G(j\omega_c)] = \tan^{-1} \frac{\sin \frac{(1-\mu)\pi}{2} + K_d\omega_c^\mu}{\cos \frac{(1-\mu)\pi}{2}}$$

$$+ \frac{\mu\pi}{2} - \frac{\pi}{2} + \tan^{-1} \frac{1 + T\omega_c^\alpha \cos \frac{\alpha\pi}{2}}{T\omega_c^\alpha \sin \frac{\alpha\pi}{2}}$$

$$= -\pi + \phi_m,$$

the relationship between K_d and μ can be established as follows:

$$K_d = \frac{1}{\omega_c^\mu} \tan\left[\phi_m - \tan^{-1} \frac{1 + T\omega_c^\alpha \cos \frac{\alpha\pi}{2}}{T\omega_c^\alpha \sin \frac{\alpha\pi}{2}} - \frac{\mu\pi}{2} - \frac{\pi}{2}\right] \cos \frac{(1-\mu)\pi}{2}$$

$$- \frac{1}{\omega_c^\mu} \sin \frac{(1-\mu)\pi}{2}. \tag{9.11}$$

According to specification (ii), we can establish the following equation for K_p

$$|G(j\omega_c)| = \frac{K_p\sqrt{\left(1 + K_d\omega_c^\mu \cos \frac{\mu\pi}{2}\right)^2 + \left(K_d\omega_c^\mu \sin \frac{\mu\pi}{2}\right)^2}}{\sqrt{\left(T\omega^{1+\alpha} \sin \frac{\alpha\pi}{2}\right)^2 + \left(\omega + T\omega^{1+\alpha} \cos \frac{\alpha\pi}{2}\right)^2}} = 1. \tag{9.12}$$

According to specification (iii) for the robustness to loop gain variations

$$\left.\frac{d(\angle(G(j\omega)))}{d\omega}\right|_{\omega=\omega_c}$$

$$= \frac{\mu K_d\omega_c^{\mu-1} \cos \frac{(1-\mu)\pi}{2}}{\cos^2 \frac{(1-\mu)\pi}{2} + \left(\sin \frac{(1-\mu)\pi}{2} + K_d\omega_c^\mu\right)^2} - \frac{\alpha T\omega^{(\alpha-1)} \sin \frac{\alpha\pi}{2}}{\left(T\omega^\alpha \sin \frac{\alpha\pi}{2}\right)^2 + \left(1 + T\omega^\alpha \cos \frac{\alpha\pi}{2}\right)^2}$$

$$= 0, \tag{9.13}$$

one can establish an equation for K_d in the following form

$$K_d = \frac{-B \pm \sqrt{B^2 - 4A^2\omega_c^{2\mu}}}{2A\omega_c^{2\mu}}, \tag{9.14}$$

where

$$A = \frac{\alpha T \omega^{(\alpha-1)} \sin \frac{\alpha\pi}{2}}{\left(T\omega^\alpha \sin \frac{\alpha\pi}{2}\right)^2 + \left(1 + T\omega^\alpha \cos \frac{\alpha\pi}{2}\right)^2},$$

$$B = 2A\omega_c^\mu \sin \frac{(1-\mu)\pi}{2} - \mu\omega_c^{\mu-1} \cos \frac{(1-\mu)\pi}{2}.$$

Clearly, we can solve equations (9.11), (9.14) and (9.12) to obtain μ, K_d and K_p.

9.3.3 Summary of Design Procedure

It can be observed from (9.11) and (9.14) that μ and K_d can be obtained jointly in theory. But it is complex to find the analytical solution of μ and K_d as the equations (9.11) and (9.14) are complicated. The graphical method can also be used as a practical and simple way to find μ and K_d. The procedure to tune the FOPD controller is as follows:

(1) Given ω_c, the gain crossover frequency.
(2) Given ϕ_m, the desired phase margin.
(3) Plot the curve 1, K_d with respect to μ, according to (9.11).
(4) Plot the curve 2, K_d with respect to μ, according to (9.14).
(5) Obtain the μ and K_d from the intersection point on the above two curves.
(6) Calculate K_p from (9.12).

9.4 Simulation Illustration

In this simulation, the plant parameter T and α in (9.11) are chosen as 0.4s and 1.4, respectively. The specifications of interest are set as $\omega_c = 10 \ rad/s$, $\Phi_m = 70^0$, and the robustness to loop gain variations is required. According to (9.11) and (9.14), two curves can be plotted in Figure 9.2. One can obtain the μ and K_d obviously from the intersection point on the two curves, that is $\mu = 1.189$ and $K_d = 0.6138$. Then K_p can be calculated from (9.12) easily, that is $K_p = 10.916$. The Bode plots of the open-loop system designed are shown in Figure 9.3. As can be seen, both the 10 rad/s gain crossover frequency specification, and the 70^0 phase margin specification, are fulfilled. The phase is forced to be flat around the gain crossover frequency.

Actually, the fractional order system and the fractional order PD controller are infinite dimensional due to the fractional order differentiator s^α and s^μ. A band-limit implementation is important in practice. As introduced in Section 6.4, the Oustaloup Recursive Algorithm [174] for the finite dimensional approximation of the fractional order operator s^γ can be utilized in a proper frequency range of practical interest in this chapter.

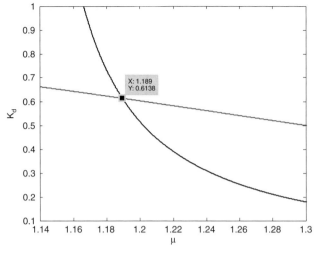

Figure 9.2 K_d versus μ

In our simulation, for the approximation of fractional order differentiator s^{μ}, the frequency range of practical interest is set to be from 0.0001Hz to 10000Hz, and N is chosen as 3 for the proper accuracy of the approximation.

Before the fractional order model is proposed to describe the real fractional order system, the approximated traditional integer order model that has always been applied for the controller design of the fractional order system is given.

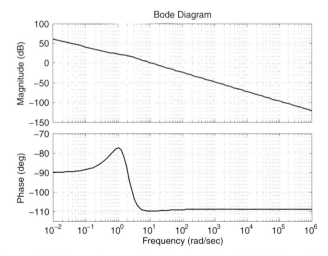

Figure 9.3 Open-loop system Bode plot with FOPD controller

For the purposes of comparison, the real fractional order systems considered can be approximated by a typical second order systems as shown in (9.15)

$$P(s)_{app} = \frac{K}{s(Ts + 1)},\qquad(9.15)$$

which can model a DC motor position system.

In order to compare the ITAE optimal proportional (P) controller and the FOPD controller designed based on the approximate integer order system (IOS) with the ones designed according to the real fractional order system (FOS) fairly, three simulation cases are presented.

- Case-1: The optimal P controller and the FOPD controller designed based on the approximate IOS simulation tests for the approximate IOS;
- Case-2: The optimal P controller and the FOPD controller designed based on the approximate IOS simulation tests for the real FOS;
- Case-3: The FOPD controller designed based on the real FOS simulation test for the FOS.

9.4.1 Case-1: IOS-Based Design for IOS

In this case, the P controller and the FOPD controller are designed and tested for the approximate integer order model (9.15). It is well known that the P controller is adopted commonly for the typical second order plants (9.15), and the ITAE optimal P controller parameter is designed as $K = 1/2T$ [68], therefore, the P controller parameter is set as 1.25 in this case. The FOPD controller can also be designed for this integer order approximation model (9.15) following the presented procedure. This integer order system is just a special case of the fractional order systems considered. The specifications of interest are also set as $\omega_c = 10\ rad/s$, $\Phi_m = 70^\circ$, and the robustness to loop gain variations is required. Following our design procedure, one can obtain that $\mu = 0.844$, $K_d = 0.368$ and $K_p = 13.860$.

In Figure 9.4, applying the ITAE optimal P controller, the unit step responses are plotted with the open-loop plant gain varying from 1 to 1.5 ($\pm 20\%$ variations from the desired value 1.25). In Figure 9.5, applying the fractional order PD controller, the unit step responses are plotted with an open-loop plant gain changing from 11.088 to 16.632 ($\pm 20\%$ variations from the desired value 13.86).

It can be seen from Figures 9.4 and 9.5 that the designed FOPD controller is effective, and performances are better than the optimum P controller for the approximate IOS (9.15).

9.4.2 Case-2: IOS-Based Design for FOS

The optimal P controller and the FOPD controller, which are designed in Case-1 based on the approximate IOS (9.15), are used to control the real FOS (9.1). From Figures 9.6 and 9.7, which also plotted the five unit step responses for the ITAE optimal P controller and the FOPD controller with the open-loop plant gain varying from the

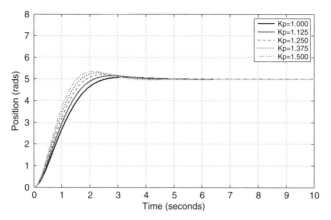

Figure 9.4 Simulation. Step responses with the ITAE optimal P controller designed based on the approximate IOS for the IOS

lower ±20% to the upper ±20% of the desired value respectively, we can see both results are not satisfactory, which confirms that these fractional order systems may require the corresponding fractional order controllers to achieve a desirable performance.

9.4.3 Case-3: FOS-Based Design for FOS

In this case, with the set of $\omega_c = 10\ rad/s$, $\Phi_m = 70^o$, following the presented design procedure, the FOPD controller parameters can be designed as $\mu = 1.189$, $K_d = 0.6138$ and $K_p = 10.916$, based on the real FOS (9.1).

Figure 9.8 shows the effectiveness of the systematic designed FOPD controller tested for the real FOS, the unit step responses are also plotted with open-loop plant

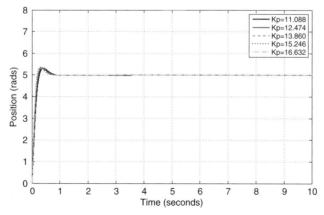

Figure 9.5 Simulation. Step responses with the FOPD controller designed based on the approximate IOS for the IOS

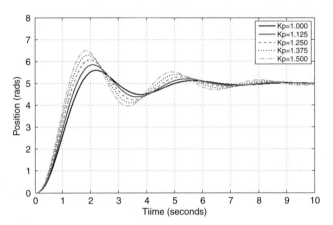

Figure 9.6 Simulation. Step responses with the ITAE optimal P controller designed based on the approximate IOS for the real FOS

gain changing from 8.733 to 14.099 (±20% variations from the desired value of 10.916). Comparing with Figures 9.6 and 9.7, the overshoots of the position step responses are much smaller and there is no oscillation. In particular, the five responses almost follow the same curve, the robustness to loop gain variations is excellent.

9.5 Experimental Validation

In this section, the practicality of the FOPD controller design scheme and the performance advantages of the FOPD controller for the real FOS presented in the simulation are also validated in a hardware-in-the-loop (HIL) experimental test bench.

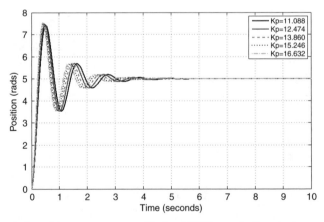

Figure 9.7 Simulation. Step responses with the FOPD controller designed based on the approximate IOS for the real FOS

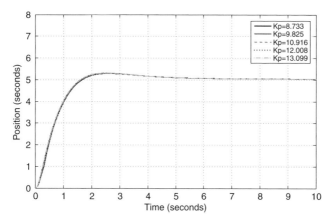

Figure 9.8 Simulation. Step responses with the FOPD controller designed based on the real FOS for the real FOS

9.5.1 HIL Experimental Setup

A fractional horsepower dynamometer was developed as a general purpose experiment platform [210]. The architecture of the dynamometer control system is shown in Figure 3.10. The setup of this experimental platform is introduced in Section 3.6.1.

Without loss of generality, consider the control system modeled by:

$$\dot{x}(t) = v(t), \tag{9.16}$$

$$\dot{v}(t) = K_u u(t) - K_b v. \tag{9.17}$$

where x is the position state, v is the velocity, and u is the control input, K_u and K_b are positive coefficients.

9.5.2 HIL Emulation of the FOS

Through simple system identification process, the dynamometer position control system can be approximately modeled by a transfer function $\frac{1.52}{s(0.4s+1)}$, which can be treated as the approximate integer order system (9.15). The fractional order system (9.1) can be emulated by modifying the dynamometer hardware-in-the-loop test bench as shown in Figure 9.9,

$$P(s) = G_m(s) \left(\frac{KA(s)}{1 + KA(s)} + \frac{KB(s)}{1 + KB(s)\frac{1}{s}} \frac{1}{K} \right) \frac{1}{s}, \tag{9.18}$$

where

$$G_m(s) = \frac{1}{Ts + 1}, \ A(s) = s^{-\alpha}, \ B(s) = s^{-\alpha+1}, \ K = \frac{1}{T},$$

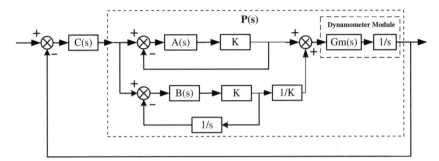

Figure 9.9 The HIL fractional order system by modifying the dynamometer

and $G_m(s)\frac{1}{s}$ is the model of the dynamometer position control system with $T = 0.4s$ and modifying the DC gain of the model as 1 without loss of the generality.

So, we can calculate the designed fractional order system from (9.18) as follows,

$$P(s) = \frac{1}{s(Ts^\alpha + 1)}. \tag{9.19}$$

The modules $A(s)$ and $B(s)$ in Figure 9.9 are also implemented using the fractional order operator module s^γ with the Oustaloup Algorithm introduced in Section 9.4.

9.5.3 Experimental Results

9.5.3.1 Simulation Results Validation

Substituting the original dynamometer bench and the modified dynamometer platform for the approximate integer order model and the fractional order system in the simulation respectively, the optimal P controllers and designed FOPD controllers in the three simulation cases are all tested in this hardware-in-the-loop manner.

First, designed based on the original dynamometer IOS, and tested for the original dynamometer setup, Figure 9.10 and Figure 9.11 show the unit step position responses using the ITAE optimal P controller and the FOPD controller. Second, designed also based on the original dynamometer IOS, but tested for the modified dynamometer FOS, Figure 9.12 and Figure 9.13 present the unit step position responses using the ITAE optimal P controller and the FOPD controller respectively. Third, designed based on and also tested for the modified dynamometer FOS, Figure 9.14 shows the unit step position responses using the FOPD controller designed by the presented tuning method in this chapter.

Comparing with Figures 9.12 and 9.13, it is obvious that, in Figure 9.14 with the properly designed FOPD controller, the overshoots of the position step responses are much smaller and there is no oscillation; in particular, the five responses almost follow the same curve, the robustness to loop gain variations is excellent. So, the simulation results are validated in our HIL experimental platform.

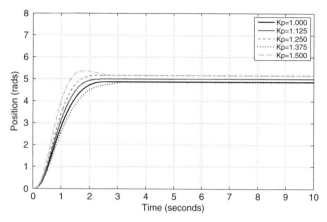

Figure 9.10 Experiment. Step responses with the ITAE optimal P controller designed based on the original dynamometer IOS for the original dynamometer IOS (For a color version of this figure, see Plate 12)

9.5.3.2 Small Unit Step Response Tests

In order to show the advantages of our presented method adequately, small unit step (0.1 rad) responses tests are performed using different controllers for the modified dynamometer FOS. As shown in Figure 9.15, the optimal P controller and the FOPD controller designed based on original dynamometer IOS, and the FOPD controller designed based on the modified dynamometer FOS, are tested for the small unit step position responses.

It can be seen that, using the optimal P controller designed based on the original dynamometer IOS, the position response has large steady-state error, which is very

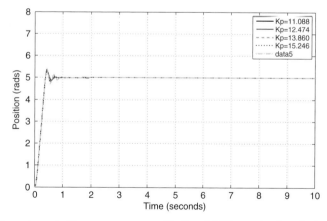

Figure 9.11 Experiment. Step responses with the FOPD controller designed based on the original dynamometer IOS for the original dynamometer IOS (For a color version of this figure, see Plate 13)

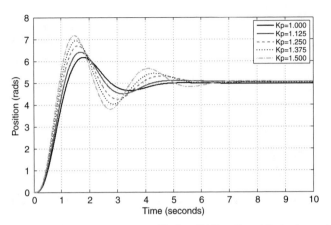

Figure 9.12 Experiment. Step responses with the ITAE optimal P designed based on the original dynamometer IOS for the modified dynamometer FOS

serious in small unit step response test, and using the FOPD controller designed based on the original dynamometer IOS, the position response has high overshoot and oscillation. But using the FOPD controller designed based on the modified dynamometer FOS, the position response does not show overshoot, the steady-state position error is almost zero, and there is no oscillation. This illustrates the advantage of the presented scheme designed in high precision control of the fractional order systems.

Furthermore, as the gain of the optimal P controller is smaller than the DC gains of the two FOPD controllers designed based on the original dynamometer IOS and the modified dynamometer FOS respectively, in order to compare them fairly, we also tested the small unit step responses using the P controller with the same DC gains

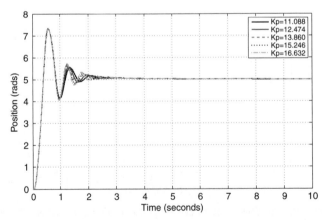

Figure 9.13 Experiment. Step responses with the FOPD controller designed based on the original dynamometer IOS for the modified dynamometer FOS

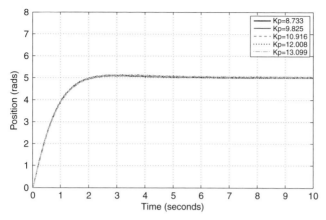

Figure 9.14 Experiment. Step responses with the FOPD controller designed based on the modified dynamometer FOS for the modified dynamometer FOS

of the two FOPD controllers as shown in Figure 9.16, where it can be seen that the systems are even unstable, when the P controllers with the gains of 13.860 and 10.916 are used respectively.

9.6 Chapter Summary

In this chapter, we presented comprehensive simulation and experimental study of fractional order proportional and derivative controller design for the fractional order position systems. In this scheme, the FOPD controller is tuned to ensure that the given

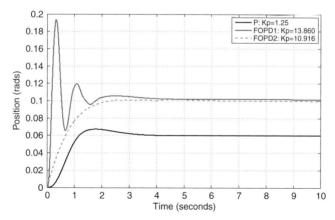

Figure 9.15 Experiment. Small step responses with the ITAE optimal P controller, the FOPD controller designed based on the original dynamometer IOS and the FOPD controller designed based on the modified dynamometer FOS

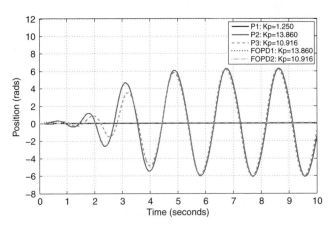

Figure 9.16 Experiment. Small step responses with three different gain P controllers and two FOPD controllers designed based on the original dynamometer IOS and the modified dynamometer FOS

gain crossover frequency and the phase margin are achieved, and the control system is robust to loop gain variations and step responses exhibit the iso-damping property. Simulation and experimental results show that the closed-loop system can achieve a favorable dynamic performance and robustness with the properly designed FOPD controller for the fractional order systems considered.

10

Fractional Order (PD) Controller Design and Comparison for Fractional Order Position Systems

10.1 Introduction

In Chapter 9, a fractional order proportional derivative (FOPD) controller was presented to control a class of fractional order position systems (FOPS) with experimental study. In this chapter, a fractional order [proportional derivative] (FO[PD]) controller is discussed for that class of FOPS with a design scheme and performance comparison. Especially, the fairness issue in comparing this FO[PD] controller with other controllers such as the traditional integer order PID (IOPID) controller and the FOPD controller has been addressed under the same number of design parameters and the same specifications. Fair comparisons of the three controllers (that is, IOPID, FOPD and FO[PD]) via the simulation tests illustrate that the IOPID controller designed may not always stabilize to achieve three design specifications while both the designed FOPD and FO[PD] controllers are stabilizing. Furthermore, the designed FO[PD] controller outperforms the designed FOPD controller for the class of FOPS [136].

10.2 Fractional Order Position Systems and Fractional Order Controllers

The FOPS discussed in this chapter is motivated by the fractional calculus-based model of membrane charging [143], [147] as shown in (9.1).

Fractional Order Motion Controls, First Edition. Ying Luo and YangQuan Chen.

As mentioned in Section 9.2, this generalized fractional capacitor membrane model does play an important role in describing the dielectric behavior of membranes, cells, tissues and a variety of biological materials [143], for example, nerve, muscle, skin and so on [147].

The IOPID, FOPD and FO[PD] controllers for fair comparison in this chapter have the transfer functions as (7.1), (7.2), and (7.3), respectively.

With the FOPS model $P(s)$ in (9.1) and $C(s)$, one of the three controllers (7.1), (7.2), and (7.3), the open-loop transfer function $G(s)$ has the form below,

$$G(s) = C(s)P(s). \tag{10.1}$$

In this chapter, the three specifications in Section 2.3 are applied to design the above three controllers for the fractional order system (9.1).

10.3 Fractional Order (PD) Controller Design

In this section, the FO[PD] controller has the transfer function form of (7.3). The phase and gain of the FOPS model described by (9.1) in frequency domain can be given as,

$$\angle[P(j\omega)] = -\tan^{-1}\frac{T\omega^{\alpha}\sin(\frac{\alpha\pi}{2})}{1+T\omega^{\alpha}\cos(\frac{\alpha\pi}{2})} - \frac{\pi}{2}, \tag{10.2}$$

$$|P(j\omega)| = \frac{1}{M}, \tag{10.3}$$

where

$$M = \sqrt{\left(T\omega^{1+\alpha}\sin\left(\frac{\alpha\pi}{2}\right)\right)^2 + \left(\omega + T\omega^{1+\alpha}\cos\left(\frac{\alpha\pi}{2}\right)\right)^2}.$$

The FO[PD] controller (7.3) can be written as,

$$C_3(j\omega) = K_{p3}[1 + K_{d3}(j\omega)]^{\mu}, \tag{10.4}$$

and the phase and gain are as follows,

$$\angle[C_3(j\omega)] = \mu\tan^{-1}(\omega K_{d3}), \tag{10.5}$$
$$|C_3(j\omega)| = K_{p3}[1 + (K_{d3}\omega)^2]^{\frac{\mu}{2}}. \tag{10.6}$$

The open-loop transfer function $G_3(s)$ is that

$$G_3(s) = C_3(s)P(s), \tag{10.7}$$

and from (10.2) and (10.5), we can obtain the phase of $G_3(s)$,

$$\angle[G_3(j\omega)] = \mu \tan^{-1}(\omega K_{d3}) - \tan^{-1}\left(\frac{T\omega^\alpha \sin \frac{\alpha\pi}{2}}{1 + T\omega^\alpha \cos \frac{\alpha\pi}{2}}\right) - \frac{\pi}{2}. \tag{10.8}$$

10.3.1 Numerical Computation Process

According to specification (i) in Section 2.3, the phase of $G_3(s)$ can be expressed as

$$\angle[G_3(j\omega_c)] = \mu \tan^{-1}(\omega_c K_{d3}) - \tan^{-1}\left(\frac{T\omega_c^\alpha \sin \frac{\alpha\pi}{2}}{1 + T\omega_c^\alpha \cos \frac{\alpha\pi}{2}}\right) - \frac{\pi}{2}$$

$$= -\pi + \phi_m, \tag{10.9}$$

so, we can establish the following relationship between K_{d3} and μ,

$$K_{d3} = \frac{1}{\omega_c} \tan\left(\frac{1}{\mu}\left(\phi_m - \frac{\pi}{2} + \tan^{-1}\left(\frac{T\omega_c^\alpha \sin \frac{\alpha\pi}{2}}{1 + T\omega_c^\alpha \cos \frac{\alpha\pi}{2}}\right)\right)\right). \tag{10.10}$$

According to specification (ii) in Section 2.3, we can establish the following equation for K_{p3}, K_{d3} and μ,

$$|G_3(j\omega_c)| = |C_3(j\omega_c)||P(j\omega_c)|$$

$$= \frac{K_{p3}(1 + (K_{d3}\omega_c)^2)^{\frac{\mu}{2}}}{N}$$

$$= 1, \tag{10.11}$$

where

$$N = \sqrt{\left(T\omega_c^{1+\alpha} \sin \frac{\alpha\pi}{2}\right)^2 + \left(\omega_c + T\omega_c^{1+\alpha} \cos \frac{\alpha\pi}{2}\right)^2}.$$

According to specification (iii) in Section 2.3 for the robustness to loop gain variations,

$$\frac{d(\angle(G_3(j\omega)))}{d\omega}\bigg|_{\omega=\omega_c} = \frac{\mu K_{d3}}{1 + (K_{d3}\omega_c)^2} - \frac{\alpha T\omega_c^{\alpha-1} \sin \frac{\alpha\pi}{2}}{\left(T\omega_c^\alpha \sin \frac{\alpha\pi}{2}\right)^2 + (1 + T\omega_c^\alpha \cos \frac{\alpha\pi}{2})^2}$$

$$= 0, \tag{10.12}$$

another relationship between K_{d3} and μ can be obtained in the following form,

$$A_3\omega_c^2 K_{d3}^2 - \mu K_{d3} + A_3 = 0, \tag{10.13}$$

that is

$$K_{d3} = \frac{\mu \pm \sqrt{\mu^2 - 4A_3^2\omega_c^2}}{2A_3\omega_c^2}, \tag{10.14}$$

where

$$A_3 = \frac{\alpha T \omega_c^{\alpha-1} \sin \frac{\alpha\pi}{2}}{\left(T\omega_c^\alpha \sin \frac{\alpha\pi}{2}\right)^2 + \left(1 + T\omega_c^\alpha \cos \frac{\alpha\pi}{2}\right)^2}.$$

Clearly, we can solve equations (10.10), (10.14) and (10.11) to obtain K_{p3}, K_{d3} and μ.

10.3.2 Design Procedure Summary

The graphical method can also be used as a practical and simple way to obtain K_{d3} and μ. The procedure to tune the parameters of the FO[PD] controller is as follows:

(1) Given parameters of the fractional order system to be controlled α and T.
(2) Given ω_c, the gain crossover frequency.
(3) Given ϕ_m, the desired phase margin.
(4) Plot the curve 1, K_{d3} with respect to μ, according to (10.10).
(5) Plot the curve 2, K_{d3} with respect to μ, according to (10.14).
(6) Obtain K_{d3} and μ from the intersection point on the above two curves.
(7) Calculate the K_{p3} from (10.11).

10.3.3 Design Example and Bode Plot Validation of FO(PD) Design

The time constant T in (9.1) is 0.4 s and the fractional order is $\alpha = 1.4$. The control design specifications of interest are set as $\omega_c = 10\,rad/s$ and $\phi_m = 70°$. According to (10.10) and (10.14), the designed values of K_{d3} and μ can be obtained obviously from the intersection point on the two curves, they are $K_{d3} = 0.9435$ and $\mu = 1.205$. Then K_{p3} can be calculated from (10.11), that is $K_{p3} = 6.3092$. The Bode plot of system designed is shown in Figure 10.1. As also can be seen, the gain crossover frequency specification, $\omega_c = 10\,rad/s$, and phase margin specification, $\phi_m = 70°$, are both fulfilled, and the phase is forced to be flat at ω_c based on the presented design.

10.4 Integer Order PID Controller and Fractional Order PD Controller Designs

The IOPID controller and FOPD controller designs also follow the same three specifications in Section 2.3 with the similar procedure as in Section 10.3.2 for the FO[PD] controller design. The FOPS to be controlled is the same with $T = 0.4$ s and $\alpha = 1.4$.

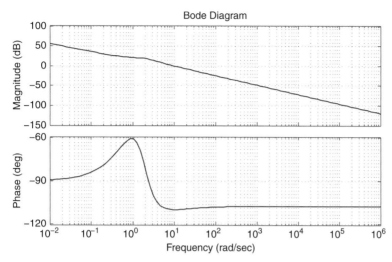

Figure 10.1 Bode plot with FO(PD) controller

Following the FO[PD] tuning formulae derivation, similar tuning formulae for both IOPID and FOPD can be obtained. The three parameters of the IOPID controller in (7.1) are designed as $K_{p1} = 18.2984$, $K_{i1} = 42.45$, and $K_{d1} = -0.0846$; and the three parameters of the FOPD controller in (7.2) are $K_{p2} = 10.916$, $K_{d2} = 0.6138$, and $\lambda = 1.189$. The Bode plots of systems designed using the IOPID controller and FOPD controller are shown in Figure 10.2 and Figure 10.3, respectively. As also can be seen, in both of the two Bode plots, the three specifications in Section 2.3 are all fulfilled.

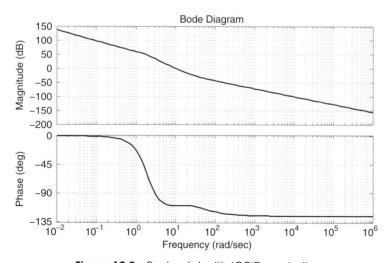

Figure 10.2 Bode plot with IOPID controller

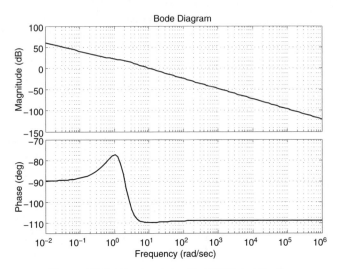

Figure 10.3   Bode plot with FOPD controller

10.5 Simulation Comparisons

The tuning procedures above for the three controllers are illustrated via numerical simulation in this section. The fractional order operators s^λ for the FOPD controller and $(1 + \tau s)^\mu$ for the FO[PD] controller are implemented by the impulse response invariant discretization (IRID) methods in time domain as presented in Section 7.5 [51], [52], [137].

The approximate finite dimensional discretized (z) transfer function of s^λ ($\lambda = 1.189$) has the following form with sampling period 0.01 s,

$$s^{1.189} \approx \frac{N_1}{D_1},$$

where

$$N_1 = z^5 - 3.423z^4 + 4.428z^3 - 2.64z^2 + 0.6928z - 0.05711,$$
$$D_1 = 0.003573z^5 - 0.007686z^4 + 0.005441z^3 - 0.00145z^2$$
$$+ 0.0001734z - 2.468 * 10^{-5},$$

and the approximate finite dimensional discretized (z) transfer function of the operator $(1 + K_{d3}s)^\mu$ ($K_{d3} = 0.9435$, $\mu = 1.205$) has the form below with the sampling period 0.01 s,

$$(1 + 0.9435s)^{1.205} \approx \frac{N_2}{D_2},$$

where

$$N_2 = z^5 - 3.392z^4 + 4.348z^3 - 2.57z^2 + 0.6685z - 0.05469$$
$$D_2 = 0.003513z^5 - 0.007414z^4 + 0.005137z^3 - 0.001348z^2$$
$$+ 0.0001721z - 2.599 * 10^{-5}.$$

The fractional order system parameters and the controller design specifications are the same as in Section 10.3.3. The parameters of the IOPID, FOPD and FO[PD] controllers are already calculated in Sections 10.3 and 10.4, respectively. However, note that, since $K_{d1} = -0.0846 < 0$, the system using the designed IOPID controller is unstable. So, we cannot obtain a properly designed IOPID controller which can guarantee the closed-loop stability and achieve the "flat phase" specification at the interested gain crossover frequency. This signifies the potential benefit of using a fractional order controller over an integer order controller.

In Figure 10.4, applying the FOPD controller, the unit step responses are plotted with the open-loop gain varying from 8.733 to 13.099 (±20% variations from the normal value 10.916). In Figure 10.5, applying the FO[PD] controller, the unit step responses are plotted with open-loop gains changing from 5.0474 to 7.5710 (±20% variations from normal value 6.3092).

It can be seen from Figure 10.4 and Figure 10.5 that both the FOPD and the FO[PD] controllers designed following the presented method in this chapter are effective. The overshoots of the step responses remain almost constant under loop gain variations, that is the iso-damping property is exhibited, that means the system is robust to loop gain changes. Furthermore, from Figure 10.6, it can be seen obviously that the overshoot of the dashed line with the designed FO[PD] controller is almost zero, and much smaller than that of the solid line with the designed FOPD controller. So, we can see that the designed FO[PD] controller outperforms the designed FOPD controller.

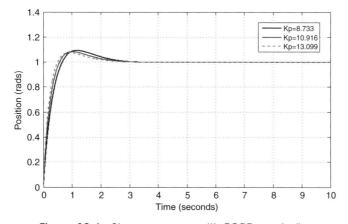

Figure 10.4 Step responses with FOPD controller

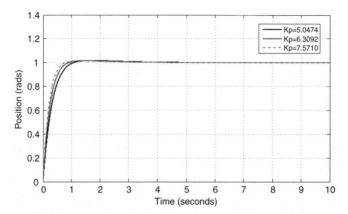

Figure 10.5 Step responses with FO(PD) controller

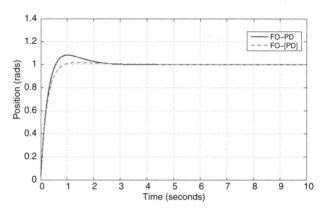

Figure 10.6 Step responses comparison with two FO controllers (For a color version of this figure, see Plate 14)

10.6 Chapter Summary

In this chapter, in order to reduce the modeling complexity, we focus on a type of simplified model for fractional order position systems. A practical and systematic design procedure of the fractional order [proportional derivative] controller is developed for this class of fractional order position systems. The fairness issue in comparing with other controllers such as the traditional integer order PID controller and the fractional order proportional derivative controller, all with three controller parameters, has been addressed under the same design specifications. Fair comparisons of the three controllers (that is, IOPID, FOPD and FO[PD]) via the simulation tests illustrate that the designed IOPID controller may not always be stabilizing to achieve all the design specifications while both designed FOPD and FO[PD] controllers are stabilizing. Furthermore, the designed FO[PD] controller outperforms the designed FOPD controller for the considered class of fractional order systems.

Part IV
Stability and Feasibility

11

Stability and Design Feasibility of Robust PID Controllers for FOPTD Systems

11.1 Introduction

11.1.1 Research Questions

11.1.1.1 Question 1

Given a stable first order plus time delay (FOPTD) system, a controller can be designed to satisfy three specifications:

(i) a specified gain crossover frequency;
(ii) a specified phase margin;
(iii) a flat phase constraint, which means that the system open-loop phase is a constant around the given gain crossover frequency, which can show the iso-damping property for the system response.

This scheme has been discussed in our previous work [136]. However, how do we know a chosen combination of gain crossover frequency and phase margin is achievable to obtain a stabilizing controller while maintaining the flat phase feature? It is desirable to create a two-dimensional figure to check the complete set of feasible gain crossover frequency and phase margin prior to the controller design and tuning, given any open-loop stable first order plus time delay system.

Fractional Order Motion Controls, First Edition. Ying Luo and YangQuan Chen.
© 2013 John Wiley & Sons, Ltd. Published 2013 by John Wiley & Sons, Ltd.

11.1.1.2 Question 2

The primary concern for the controller design and tuning is to maintain the stability of the control system. Stability is the minimum requirement for the controller design. The desirable approach for the controller design in modern control theory is to find the complete set of stabilizing controllers, and then obtain a certain unique controller solution from this complete stabilizing set, according to some controller design constraints. The question is, how to find this complete stabilizing region for the controller design first, and how to search for the optimal, unique controller in this region given a specific set of controller design specifications?

11.1.2 Previous Work

In the past few decades, many techniques of design and tuning of PID controllers have been proposed. Some of the most popular methods are the Ziegler-Nichols method [249], the Cohen-Coon rule [59], the modified Ziegler-Nichols scheme [249], the integral performance criteria [202], the Åstrom-Hagglund method [13], and so on. Meanwhile, in order to improve the feedback control performance, various PID controllers have been proposed, as typical examples, we mention the PID-deadtime controller [57], [97], the IMC-PID controller [97], [118], the Smith predictor-PID controller [77], [203], etc. Some synthesis schemes for the fractional order PID controller in feedback control systems are presented in [87], [116], [136], [160]. These results showed the potential of the fractional order controllers to improve both the stability and robustness of the feedback control systems.

For the PID controllers' design, stability is a fundamental issue. Guaranteeing the system stabilization is the minimum requirement for any PID controller design and tuning technique. Recently, there has been a trend of calculating the stabilizing region of the parameters for the PID controller design [1], [200], [201], [204], [207]. In [200], a version of the Hermite-Biehler Theorem derived by Pontryagin [191] was investigated to determine the entire stabilizing region of the PID parameters for the FOPTD models; and the similar result for the stabilizing region of the PID controllers was achieved in [151] by an alternative way using the classical Nyquist stability criterion. In [87], the formulation and numerical scheme of the stabilizing fractional order PID controller design for the fractional order time delay systems are presented.

11.1.3 Contributions in This Chapter

Given any open-loop stable first order plus time delay (FOPTD) model, all stabilizing PID controllers can be found in terms of a 3D plot of three gains, which is a solved problem [87], [151], [191]. However, how to determine a unique set of integer order or fractional order PID gains depends on the controller design specifications used. Furthermore, another practical question is how to know the design specifications used are achievable or feasible for a given FOPTD plant. It is desirable to know the ranges of the feasible design specifications given for an FOPTD plant.

In order to find the solution of the above issue for fractional order PID controllers design, the integer order PID controllers are studied in this chapter as a baseline. First, we focus on three design specifications: gain crossover frequency, phase margin, and the "flat phase" condition, that is, the phase slope at the gain crossover frequency is zero, guaranteeing robustness to loop gain variations. The motivation of this robustness specification can accommodate both controller and plant gain variations since controller implementation may be subject to the imprecision inherent in analog-digital and digital-analog conversions, the resolution limitation of the measuring instruments and the approximation errors in numerical computations [107], meaning that the coefficients of the PID controller may be changed in the course of the control implementation. This tuning constraint makes the open-loop phase flat around the pre-specified gain crossover frequency, which means the derivative of the phase with respect to the frequency is zero at the gain crossover frequency point, for example, the system step response shows "iso-damping property" and is robust to the loop gain variations which include the DC gain uncertainties of the plant and the implementation imprecision of the PID controller. Thus, the control performance with this designed PID controller degrades "gracefully" as the loop gain changes.

This chapter provides a new approach for stable PID controller design based on the FOPTD models. With this designed PID controller, it can be guaranteed that the system is stable, two specifications of phase margin and gain crossover frequency as design specifications are satisfied, and the flat phase constraint is also fulfilled as the third specification [254]. The stability region of the PID controller parameters is first determined according to a graphical stability criterion. Then, a three-dimensional surface satisfying the pre-specified phase margin, and a three-dimensional curve on this surface satisfying the phase margin and gain crossover frequency, can be drawn in the stabilizing PID parameter space. Then, by defining a relative function according to the flat phase constraint, a specific point on the three-dimensional curve to determine the PID controller can be found.

Since the PID controller design scheme in this chapter is based on the three specifications which include two performance specifications and a flat phase constraint, how to select the two design specifications is a valuable and instructional question. The key point of the contributions in this chapter is that, not only the PID controller can be determined by three specifications, but also the achievable ranges of the two specifications in the specifications can be found to guarantee the existence of the stable PID solution and satisfying three specifications. Therefore, the feasibility of two design specifications can be verified from the achievable region in advance, prior to the controller design. Furthermore, from the normalization of the FOPTD system, the efforts of the PID controller design are not even required for any open-loop stable FOPTD system. The desired PID can be easily looked up in the constructed table satisfying the given specifications and robustness on the loop gain variations. A simulation example is presented to illustrate the advantages of the designed PID controller over the traditional Ziegler-Nichols PID controller based on the FOPTD model.

In summary, the significance of contributions in this work is that, under the above three controller design specifications, the robust PID synthesis problem for all FOPTD models can be considered as *completely* solved in the sense that a lookup table can be constructed for a practical design scenario: Given the delay to time constant ratio, the feasible ranges of specifications can be determined first; given (feasible) specifications, unique PID parameters can be retrieved from the table, with the "flat phase" property embedded.

11.2 Stability Region and Flat Phase Tuning Rule for the Robust PID Controller Design

11.2.1 Preliminary

The PID controller design synthesis in this chapter is based on the first order plus time delay (FOPTD) models characterized by the following transfer function,

$$P(s) = \frac{K}{Ts+1}e^{-Ls}, \tag{11.1}$$

where K represents the steady-state gain of the plant, T is the apparent time constant of the plant, and L represents the apparent time delay.

Considering the feedback control system shown in Figure 11.1, where r is the command reference signal, y is the output signal of the plant, $P(s)$ given by (11.1) is the plant to be controlled, and $C(s)$ is the designed controller. In this chapter, we focus on the following proportional-integral-derivative (PID) controller,

$$C(s) = K_p + \frac{K_i}{s} + K_d s, \tag{11.2}$$

where K_p is the proportional gain, K_i is the integral gain, and K_d is the derivative gain.

Definition 11.2.1   *In Figure 11.1, M_T is a gain-phase margin tester [39], which provides information for plotting the boundaries of constant gain margin and phase margin in the parameter plane [87]. The transfer function of M_T is given in the form below,*

$$M_T(A, \phi) = Ae^{-j\phi}. \tag{11.3}$$

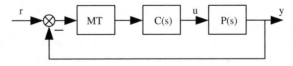

Figure 11.1   The feedback control system

Property 11.2.1 *Assuming $\phi = 0$ for M_T in (11.3), the controller parameters can be obtained satisfying a given gain margin A of the control system as shown in Figure 11.1. Meanwhile, assuming $A = 1$ for M_T in (11.3), one can obtain the controller parameters for a given phase margin ϕ [87].*

11.2.2 Stability Region of PID Controller for FOPTD Plants

From equations (11.1), (11.2) and (11.3), the open-loop transfer function of the feedback control system in Figure 11.1 is,

$$G(s) = M_T(A, \phi)C(s)P(s). \tag{11.4}$$

The closed-loop transfer function can be expressed as,

$$\Phi(s) = \frac{M_T(A, \phi)C(s)P(s)}{1 + M_T(A, \phi)C(s)P(s)}. \tag{11.5}$$

Replacing (11.1), (11.2) and (11.3) into (11.5) yields,

$$\Phi(s) = \frac{Ae^{-j\phi}Ke^{-Ls}(K_d s^2 + K_p s + K_i)}{s(Ts + 1) + Ae^{-j\phi}Ke^{-Ls}(K_d s^2 + K_p s + K_i)}. \tag{11.6}$$

Hence, the characteristic equation of the closed-loop system $\Phi(s)$ is as follows,

$$D(K_p, K_i, K_d, A, \phi; s) = s(Ts + 1) + Ae^{-j\phi}Ke^{-Ls}(K_d s^2 + K_p s + K_i) = 0. \tag{11.7}$$

With the given FOPTD model in (11.1), the initial target is to find the complete set of PID controllers stabilizing the system. The system stability is depending on the roots' locations of the characteristic equation (11.7) with $A = 1$ and $\phi = 0°$. If all the roots of the polynomial (11.7) are located in the left half of the s-plane, the closed-loop system (11.5) is bounded-input bounded-output stable. There are three parameters K_p, K_i and K_d in the PID controller.

Definition 11.2.2 *If $(K_p, K_i, K_d) \in Q$, and all the roots of $D(K_p, K_i, K_d, A, \phi; s) = 0$ line in the left half of the s-plane, then Q is the stability region of these three controller parameters K_p, K_i and K_d.*

In order to achieve the Property 11.2.2, the following definitions are explained:

Definition 11.2.3 *Infinity root boundary (IRB): The infinity root boundary is defined by the equation*

$$D(K_p, K_i, K_d, A, \phi; s = \infty) = 0,$$

and the boundary can be obtained [87] as follows,

$$K_d = \pm \frac{T}{AK}, \tag{11.8}$$

where it can be seen that the stabilizing interval of K_d is not concerned with the setting of phase margin ϕ, which means that K_d has nothing to do with the phase margin settings for the PID controller design based on the FOPTD plants.

Although the theoretical method of calculating the IRB is difficult because of the time delay in the FOPTD system, which can generate an infinite number of roots, the asymptotical location of roots far from the origin [26] can be used to figure out the IRB.

Definition 11.2.4 *Real root boundary (RRB): The real root boundary is defined by the equation*

$$D(K_p, K_i, K_d, A, \phi; s = 0) = 0,$$

so, one can obtain the boundary as,

$$K_i = 0,$$

where we can see the boundary of K_i always stays the same in spite of different phase margins and gain margins.

Definition 11.2.5 *Complex root boundary (CRB): Substituting s with $j\omega$ in (11.7), the complex root boundary can be defined from*

$$D(K_p, K_i, K_d, A, \phi; s = j\omega) = 0, \tag{11.9}$$

as follows,

$$
\begin{aligned}
&D(K_p, K_i, K_d, A, \phi; j\omega) \\
&= j\omega(j\omega T + 1) + A e^{-j\phi} K e^{-j\omega L}((j\omega)^2 K_d + j\omega K_p + K_i) \\
&= (-T\omega^2 + j\omega) + K A e^{-j(\phi+\omega L)}(-\omega^2 K_d + K_i + j\omega K_p) \\
&= (-T\omega^2 + j\omega) \\
&\quad + KA(\cos(\phi + \omega L) - j\sin(\phi + \omega L))(-\omega^2 K_d + K_i + j\omega K_p) \\
&= -T\omega^2 + KA((-\omega^2 K_d + K_i)\cos(\phi + \omega L) + \omega K_p \sin(\phi + \omega L)) \\
&\quad + j(\omega + KA(\omega K_p \cos(\phi + \omega L) - (-\omega^2 K_d + K_i)\sin(\phi + \omega L))) \\
&= 0.
\end{aligned}
\tag{11.10}
$$

Considering the real part and the imaginary part of (4.10) respectively, one can obtain,

$$- T\omega^2 + KA((-\omega^2 K_d + K_i) \cos(\phi + \omega L) + \omega K_p \sin(\phi + \omega L)) = 0; \quad (11.11)$$

$$\omega + KA(\omega K_p \cos(\phi + \omega L) - (-\omega^2 K_d + K_i) \sin(\phi + \omega L)) = 0. \quad (11.12)$$

From (11.12),

$$K_d = \frac{KA \sin(\phi + \omega L) K_i - KA\omega K_p \cos(\phi + \omega L) - \omega}{KA\omega^2 \sin(\phi + \omega L)}, \quad (11.13)$$

substituting (11.13) into (11.11) yields,

$$K_p = \frac{T\omega \sin(\phi + \omega L) - \cos(\phi + \omega L)}{KA}, \quad (11.14)$$

so, one can obtain,

$$K_i = \frac{\omega \sin(\phi + \omega L) + T\omega^2 \cos(\phi + \omega L)}{KA} + \omega^2 K_d. \quad (11.15)$$

Hence, given K_d, the curve of K_i with respect to K_p can be plotted with $\omega \to +\infty$ from zero.

Property 11.2.2   *The controller parameter boundaries of the stability region Q can be determined by the infinity root boundary (IRB), the real root boundary (RRB), and the complex root boundary (CRB)* [1], [86], [87], [93].

Proof: With $A = 1$, $\phi = 0°$ and a given K_d, the parameter-plane (K_p, K_i) can be divided into stable and unstable regions by the IRB, RRB and CRB presented above. The stable region can be detected by testing one arbitrary point in each region. The system characteristic equation with a PID controller whose parameters are chosen from the stable region has no root locating in the right half of the s-plane. Conversely, if a PID controller is chosen from the unstable region, the system characteristic equation must have some roots in the right half of the s-plane. Thus, the stability region of the parameters K_i and K_p can be fixed by the RRB and CRB conditions with a given K_d in the boundary (4.8) determined by the IRB condition. Sweeping over all the stabilizing $K_d \in [-T/K, T/K]$, the three-dimensional stability region in the parameter-space for the three PID parameters can be determined, which is called the *complete stability region*. ∎

11.3 PID Controller Design with Pre-Specifications on ϕ_M and ω_C

11.3.1 Design Scheme

Since the *complete stability region* is determined, a special surface in the *complete stability region* of the parameter-space can be drawn to satisfy the designed phase margin ϕ_m with the set of $A = 1$ and $\phi = \phi_m$ in (4.10), or the designed gain margin A_m with the set of $\phi = 0°$ and $A = A_m$ in (4.10).

Definition 11.3.1   *Relative stability line and surface: Given one specification – phase margin $\phi = \phi_m$ ($A = 1$), and from (11.9) in CRB, a relative stability line can be drawn in the (K_p, K_i) parameter-plane as $\omega \to \omega_0$ from zero with a certain fixed $K_{d1} \in [-T/K, T/K]$. ω_0 is the maximum frequency guaranteing the pre-specified phase margin with the fixed K_{d1} on the relative stability line. Sweeping all the K_d in $[-T/K, T/K]$, a surface in the three-dimensional parameter-space can be generated satisfying the pre-specified phase margin ϕ_m, which is called the relative stability surface. Since there exists a maximum frequency ω_0 for every relative stability line, the frequency boundary of the relative stability surface can be found.*

Property 11.3.1   *The frequency ω of every point on the relative stability line can be used as the gain crossover frequency with the corresponding PID controller parameters.*

Proof: From equation (11.5) and the characteristic equation (11.7), one can obtain,

$$D(K_p, K_i, K_d, A, \phi; s)|_{s=j\omega} = 1 + M_T(A, \phi)C(s)P(s)|_{s=j\omega} = 0, \qquad (11.16)$$

which means that the open-loop transfer function $G(s)$ is equal to -1, that is,

$$G(s) = M_T(A, \phi)C(s)P(s)|_{s=j\omega} = -1,$$

so, one can obtain,

$$|M_T(A, \phi)C(s)P(s)|_{s=j\omega}| = 1, \angle(M_T(A, \phi)C(s)P(s)|_{s=j\omega}) = -\pi.$$

If $A = 1$ and $\phi = \phi_m$, all the ω satisfying equation (11.16) can be used as the gain crossover frequencies with phase margin ϕ_m for the plant (11.1) with a proper PID controller. Actually, every point with $\omega \in (0, \omega_0]$ on the *relative stability line* corresponds to a proper PID controller which can satisfy (11.16), namely, all the $\omega \in (0, \omega_0]$ can be used as the gain crossover frequencies with the corresponding PID controller parameters chosen from the points on the *relative stability line*.   ∎

Therefore, on a certain *relative stability line* corresponding to a fixed K_{d1}, a point with a pair of the parameters K_p and K_i can be determined with the other specification – gain crossover frequency $\omega = \omega_c$.

Definition 11.3.2   *Relative stability curve: With the pre-specified ω_c, ϕ_m and a certain K_{d1}, the other two PID parameters K_p and K_i can be determined from a point on the relative stability line. Therefore, one $K_d \in [-T/K, T/K]$ corresponds to one point satisfying the two specifications ω_c and ϕ_m on the relative stability line. In the same way, sweeping all the K_d in $[-T/K, T/K]$, a relative stability curve in the three-dimensional parameter-space can be found to guarantee that, all the points on this curve can meet the pre-specified ω_c and ϕ_m.*

11.3.2 Flat Phase Tuning Rule for the Robust PID Controller Design

To achieve the robust property for the designed PID controller, the "flat phase" rule is applied to select the unique set of PID parameters.

From (11.11) and (11.12), one can obtain,

$$\phi = \arctan \frac{T\omega^2 K_p + K_i - \omega^2 K_d}{-\omega K_p + \omega T K_i - \omega^3 T K_d} - \omega L + n\pi, \tag{11.17}$$

where n is an integer which guarantees,

$$\phi + \omega L - n\pi = \arctan \frac{T\omega^2 K_p + K_i - \omega^2 K_d}{-\omega K_p + \omega T K_i - \omega^3 T K_d} \in (-\pi/2, \pi/2).$$

In order to enhance the system robustness to loop gain variations, which include the uncertainty of the plant steady-state gain and the entire change of the PID controller coefficients due to the finite work length effect, the "flat phase" constraint as shown in Definition 11.3.3 as an additional tuning specification is proposed to select the unique PID controller in the three-dimensional parameter-space.

Definition 11.3.3   *Flat phase constraint: The flat phase means the phase of the open-loop system is flat around the gain crossover frequency point in the Bode plot.*

Property 11.3.2   *With this flat phase constraint, the open-loop system phase can maintain almost the same value when the loop gain changes in a certain interval, namely, the system with this designed PID is robust to the loop gain variations. The overshoots of the step responses are almost the same with the variations of loop gain in certain range. Hence, the control performance of the system with the designed PID controller degrades gracefully when the steady-state gains of the plant and the PID controller change.*

To satisfy the flat phase tuning constraint, the derivative of the open-loop system phase θ with respect to the frequency ω is forced to be zero at the gain crossover frequency point, that is,

$$\frac{d\theta}{d\omega} = 0.$$

As mentioned in Section 11.3.4, ϕ can be used as the phase margin with $A = 1$ in (11.10). Thus, open-loop phase $\theta = \phi - \pi$, and,

$$\frac{d\theta}{d\omega} = \frac{d\phi}{d\omega} = 0.$$

From (11.17), one can obtain,

$$\frac{d\phi}{d\omega} = \frac{E}{B^2 + C^2} - L = 0, \tag{11.18}$$

where

$$
\begin{aligned}
E &= -T\omega^2 K_p^2 + (T^2\omega^2 + 1)K_p K_i + (T^2\omega^4 + \omega^2)K_p K_d + 2T\omega^2 K_d K_i \\
&\quad - T\omega^4 K_d^2 - T K_i^2, \\
B &= -\omega K_p + \omega T K_i - \omega^3 T K_d, \\
C &= T\omega^2 K_p + K_i - \omega^2 K_d.
\end{aligned}
$$

Definition 11.3.4   *Flat phase stable point: From Section 11.3.2, all the points with PID parameters on the relative stability curve can be tested by the equation (11.18). If a certain point with the PID parameters (K_p, K_i, K_d) can be found to guarantee the relationship (11.18), this point is called the flat phase stable point.*

Thus, the PID controller whose parameters are selected from this *flat phase stable point* can achieve the desired control performance introduced by the two specifications ω_c and ϕ_m, and the robustness to loop gain variations.

11.3.3 Design Procedures Summary with An Example

In this section, we present an examples to summarize the procedures of the proposed PID controller synthesis.

Step 1: Given the FOPTD plant with $K = 1$, $T = 1s$ and $L = 0.1s$, pre-specify the phase margin $\phi_m = 50°$ and gain crossover frequency $\omega_c = 10\,rad/s$.

Step 2: Find the stability interval of $K_d \in [-1, 1]$ from (11.8) according to IRB. Draw the straight line $K_i = 0$ in the parameter-plane of K_p and K_i following RRB. Choose $K_d = 0.5$ and draw the line of K_p with respect to K_i in the (K_p, K_i)-plane according to the relationships in (11.14) and (11.15) from CRB. Detect the stable region with a random point test as shown as the section surrounded by the solid line in Figure 11.2a. Obtain the *complete stability region* as shown in Figure 11.2(b) by sweeping the $K_d \in [-1, 1]$, following the scheme introduced in Section 11.3.1. In order to show the boundary clearly of the three-dimensional *complete stability region*, the boundary plane $K_i = 0$ is clearly presented in Figure 11.2(b).

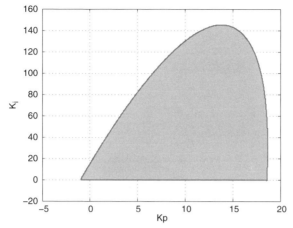

(a) Stability region of K_i with respect to K_p with $K_d = 0.5$

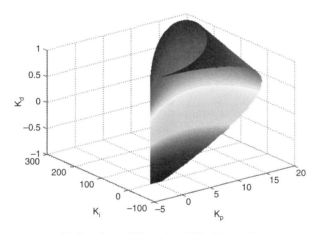

(b) Complete stability region of K_i, K_p and K_d

Figure 11.2 Stability region of K_i with respect to K_p with $K_d = 0.5$ and complete stability region

Step 3: With the fixed $K_d = 0.5$ and a pre-specified phase margin $\phi_m = 50°$, the *relative stability line* in the (K_p, K_i)-plane can be drawn as shown as the dashed line with $\omega \in (0, \omega_0]$ in Figure 11.3(a), and the relative stability region surrounded by the solid line can be compared with the complete stability region as shown as the solid line and $K_i = 0$ surrounded section. Find the *relative stability surface* by sweeping $K_d \in [-1, 1]$ in the three-dimensional parameter-space as shown in Figure 11.3(b), so the PID controllers whose parameters are chosen from the points on this *relative stability surface* can all satisfy the pre-specified phase margin $\phi_m = 50°$ for the given FOPTD system.

Step 4: According to the *relative stability surface* in Figure 11.3(b), the relation curve of the maximum frequency ω_0 on a *relative stability line* with respect to K_d can be drawn

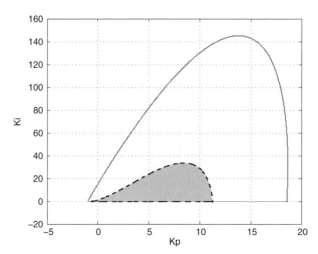

(a) Stability region comparison of K_i and K_p with $\phi_m = 0°$ and $\phi_m = 50°$ ($K_d = 0.5$)

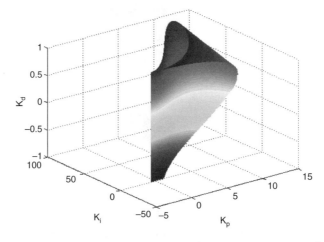

(b) Three-dimensional stability region of K_i, K_p and K_d with $\phi_m = 50°$.

Figure 11.3 Stability region comparison of K_i and K_p and three-dimensional stability region with $\phi_m = 50°$

in Figure 11.4. It can be seen that the maximum value of ω_0 on all the *relative stability line* is $\omega_{0\,max} = 22.73\,rad/s$. On the dashed *relative stability line* in Figure 11.3(a) with $K_d = 0.5$, give the other design specification, gain crossover frequency $\omega_c = 5\,rad/s$, the other two parameters K_p and K_i of the PID controller can be determined, as shown as the intersection point with $\omega = \omega_c = 5\,rad/s$ in Figure 11.5a which is zoomed in from Figure 11.3(a). Sweeping all the $K_d \in [-1, 1]$ as shown in Figure 11.5(b), one can find the *relative stability curve* presented by the dashed line in Figure 11.6a on the *relative stability surface* with $\phi_m = 50°$. All the points on this *relative stability curve* can satisfy the two specifications $\phi_m = 50°$ and $\omega_c = 5\,rad/s$ simultaneously.

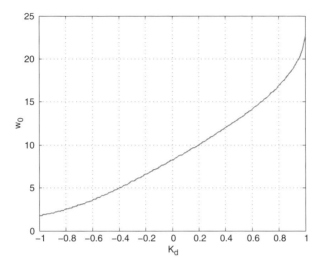

Figure 11.4   Relation curve of ω_0 vs. K_d, $\omega_{0\,max} = 22.73\,rad/s$

Step 5: Test all the points on the the *relative stability curve* to find a unique solution of equation (11.18) from the "flat phase" constraint. This solution point is illustrated as the star on the *relative stability curve* in the three-dimensional parameter-space of Figure 11.6(b). Find the corresponding PID parameters at this *flat phase stable point* to fix the unique PID controller satisfying the pre-specified phase margin $\phi = 50°$, gain crossover frequency $\omega_c = 5\,rad/s$, and the flat phase constraint with the given FOPTD plant.

11.3.4 How to Find the Achievable Region of the Two Specifications?

Since the PID controller design scheme in this chapter is based on the three specifications which include two pre-specified design specifications and a flat phase constraint, how to select the two design specifications is a valuable and instructional question. In this section, the achievable region of two design specifications can be found to guarantee the existence of the PID solution which is stable and can satisfy the three specifications. Therefore, the feasibility of the chosen two design specifications can be verified or checked from the achievable ranges in advance, prior to the controller design. Moreover, after the FOPTD system is normalized, the desired PID can be easily looked up in the table satisfying the given specifications and robustness on the loop gain variations for any open-loop stable FOPTD systems.

11.3.4.1 Collection Scheme for Achievable Region

From Section 11.3.4, with a pre-specified phase margin ϕ_m, there exists a maximum value $\omega_{0\,max}$ of the frequency ω_0 on all the *relative stability lines* from the relation curve of ω_0 with respect to K_d in Figure 11.4. Therefore, with different gain crossover frequency

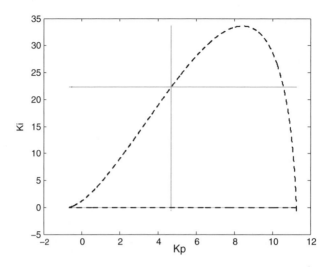

(a) The designed K_i and K_p satisfying $\omega_c = 5\ rad.s$, $\phi_m = 50°$ with $K_d = 0.5$

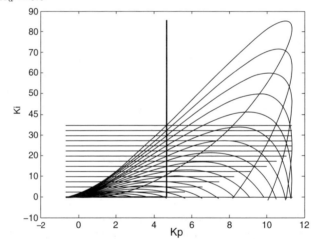

(b) The designed K_i and K_p satisfying $\omega_c = 5\ rad.s$, $\phi_m = 50°$ sweeping all the optional K_d

Figure 11.5 The designed K_i and K_p with $K_d = 0.5$ and those with sweeping all the optional K_d from the black line intersections

$\omega_c \in (0, \omega_{0\,max}]$, the existence of all the flat phase stable points can be detected for the pre-specified phase margin ϕ_m.

So, with a given phase margin, the region for choosing the gain crossover frequency to find the desired PID controller satisfying three specifications, can be decided by searching the frequency in between $(0, \omega_{0\,max}]$. Whereafter, sweeping the phase margin design specification from 0 to $2\pi\ rad$, the complete information about the achievable region of the phase margin and gain crossover frequency specifications

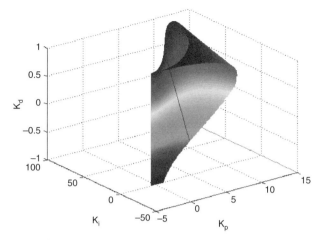

(a) The relative stability curve on the relative stability surface

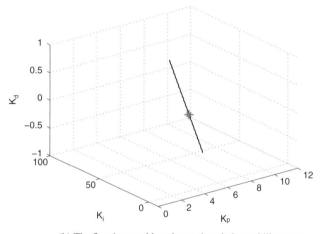

(b) The flat phase stable point on the relative stability curve

Figure 11.6 The relative stability curve and the flat phase stable point in the three-dimensional parameter-space (For a color version of this figure, see Plate 15)

can be collected to guarantee the existence of a PID controller solution satisfying the pre-specified specifications and the flat phase tuning constraint.

According to this instructional achievable region information, choosing a phase margin ϕ_m and a gain crossover frequency ω_c properly, the desired stable and robust PID controller can be obtained following the proposed design synthesis in this chapter.

11.3.4.2 Example Illustration and Comparison

Step 6: With the maximum optional frequency $\omega_{0\,max}$ from *Step 4*, search all the gain crossover frequency specifications the ω_c in $(0, \omega_{0\,max}]$ which can lead to a flat phase

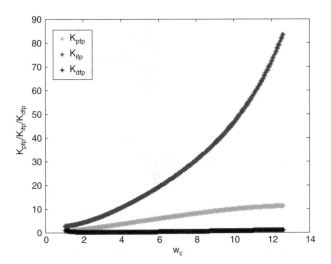

Figure 11.7 K_{pfp}, K_{ifp} and K_{dfp} v.s. ω_c with $\phi_m = 50°$

stable point, with the pre-specified phase margin $\phi_m = 50°$. So, the achievable region information of the gain crossover frequency ω_c can be collected from Figure 11.7 with $\phi_m = 50°$, and the achievable interval of ω_c corresponds to the nonzero solutions of the flat phase stable point (K_{pfp}, K_{ifp}, K_{dfp}). Sweeping the different phase margin $\phi_m \in [0, 2\pi]$, the complete achievable region of the phase margin and gain crossover frequency can be drawn as shown as the dark region in Figure 11.8a.

Remark 11.3.1   *The FOPTD system (11.1) can be normalized as follows:*

$$P_n(s) = \frac{1}{Ts+1}e^{-(L/T)Ts} = \frac{1}{s'+1}e^{-L's'}, \tag{11.19}$$

where $s' = Ts$ and $L' = L/T$. The parameter K in (11.1) can be normalized as 1, since the steady-state gain of the plant can always be used as part of the gain of the PID controller. So, if the complete information of the achievable set of specifications ϕ_m and ω_c is collected for the standard form of the control system plant below,

$$P_0(s) = \frac{1}{s+1}e^{-Ls}, \tag{11.20}$$

where L is equal to L' in (11.19), then the complete achievable region of the specifications ϕ_m and ω_c can be easily found for the normalized FOPTD system (11.19), with the proportional change of the ω_c axis, $\omega_c = \omega_c'/T$ as $s' = Ts$.

Following the example in Section 11.3.3, the complete information for the achievable ϕ_m and ω_c is collected for the standard control system (11.20) with $L = 0.1s$. In order

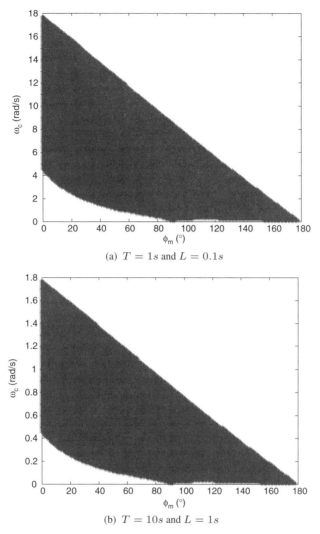

(a) $T = 1s$ and $L = 0.1s$

(b) $T = 10s$ and $L = 1s$

Figure 11.8 The achievable regions of ω_C vs. ϕ_m with $T = 1s$ and $L = 0.1s$, $T = 10s$ and $L = 1s$

to validate the content in Remark 11.3.1, by repeating the procedures $Step1$ to $Step\ 6$ in Section 11.3.3, the achievable region of ϕ_m and ω_c is also collected for the normalized control plant (11.19) with $T = 10s$ and $L = 1s$, where $L' = L/T = 0.1s$ which is equal to the time delay L in (11.20) above. As shown in Figure 11.8(b), the figure is the same with Figure 11.8a except the $1/T = 0.1$ times proportional relationship of the ω_c axis. For the standard form (11.20) of the control systems, different achievable regions of the ϕ_m and ω_c can be found with different time delay L. Following the procedures in Section 11.3.3, the region of achievable specifications is collected with $L = 1s$ and $L = 10s$ in Figure 11.9(a) and Figure 11.9(b), respectively.

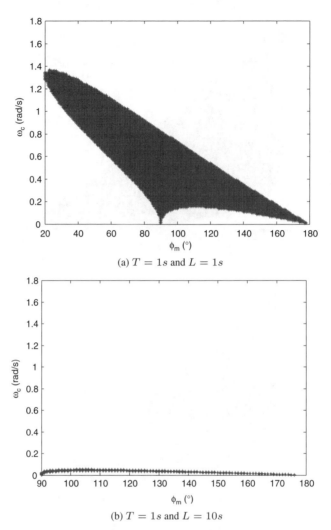

Figure 11.9 The achievable regions of ω_c vs. ϕ_m with $T = 1s$ and $L = 1s$, $T = 1s$ and $L = 10s$ (For a color version of this figure, see Plate 16)

With the achievable specification region information based on the standard form plants (11.20), an arbitrary FOPTD system can obtain the complete information of the achievable specifications directly by this normalization scheme without re-design and calculation. This normalization strategy can obviously simplify the stable and robust PID controller design and the achievable region information collection of the two specifications.

Remark 11.3.2 *In this chapter, the proposed PID controller design synthesis is applied to the common FOPTD plants in process control. The motivation for this special application* *demonstration is to show the proposed design synthesis more clearly with the normalization scheme in Remark 11.3.1. Actually, this proposed PID design synthesis can be applied to a general control system model $P_g(s)$ below, following the summarized procedures Step 1 to Step 6 above,*

$$P_g(s) = \frac{N(s)}{D(s)} = \frac{b_i s^i + b_{i-1} s^{i-1} + \cdots + b_1 s + b_0}{a_i s^i + a_{i-1} s^{i-1} + \cdots + a_1 s + a_0} e^{Ls}, \tag{11.21}$$

where a_i, b_i are coefficients, i is a positive integer, and L is the time delay.

11.4 Simulation Illustration

In this section, the proposed PID controller design synthesis is validated by the numerical simulation illustration. In order to validate the advantages of the designed PID controller, the traditional Ziegler-Nichols PID (ZNPID) controller is compared with the designed PID controller for the given FOPTD simulation model following the proposed tuning synthesis in this chapter.

The FOPTD simulation model (11.1) is chosen with $K = 1$, $T = 1s$ and $L = 0.1s$. According to the Ziegler-Nichols tuning rule [249] for the FOPTD systems, the parameters of the ZNPID controller can be decided by the following formulas,

$$K_{pzn} = \frac{1.2T}{KL}, \quad K_{izn} = \frac{K_{pzn}}{2L}, \quad K_{dzn} = \frac{K_{pzn}L}{2}.$$

Thus, the ZNPID controller can be determined as,

$$C_{zn}(s) = K_{pzn} + K_{izn}\frac{1}{s} + K_{dzn}s = 12 + 60\frac{1}{s} + 0.6s.$$

With this ZNPID controller and the considered FOPTD simulation model, the Bode plot of the open-loop transfer function can be drawn in Figure 11.10a. It can be seen that, if the magnitude of the system increases 3dB from the gain crossover point, the phase increases to $-141°$ from $-144°$ with $3°$ phase change; conversely, if the magnitude decreases 3dB from the gain crossover point, the phase decreases to $-178°$ from $-144°$ with $34°$ phase change. So, this system is not robust to the system gain variations from the Bode plot.

Following the PID controller design synthesis in this chapter, the PID controller satisfying two specifications (phase margin $\phi_m = 50°$ and gain crossover frequency $\omega_c = 5\,rad/s$) and the flat phase tuning constraint, can be designed based on the given FOPTD simulation model,

$$C_{fp}(s) = K_{pfp} + K_{ifp}\frac{1}{s} + K_{dfp}s = 4.71 + 14.48\frac{1}{s} + 0.19s.$$

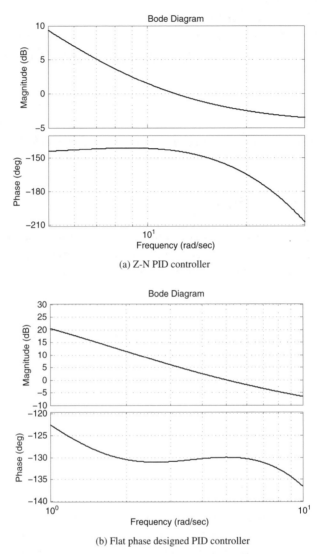

(a) Z-N PID controller

(b) Flat phase designed PID controller

Figure 11.10 Open-loop system Bode plot with Z-N PID controller and flat phase designed PID controller

The Bode plot of the open-loop system with the designed PID controller and the given simulation model can be drawn in Figure 11.10(b). One can observe that, if the magnitude of the system increases 3dB from the gain crossover point, the phase increases to $-130.4°$ from $-130°$ with $0.4°$ phase change; conversely, if the magnitude decreases 3dB from the gain crossover point, the phase decreases to $-130.9°$ from $-130°$ with $0.9°$ phase change. So, it is obvious that this system is much more robust to the system gain variations from the Bode plot than the system with the ZNPID controller.

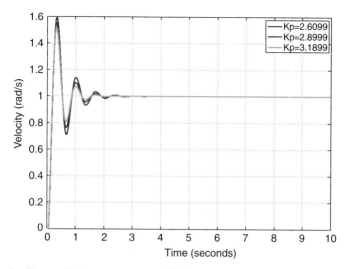

Plate 1 The unit step responses with the fractional order (PI) controller

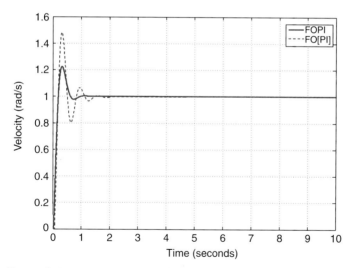

Plate 2 The unit step responses with the fractional order PI and (PI) controllers

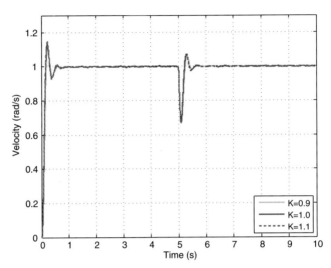

Plate 3 Experiment. Step responses and disturbance rejections using FO(PI) with loop gain variations ($\omega_c = 15\ rad/s$, $\phi_m = 65°$)

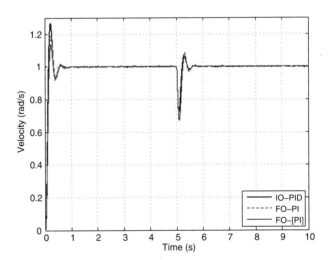

Plate 4 Experiment. Step response and disturbance rejection comparison ($\omega_c = 15\ rad/s$, $\phi_m = 65°$)

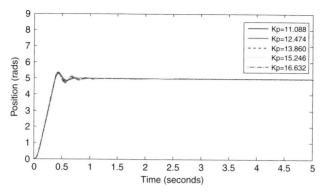

Plate 5 Dynamometer real-time experiment. Step position responses with FOPD controller

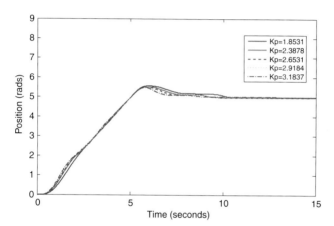

Plate 6 Dynamometer real-time experiment. Ramp position responses with ITAE optimal PI controller

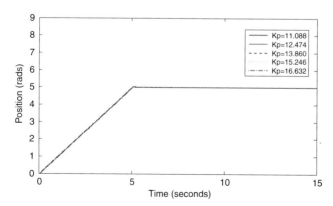

Plate 7 Dynamometer real-time experiment. Ramp position response with FOPD controller

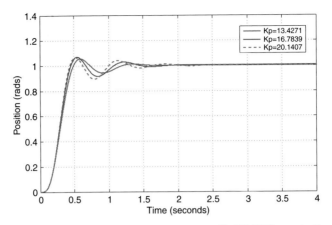

Plate 8 Experiment. Step responses with FO(PD) controller

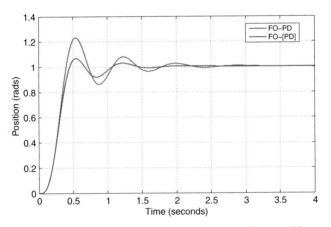

Plate 9 Experiment. Step responses comparison with two FO controllers

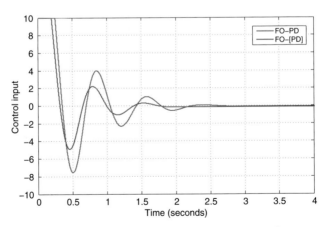

Plate 10 Experiment. Control input signals with two FO controllers

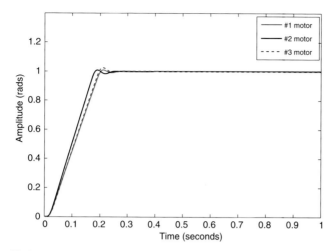

Plate 11 Step response comparison of the three motors

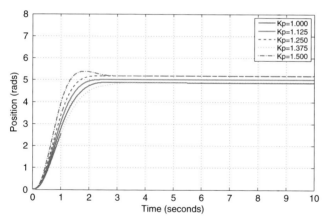

Plate 12 Experiment. Step responses with the ITAE optimal P controller designed based on the original dynamometer IOS for the original dynamometer IOS

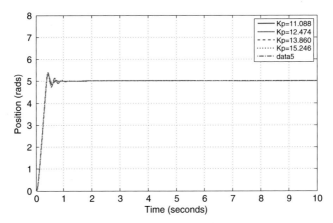

Plate 13 Experiment. Step responses with the FOPD controller designed based on the original dynamometer IOS for the original dynamometer IOS

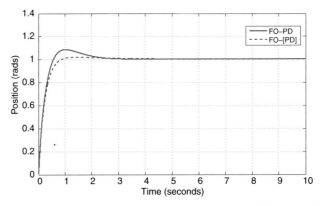

Plate 14 Step responses comparison with two FO controllers

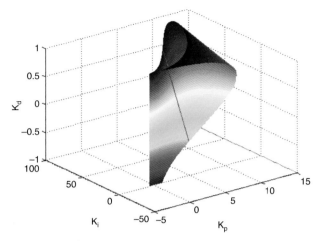

(a) The relative stability curve on the relative stability surface

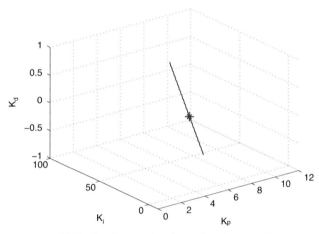

(b) The flat phase stable point on the relative stability curve

Plate 15 The relative stability curve and the flat phase stable point in the three-dimensional parameter-space

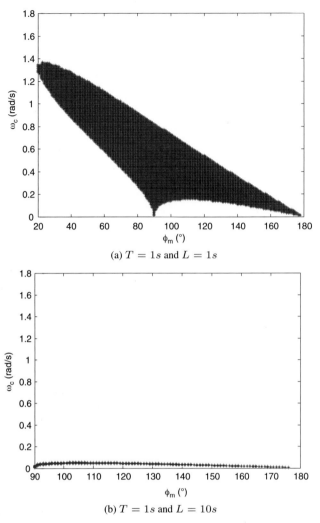

(a) $T = 1s$ and $L = 1s$

(b) $T = 1s$ and $L = 10s$

Plate 16 The achievable regions of ω_c vs. ϕ_m with $T = 1s$ and $L = 1s$, $T = 1s$ and $L = 10s$

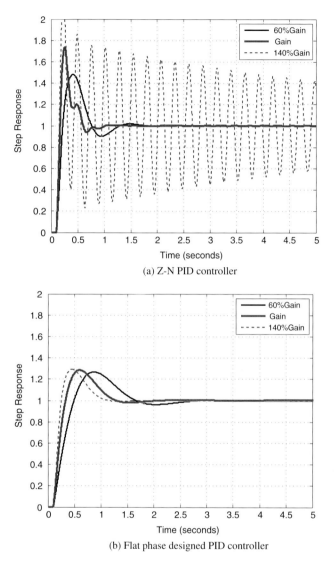

(a) Z-N PID controller

(b) Flat phase designed PID controller

Plate 17 Step responses using the Z-N PID controller and the flat phase designed PID controller with system gain variations

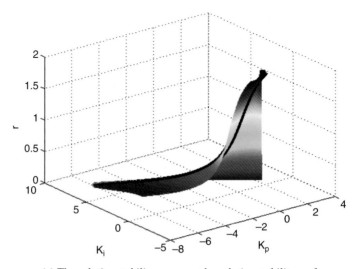

(a) The *relative stability curve* on the *relative stability surface*

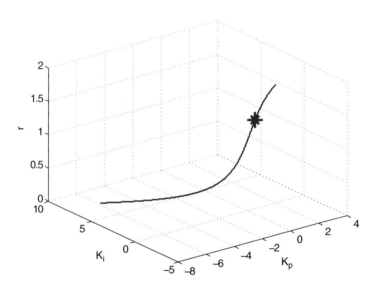

(b) The *flat phase stable point* on the *relative stability curve*

Plate 18 The *relative stability curve* and the *flat phase stable point* in the three-dimensional parameter-space

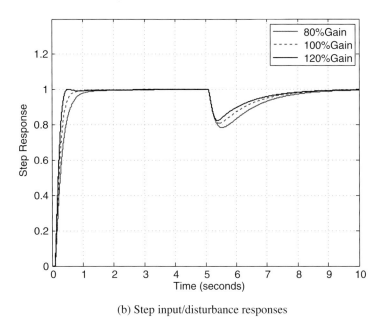

(b) Step input/disturbance responses

Plate 19 (b) Step input/disturbance responses with loop gain variations using designed #2 FOPI controller $C_{fopi2}(s)$

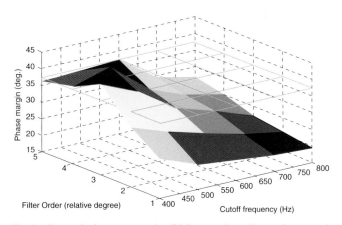

Plate 20 Illustration of phase margin (PM) as a function of n_Q and ω_Q in DOB

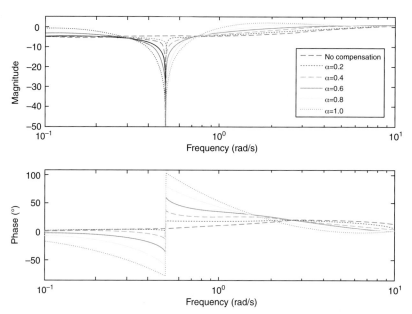

Plate 21 Experiment: Bode plots of the sensitivity function with $\omega_1 = 0.5$ and $\alpha \in (0, 1)$

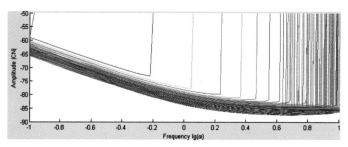

(a) Enlarged 2D Bode plot of the amplitude w. r. t. ω using IOPI with $T_{0IOPI} = 0.577\ Nm$

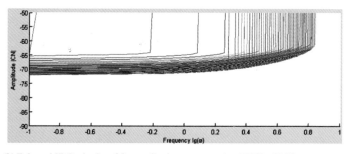

(b) Enlarged 2D Bode plot of the amplitude w. r. t. ω using FOPD with $T_{0FOPD} = 0.622\ Nm$

Plate 22 Enlarged 2D Bode plot of the amplitude with respect to ω

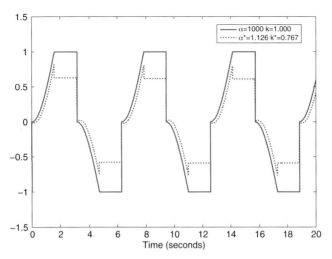

Plate 23 Output comparison of the *ICI* and the *OFOCI*$_3$

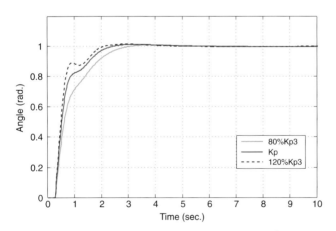

Plate 24 Simulation: Step responses using the designed $(PI)^\lambda$ (80°) controller (20.24) with system gain variations

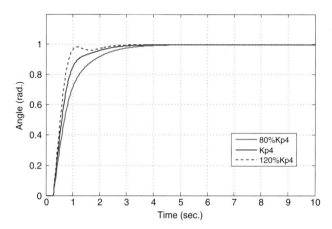

Plate 25 Simulation: Step responses using the designed $(PI)^\lambda$ controller (20.22) with system gain variations

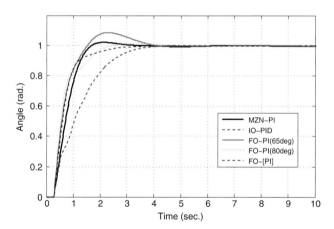

Plate 26 Simulation: Step responses comparison using the designed stabilizing fractional/integer order controllers

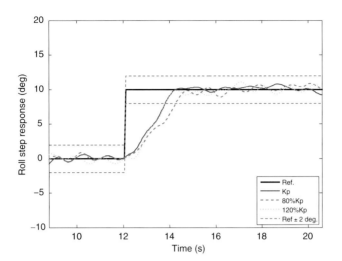

Plate 27 Experiment: Step responses using the designed $(PI)^\lambda$ controller (20.22) with system gain variations

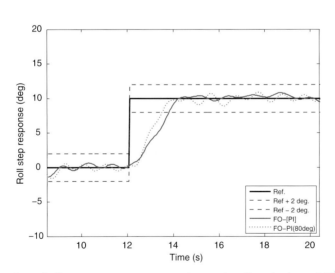

Plate 28 Experiment: Step responses comparison using the designed PI^λ (80°) (20.24) and $(PI)^\lambda$ (20.22) controllers

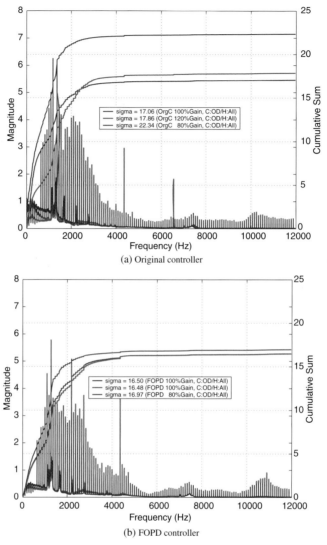

(a) Original controller

(b) FOPD controller

Plate 29 TMR comparison with original controller and designed FOPD controller

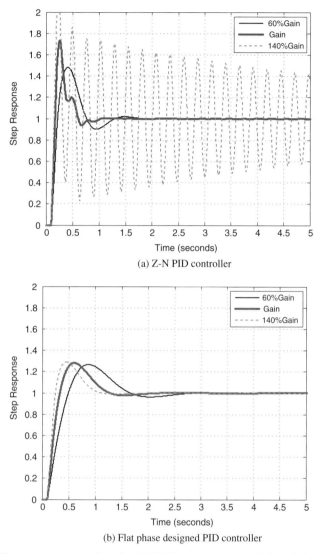

(a) Z-N PID controller

(b) Flat phase designed PID controller

Figure 11.11 Step responses using the Z-N PID controller and the flat phase designed PID controller with system gain variations (For a color version of this figure, see Plate 17)

In order to validate the robustness comparison on the Bode plots of the system with ZNPID and the system with the designed PID, the step responses are tested using these two PID controllers with the same given simulation FOPTD model. Since the analysis of the phase margin changing under the ±3dB magnitude variations on the Bode plots is presented above, the ±40% system gain variations are performed in the step response tests as ±3dB corresponds to 1.4125/0.7079 times change of the system gain. Figure 11.11(a) shows the unit step responses using the ZNPID controller with

±40% system gain variations. Correspondingly, Figure 11.11(b) shows the unit step responses using our designed PID controller with ±40% system gain variations. It can be seen clearly that the system using our designed PID is much more robust to the system gain variations than that using the ZNPID controller.

11.5 Chapter Summary

In this chapter, given any open-loop stable FOPTD model, all stabilizing traditional integer order PID controllers can be found in terms of a 3D plot of three gains. Given specifications such as gain crossover frequency and phase margin, the three PID gains are not unique. To make three gains unique, an additional constraint "flat phase" condition is added, that is, the phase slope at the gain crossover frequency is zero, guaranteeing robustness to loop gain variations. This chapter offers an analytical synthesis method to obtain unique PID gains. Furthermore, a procedure is suggested to find out all feasible specifications. With a scaling idea, the FOPTD model can be reduced to having only one plant parameter (delay to time constant ratio), therefore, a lookup table can be constructed for a practical design scenario: Given the delay to time constant ratio, the feasible ranges of specifications can be determined; given (feasible) specifications, unique PID parameters can be retrieved from the constructed table, with the "flat phase" property embedded. This scheme serves as an example that we can claim to have completely solved the robust PID synthesis problem for FOPTD models widely used in the process industry which we feel very excited about.

12

Stability and Design Feasibility of Robust FOPI Controllers for FOPTD Systems

12.1 Introduction

As presented in Chapter 11, for all open-loop stable first order plus time delay (FOPTD) systems, a traditional integer order PID controller can be designed to satisfy the given crossover frequency, phase margin and a flat phase constraint. This flat phase means that the system open-loop phase is constant around the given gain crossover frequency, which can show the iso-damping property for the system responses. Meanwhile, the questions about how we can know a selected combination of gain crossover frequency and phase margin is achievable for a proper integer order PID controller while maintaining the flat phase feature, and if this designed integer order PID controller is stable, are also addressed. In this chapter, in order to find a guideline for choosing the proper gain crossover frequency and phase margin to ensure the existence of the stabilizing and desired fractional order controllers, a fractional order proportional integral (FOPI) controller design synthesis is proposed for all stable FOPTD systems.

Using the design scheme in this chapter, a two-dimensional figure for the complete set of feasible gain crossover frequency and phase margin can be drawn for any given stable FOPTD system. With this complete set as the prior knowledge, all the combinations of phase margin and gain crossover frequency can be verified prior to the controller design. Only if the combination is chosen from this achievable region can the existence of the stabilizing and desired FOPI be guaranteed. In particular, it is interesting to compare the areas of the feasible region for the FOPI controller and that for the IOPID controller from Chapter 11.

As a starting point, a scheme for finding the complete stability region of the FOPI controller for the given FOPTD system is presented, and then a scheme for designing a stabilizing FOPI controller satisfying the given gain crossover frequency, phase margin and flat phase constraint is presented in detail. After that, the complete feasible region of the gain crossover frequency and the phase margin is constructed. This feasible region for the FOPI controller is compared with that for the traditional IOPID controller. This area comparison shows the advantage of the FOPI over the traditional IOPID clearly. A simulation illustration is presented to show the effectiveness and the performance of the designed FOPI controller compared with the optimized integer order proportional integer (IOPI) controller, and the designed IOPID controller following the scheme in Chapter 11 [222].

12.2 Stabilizing and Robust FOPI Controller Design for FOPTD Systems

12.2.1 The Plant and Controller Considered

Considering the feedback control system as shown in Figure 11.1, $P(s)$ is the control plant, and $C(s)$ is the designed controller. The considered plant $P(s)$ in this chapter represents all the open-loop stable FOPTD systems, which are characterized as the transfer function in (11.1).

In this chapter, we focus on the fractional order proportional integral (FOPI) controller $C(s)$ as follows,

$$C(s) = K_p + \frac{K_i}{s^r}, \tag{12.1}$$

where K_p is the proportional gain, K_i is the integral gain, and the real number $r \in (0, 2)$ is the fractional order [188].

In Figure 11.1, M_T is a gain-phase margin tester [39], which provides information for plotting the boundaries of the constant gain margin and the phase margin in the parameter plane [87]. The transfer function of M_T is given as (11.3). Setting $\phi = 0$ in (11.3), the controller parameter boundaries can be obtained satisfying a given gain margin A of the control system as shown in Figure 11.1. Meanwhile, setting $A = 1$ in (11.3), one can obtain the controller parameter boundaries for a given phase margin ϕ.

12.2.2 Stability Region Analysis of the FOPI Controller

The open-loop transfer function of the feedback control system in Figure 11.1 can be derived from (11.1), (12.1) and (11.3),

$$G(s) = M_T(A, \phi)C(s)P(s). \tag{12.2}$$

The closed-loop transfer function can be expressed as,

$$T(s) = \frac{M_T(A, \phi)C(s)P(s)}{1 + M_T(A, \phi)C(s)P(s)}. \tag{12.3}$$

Substituting (11.1), (12.1), and (11.3) into (12.3) yields,

$$T(s) = \frac{Ae^{-j\phi}e^{-Ls}K(K_ps^r + K_i)}{s^r(Ts + 1) + Ae^{-j\phi}e^{-Ls}K(K_ps^r + K_i)}. \tag{12.4}$$

Hence, the characteristic equation of the closed-loop system (12.3) is,

$$D(K_p, K_i, r, A, \phi; s) = s^r(Ts + 1) + Ae^{-j\phi}e^{-Ls}K(K_ps^r + K_i) = 0. \tag{12.5}$$

The primary concern for the controller design is the complete set of controllers which can stabilize the system. On the FOPI controller design for the stable FOPTD plants, the system stability is depens on the root locations of the characteristic equation (12.5) with $A = 1$ and $\phi = 0°$. If all roots of the polynomial (12.5) are located in the left half of the s-plane, the closed-loop system (12.3) is bounded-input bounded-output stable. There are three parameters K_p, K_i and r for the FOPI controller. The stability region Q of these three controller parameters is defined as that, if $(K_p, K_i, r) \in Q$, all the roots of $D(K_p, K_i, r, A, \phi; s)$ lie in the left half of the s-plane. The boundaries of the controller parameters stability region Q can be determined by the real root boundary (RRB) and the complex root boundary (CRB) [1], [86], [87], [93].

- Region of r: For the fractional order r in the FOPI controller, the chosen range is defined as $r \in (0, 2)$.
- RRB: The real root boundary is defined by the equation $D(K_p, K_i, r, A, \phi; s = 0) = 0$, so one can obtain the boundary as,

$$K_i = 0.$$

- CRB: Substituting $j\omega$ for s in (12.5), the complex root boundary can be defined from $D(K_p, K_i, r, A, \phi; s = j\omega) = 0$ as follows,

$$\begin{aligned}
D(K_p, K_i, r, A, \phi; j\omega) &= (j\omega)^r(jT\omega + 1) + Ae^{-j\phi}e^{-j\omega L}K(K_p(j\omega)^r + K_i) \\
&= \omega^r \cos\frac{r\pi}{2} - T\omega^{1+r}\sin\frac{r\pi}{2} \\
&\quad + AK\cos(\phi + \omega L)(K_i + K_p\omega^r \cos r\pi 2) + AK\sin(\phi + \omega L)K_p\omega^r \sin\frac{r\pi}{2} \\
&\quad + j(T\omega^{1+r}\cos\frac{r\pi}{2} + \omega^r \sin\frac{r\pi}{2} + AK\cos(\phi + \omega L)K_p\omega^r \sin\frac{r\pi}{2} \\
&\quad - AK\sin(\phi + \omega L)(K_i + K_p\omega^r \cos r\pi 2)) \\
&= 0,
\end{aligned} \tag{12.6}$$

where

$$j^r = e^{\frac{r\pi}{2}} = \cos\frac{r\pi}{2} + j\sin\frac{r\pi}{2}.$$

Considering the real part and the imaginary part of (12.6) respectively, one can obtain,

$$\omega^r \cos\frac{r\pi}{2} - T\omega^{1+r}\sin\frac{r\pi}{2}$$
$$+ AK\cos(\phi + \omega L)(K_i + K_p\omega^r \cos r\pi 2)$$
$$+ AK\sin(\phi + \omega L)K_p\omega^r \sin\frac{r\pi}{2}$$
$$= 0; \tag{12.7}$$
$$T\omega^{1+r}\cos\frac{r\pi}{2} + \omega^r \sin\frac{r\pi}{2}$$
$$+ AK\cos(\phi + \omega L)K_p\omega^r \sin\frac{r\pi}{2}$$
$$- AK\sin(\phi + \omega L)(K_i + K_p\omega^r \cos r\pi 2)$$
$$= 0. \tag{12.8}$$

From (12.7) and (12.8), we get,

$$B_1 + AKC_1E + AKS_1F = 0, \tag{12.9}$$

$$B_2 + AKC_1F - AKS_1E = 0, \tag{12.10}$$

where,

$$B_1 = \omega^r C_2 - T\omega^{1+r}S_2,$$
$$B_2 = T\omega^{1+r}C_2 + \omega^r S_2,$$
$$C_1 = \cos(\phi + \omega L), \quad S_1 = \sin(\phi + \omega L),$$
$$C_2 = \cos\frac{r\pi}{2}, \quad S_2 = \sin\frac{r\pi}{2},$$
$$E = K_i + K_p\omega^r C_2, \quad F = K_p\omega^r S_2.$$

From (12.9) and (12.10),

$$K_p = \frac{-(B_1S_1 + B_2C_1)}{AKS_2\omega^r}, \tag{12.11}$$

$$K_i = \frac{B - B_1S_1C_1 - B_2C_1^2}{AKS_1} + \frac{B_1S_1C_2 + B_2C_1C_2}{AKS_2}. \tag{12.12}$$

Hence, given r, the curve of K_i versus K_p can be plotted with $\omega \to +\infty$ from zero.

So, with $A = 1$, $\phi = 0°$ and a fixed fractional order r, the parameter plane (K_p, K_i) can be divided into stable and unstable regions by the RRB and CRB presented above. The stable region can be detected by testing one random point in each region [238]. Thus, the stability region of the parameters K_i and K_p can be fixed by the RRB and CRB conditions with a fixed r in the interval $(0, 2)$. By sweeping over all the $r \in (0, 2)$, the three dimensional stability region in the parameter space for the three parameters of FOPI can be determined, which is called the *complete stability region*.

12.2.3 FOPI Parameters Design with Two Specifications

Since the *complete stability region* is determined, the special surface in the *complete stability region* can be drawn to satisfy the specified phase margin ϕ_m when we set $A = 1$ and $\phi = \phi_m$ in (12.6), or satisfy the specified gain margin A_m with $\phi = 0°$ and $A = A_m$ in (12.6).

Given a specification of phase margin ϕ_m–a *relative stability line* can be drawn in the (K_p, K_i)-plane as $\omega \to \omega_0$ from zero with a certain fixed $r_1 \in (0, 2)$, by setting $A = 1$ and $\phi = \phi_m$ in (12.6). ω_0 is the maximum frequency on the *relative stability line* and in the *complete stability region*. Sweeping all the r in $(0, 2)$, a surface in the three-dimensional parameter space can be generated satisfying the pre-specified ϕ_m. This surface is called as the *relative stability surface*. The maximum frequency ω_{0max} in all ω_0 on all *relative stability lines* with $r_1 \in (0, 2)$ is the gain crossover frequency upper boundary of the *relative stability surface*.

Given another specification of gain crossover frequency ω_c–a point corresponding to the parameters K_p and K_i on the *relative stability line* can be determined with a fixed order r_1.

Actually, from (12.3) and (12.5), one can obtain the characteristic equation of the closed-loop system (12.3),

$$1 + M_T(A, \phi)C(s)P(s)|_{s=j\omega} = 0,$$

which means the open-loop transfer function $G(s)$ is equal to -1 with $s = j\omega$,

$$G(s)|_{s=j\omega} = M_T(A, \phi)C(s)P(s)|_{s=j\omega} = -1, \tag{12.13}$$

so one can find the magnitude equation,

$$|M_T(A, \phi)C(s)P(s)|_{s=j\omega}| = 1,$$

and the phase equation,

$$\angle(M_T(A, \phi)C(s)P(s)|_{s=j\omega}) = -\pi.$$

If setting $A = 1$ and $\phi = \phi_m$, all the $\omega \in (0, \omega_0]$ satisfying equation (12.13) can be treated as the gain crossover frequencies for the control system (12.3) in Figure 11.1. Since the *relative stability lines* are generated from the equation (12.5) which is equal to the equation (12.13), all the frequency ω corresponding to the points on the *relative stability lines* can be treated as the gain crossover frequency.

So, with the pre-specified ω_c, ϕ_m and a fixed r_1, the other two FOPI parameters K_p and K_i can be determined on a point of the *relative stability lines*. In the same way, sweeping all the r in $(0, 2)$, a curve in the three dimensional parameter space can be determined, which is the *relative stability curve*. All the points on this curve can guarantee the two specifications ω_c and ϕ_m.

To make the FOPI controller parameter setting unique, we need one more condition or constraint.

12.2.4 FOPI Parameters Design with an Additional Flat Phase Constraint

In this section, an additional flat phase tuning constraint is presented to make the parameters of the FOPI controller unique.

From (12.9) and (12.10),

$$S_1 = \frac{B_2 E - B_1 F}{AK(E^2 + F^2)}, \tag{12.14}$$

$$C_1 = \frac{-B_1 E - B_2 F}{AK(E^2 + F^2)}, \tag{12.15}$$

so, one can obtain,

$$\phi = \arctan \frac{B_1 F - B_2 E}{B_1 E + B_2 F} - \omega L + n\pi, \tag{12.16}$$

where n is an integer which guarantees,

$$\phi + \omega L - n\pi = \arctan \frac{B_1 F - B_2 E}{B_1 E + B_2 F} \in (-\pi/2, \pi/2).$$

In order to make the system robust to the loop gain variations, the flat phase constraint as an additional specification is presented to design the FOPI controllers. The flat phase means the phase of the open-loop system is flat around the gain crossover frequency point in the Bode plot. With this constraint, the system phase can retain almost the same value when the loop gain changes in a certain interval, namely, the system with this designed FOPI is robust to the loop gain variations, and the overshoots of the step responses are almost the same with the loop gain variations in a certain range.

In order to satisfy the flat phase tuning constraint, the derivative of the open-loop phase θ with respect to the frequency ω is forced to be zero at the gain crossover frequency point, for example,

$$\frac{d\theta}{d\omega} = 0.$$

As mentioned in Section 12.2.3, ϕ can be treated as the phase margin by setting $A = 1$ in (12.6). Thus, $\theta = \phi - \pi$, and,

$$\frac{d\theta}{d\omega} = \frac{d\phi}{d\omega} = 0.$$

From (12.16), one can obtain,

$$\frac{d\phi}{d\omega} = \frac{(B_1^2 + B_2^2)(EF' - E'F) + (B_1' B_2 - B_1 B_2')(E^2 + F^2)}{(B_1 E + B_2 F)^2 + (B_1 F - B_2 E)^2} - L$$

$$= 0, \tag{12.17}$$

where,

$$E' = C_2 K_p r \omega^{r-1},$$
$$F' = S_2 K_p r \omega^{r-1},$$
$$B_1' = C_2 r \omega^{r-1} - S_2 T(1+r)\omega^r,$$
$$B_2' = S_2 r \omega^{r-1} + C_2 T(1+r)\omega^r. \tag{12.18}$$

From Section 12.3.3, all FOPI parameter points on the *relative stability curve* in the three dimensional parameter space satisfying both the pre-specified design specifications ϕ_m and ω_c, can be tested by the equation (12.17). If a certain point with the FOPI parameters (K_p, K_i, r) on the *relative stability curve* can be found to satisfy the relationship (12.17), this point is called the *flat phase stable point*, which is the goal of this FOPI controller design for the given FOPTD system.

Thus, the FOPI controller from this *flat phase stable point* can achieve the desired control performance introduced by two specifications ϕ_m and ω_c, and the robustness to the loop gain changes due to the flat phase tuning constraint.

12.2.5 Achievable Region of Two Design Specifications for the FOPI Controller Design

From Section 12.2.3, the upper boundary ω_{0max} of the gain crossover frequency specification can be found to satisfy the given phase margin ϕ_m with a fixed r_1 in $(0, 2)$. Pre-specifying a phase margin ϕ_m, one *relative stability curve* in the three dimensional parameter space can be detected for one $\omega_c \in (0, \omega_{0max}]$. The existence of the *flat phase stable point* can be detected on each *relative stability curve* with each ω_c.

So, with a fixed phase margin, the region for choosing the gain crossover frequency to find the desired FOPI controller can be decided by searching the frequencies in between $(0, \omega_{0max}]$. Whereafter, sweeping the phase margin specification from 0 to 2π, the complete achievable region for the phase margin and gain crossover frequency, guaranteeing the existence of FOPI controllers that meet the flat phase tuning constraint, can be found.

According to this instructional information of the achievable region, a feasible combination of phase margin ϕ_m and gain crossover frequency ω_c can be checked in advance before any FOPI controller is designed, and the desired stabilizing and robust FOPI controller can be designed following the proposed synthesis in this chapter.

12.3 Design Procedures Summary with an Illustrative Example

In this section, the procedures are summarized with an illustrative example for the proposed FOPI controller design.

Step 1: Given the stable FOPTD plant with $K = 1, T = 1s$ and $L = 1s$, and two specifications on the phase margin $\phi_m = 50°$ and gain crossover frequency $\omega_c = 0.5 \ rad/s$.[1]

Step 2: With the range of fractional order $r \in (0, 2)$, choose $r = 0.5$ and draw the line of K_p with respect to K_i in the (K_p, K_i)-plane according to the equations (12.11) and (12.12) of CRB in Section 12.2.2. Draw the line $K_i = 0$ according to RRB in Section 12.2.2. Detect the stabilizing region with a random point test [238] as shown as the dark line boundary surrounded section in Figure 12.1(a). Obtain the *complete stability region* as shown in Figure 12.1(b) by sweeping all the r in $(0, 2)$, following the scheme introduced in Section 12.2.2.

Step 3: With the pre-specified $\phi_m = 50°$ and fixed $r = 0.5$, the *relative stability line* in the (K_p, K_i)-plane can be drawn as the dashed line in Figure 12.2(a), which can be compared with the complete stability boundary as shown as the dark line. Find the *relative stability surface* by sweeping all r in $(0, 2)$ in Figure 12.2(b) in the three-dimensional parameter space, satisfying the pre-specified phase margin $\phi_m = 50°$. According to the *relative stability surface* in Figure 12.2(b), the maximum value ω_{0max} of ω_0 which corresponds to the biggest frequency point on each *relative stability line* for different r, is $\omega_{0max} = 1.38 \ rad/s$.

Step 4: Given the gain crossover frequency $\omega_c = 0.5 \ rad/s$, phase margin $\phi_m = 50°$ and $r = 0.5$, the other two parameters K_p and K_i can be determined from the intersection point in Figure 12.3(a) with $\omega = \omega_c = 0.5 \ rad/s$. Sweeping all the $r \in (0, 2)$ as shown in Figure 12.3(b), one can find the *relative stability curve* presented as the black curve in Figure 12.4(a) on the *relative stability surface* with $\phi_m = 50°$. All points on this *relative stability curve* can satisfy the two specifications $\phi_m = 50°$ and $\omega_c = 0.5 \ rad/s$ simultaneously.

---

[1]Previously, there was no way to know in advance if this combination of ϕ_m and ω_c is feasible to obtain the desired controller. This chapter proposed the synthesis for the FOPI controller design with the complete achievable set for the combinations.

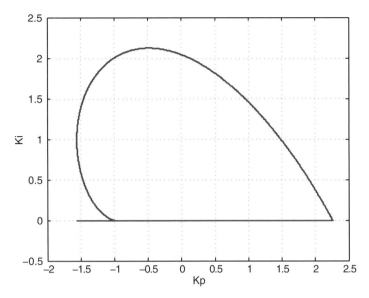

(a) Stability boundary of K_i vs. K_p with $r = 0.5$

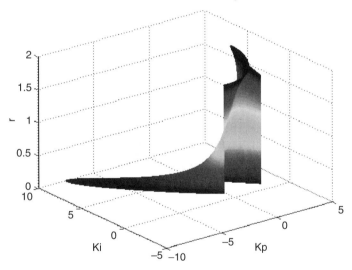

(b) *Complete stability region* of K_i, K_p and r

Figure 12.1 Stability boundary of K_i vs. K_p with $r = 0.5$ and the *complete stability region*

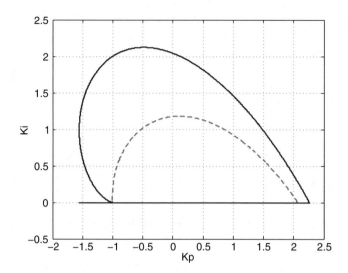

(a) Stability region comparison of K_i vs. K_p with $\phi_m = 0°$ in dark line and $\phi_m = 50°$ in dashed line ($r = 0.5$)

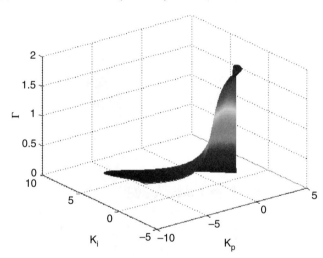

(b) *Three dimensional relative stability surface* of K_i, K_p and r with $\phi_m = 50°$

Figure 12.2 Stability boundary comparison of K_i vs. K_p and three dimensional *relative stability surface* with $\phi_m = 50°$

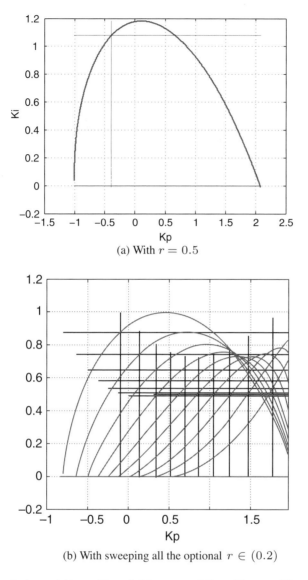

(a) With $r = 0.5$

(b) With sweeping all the optional $r \in (0.2)$

Figure 12.3 The designed K_i vs. K_p satisfying $\omega_c = 0.5$ rad/s, $\phi_m = 50°$ with $r = 0.5$ and those with sweeping all the optional $r \in (0, 2)$

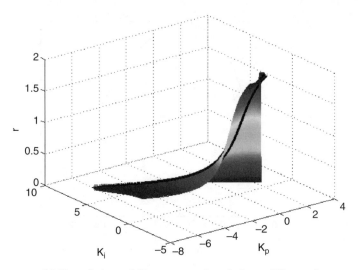

(a) The *relative stability curve* on the *relative stability surface*

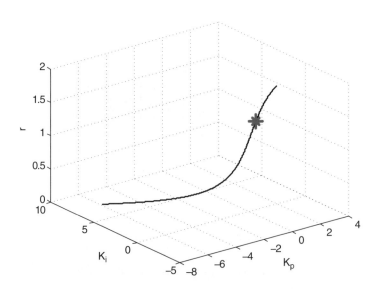

(b) The *flat phase stable point* on the *relative stability curve*

Figure 12.4 The *relative stability curve* and the *flat phase stable point* in the three-dimensional parameter space (For a color version of this figure, see Plate 18)

Step 5: Test all points on the *relative stability curve* to find a solution to equation (12.17). This point is illustrated as a dark star on the *relative stability curve* in the three dimensional parameter space of Figure 12.4(b). This point is called the *flat phase stable point*. Find the corresponding parameters (K_p, K_i, r) on this *flat phase stable point* to obtain the unique FOPI controller satisfying the pre-specified phase margin, gain crossover frequency and the flat phase constraint.

12.4 Complete Information Collection for Achievable Region of ω_c and ϕ_m

In order to highlight the scheme of complete achievable region collection for the ω_c and ϕ_m, this procedure is separate from the procedure summary in Section 12.3.

Step 6: With the frequency boundary ω_{0max} from *Step 3*, obtain the optional range of the gain crossover frequency $\omega_c \in (0, \omega_{0max}]$ under the pre-specified phase margin $\phi_m = 50°$. One *relative stability curve* can be drawn with one $\omega_c \in (0, \omega_{0max}]$ under $\phi_m = 50°$, and the existence of the *flat phase stable point* can be tested on each *relative stability curve*. So, the achievable gain crossover frequency ω_c to guarantee the existence of the *flat phase stable point* with $\phi_m = 50°$, can be obtained from Figure 12.5. The achievable ω_c interval corresponds to the nonzero solution of the *flat phase stable point* $(K_{pfp}, K_{ifp}, r_{fp})$. Testing a different phase margin $\phi_m \in (0, 2\pi)$, the complete information for the achievable region of the design specificaitons, namely, the phase margin and gain crossover frequency, can be obtained as shown as the dark region in Figure 12.6(a).

Note that the normalized plant idea in Remark 11.3.1 is also applied here.

For the normalized standard form (11.20) of the open-loop stable FOPTD systems, the different achievable region of ϕ_m and ω_m can be constructed with a different time delay L. Following the procedures in Section 12.3 and Section 12.4, the achievable regions are shown with $L = 10s$ and $L = 0.1s$ in Figure 12.7(a) and Figure 12.8(a), respectively.

With the achievable specification information based on the standard form plants (11.20), any open-loop stable FOPTD system, we can find the complete achievable region of the two design specifications directly by this normalization scheme without any recalculation.

Substituting the derivative gain K_d of the IOPID controller for the fractional order r of the FOPI controller, the IOPID controller can also be designed to guarantee the pre-specified ω_c, ϕ_m, and the flat phase constraint by following the synthesis in this Chapter 11. The complete achievable region of the design specifications can be constructed as well.

Based on the normalized form (11.20) of the FOPTD systems, the complete achievable regions of the ϕ_m and ω_c are constructed for IOPID with different time delays $L = 1s$, $L = 10s$, and $L = 0.1s$, as shown in Figure 12.6(b), Figure 12.7(b), and Figure 12.8(b), respectively.

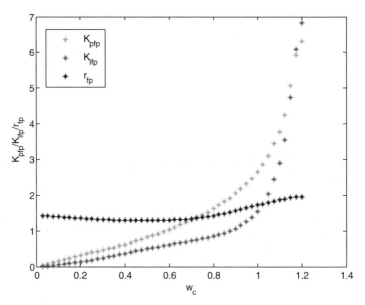

Figure 12.5 K_{pfp}, K_{ifp} and r_{fp} v.s. ω_c with $\phi_m = 50°$

Remark 12.4.1   *From the comparisons of Figure 12.6 and Figure 12.7 , it can be seen clearly that, for the FOPI controller, the achievable region of ϕ_m and ω_c is much bigger than that for the IOPID controller with the time delays $L = 1s$ and $L = 10$. Especially, when the system (11.20) is a delay dominant case, for example, $L = 10 >> T = 1$, the feasible region of ϕ_m and ω_c for FOPI is significantly bigger than that for IOPID. This larger achievable region for FOPI over IOPID gives the users more capability and flexibility to design the proper controller and gain the desired control performance. The advantage of the proposed FOPI controller over the traditional IOPID controller is illustrated clearly through these comparisons of the achievable regions of the two design specifications.*

Comparing Figure 12.8(a) and Figure 12.8(b), the achievable region of the two specifications for FOPI is smaller than that for IOPID. However, the achievable set for FOPI covers the blank part of the achievable set for IOPID with small values of ϕ_m and ω_c.

In summary:

When $L/(L + T) \to 1$, FOPI is more needed than IOPID.

When $L/(L + T) \to 0$, FOPI is an extended option of IOPID.

12.5 Simulation Illustration

In this section, the designed FOPI controller is validated by the numerical simulation illustration. In order to verify the effectiveness of this proposed controller design synthesis and show the advantages of the designed FOPI controller, an IOPI controller is optimized following the recognized method in [13]. This optimized integer order

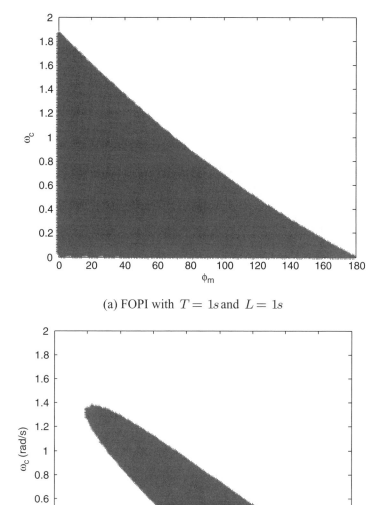

(a) FOPI with $T = 1s$ and $L = 1s$

(b) IOPID with $T = 1s$ and $L = 1s$

Figure 12.6 The achievable region of ω_c vs. ϕ_m for FOPI and IOPID design with $T = 1s$ and $L = 1s$

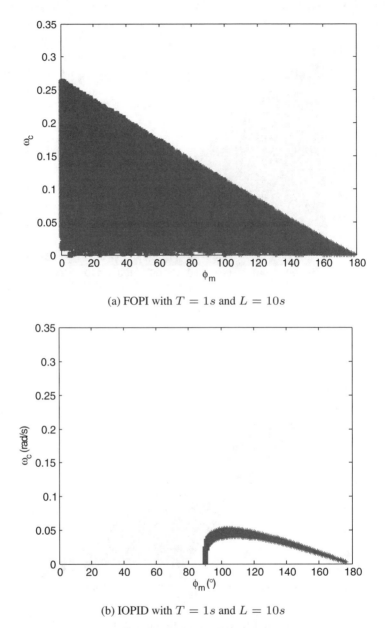

(a) FOPI with $T = 1s$ and $L = 10s$

(b) IOPID with $T = 1s$ and $L = 10s$

Figure 12.7 The achievable region of ω_c and ϕ_m for FOPI and IOPID design with $T = 1s$ and $L = 10s$

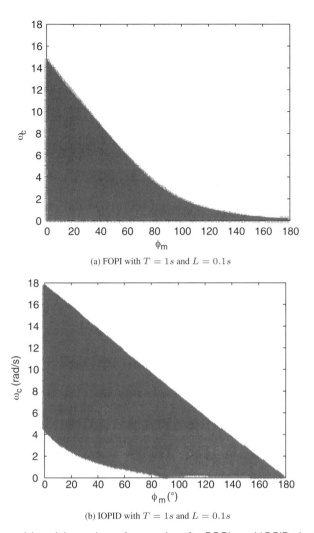

(a) FOPI with $T = 1s$ and $L = 0.1s$

(b) IOPID with $T = 1s$ and $L = 0.1s$

Figure 12.8 The achievable region of ω_c and ϕ_m for FOPI and IOPID design with $T = 1s$ and $L = 0.1s$

controller is compared with the designed FOPI controller for the given FOPTD model in the simulation.

Given the FOPTD system (11.1) with $K = 1$, $T = 1s$ and $L = 0.1s$, according to the "AMIGO tuning rule" for the FOPTD systems (page 72, [167]) [13], the parameters of the IOPI controller can be optimally designed by the following formulas,

$$K_{iop1} = \frac{0.15}{K} + \left[0.35 - \frac{LT}{(L+T)^2}\right]\frac{T}{KL} = 2.8236,$$

$$K_{ioi1} = K_{iop}\left(0.35L + \frac{13LT^2}{T^2 + 12TL + 7L^2}\right) = 4.6464.$$

Thus, the IOPI controller can be optimized as,

$$C_{iopi}(s) = K_{iop1} + K_{ioi1}\frac{1}{s} = 2.8236 + 4.6464\frac{1}{s}.$$

Using this optimal IOPI controller, the open-loop Bode can be plotted as shown in Figure 12.9(a). The gain crossover frequency and phase margin can be found as $0.485Hz$ and $62.3°$ respectively with this IOPI controller in Figure 12.9(a). In order to make a fair comparison, this specification combination of gain crossover frequency $0.485Hz$ and phase margin $62.3°$ is used to design the integer order PID controller following the method in Chapter 11, and the FOPI controller following the scheme proposed in this chapter. Therefore, the IOPID and FOPI controllers can be designed, respectively, as follows,

$$C_{iopid1}(s) = 2.8210 + 5.6638\frac{1}{s} + \frac{0.1092s}{T_{d1}s + 1},$$

where $1/(T_ds + 1)$ is a low pass filter for the derivative item; T_{d1} is set as K_d/N with $N = 10$, according to the choosing range $N \in [8, 20]$ in [12].

$$C_{fopi1}(s) = 3.3367 + \frac{6.1977}{s^{1.4433}}.$$

Using the designed IOPID and FOPI controllers $C_{iopid1}(s)$ and $C_{fopi1}(s)$, the open-loop Bode plots are shown in Figure 12.10(a) and Figure 12.11(a) with the given FOPTD model. It can be seen clearly the flat phase constraint is satisfied in these two sets of Bode plots. In practice, the controllers are always implemented with discretization. The sampling frequency is chosen as 2kHz. The step input responses and disturbance (magnitude 1 step) rejection responses using optimized IOPI controller $C_{iopi}(s)$ are presented in Figure 12.9(b). The robustness performances using this optimized IOPI controller are illustrated with ±20% loop gain variations. Meanwhile, the step input responses and disturbance (magnitude 1 step) rejection responses using the designed IOPID controller $C_{iopid1}(s)$ are presented in Figure 12.10(b). The robustness performances using this designed IOPID controller are also illustrated with ±20% loop gain variations.

For the effectiveness verification of the designed FOPI controller C_{fopi1}, the fractional order operator s^r is implemented by the impulse response invariant discretization (IRID) method in time domain [51]. The step input responses and disturbance rejection responses with ±20% loop gain variations are presented in Figure 12.11(b). This performance shows the system robustness to loop gain variations using the designed FOPI controller. Comparing with the step input and disturbance responses using the optimized IOPI and the designed IOPID controller in Figures 12.9(b) and 12.10(b), the overshoots using the FOPI controller in Figure 12.11(b) are smallest and with similar rising time.

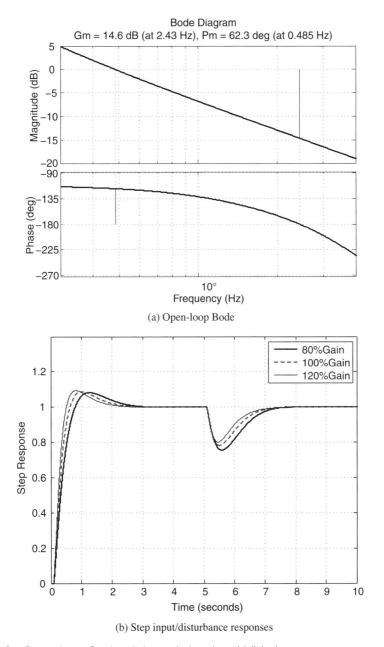

(a) Open-loop Bode

(b) Step input/disturbance responses

Figure 12.9 Open-loop Bode plots and step input/disturbance responses with loop gain variations using optimized IOPI controller $C_{iopi}(s)$

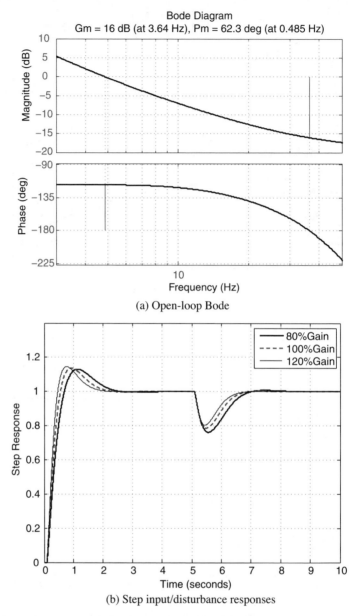

(a) Open-loop Bode

(b) Step input/disturbance responses

Figure 12.10 Open-loop Bode plots and step input/disturbance responses with loop gain variations using designed IOPID controller $C_{iopid1}(s)$

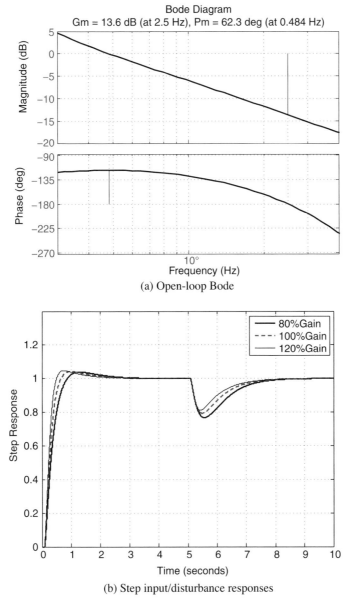

Figure 12.11 Open-loop Bode plots and step input/disturbance responses with loop gain variations using designed #1 FOPI controller $C_{fopi1}(s)$

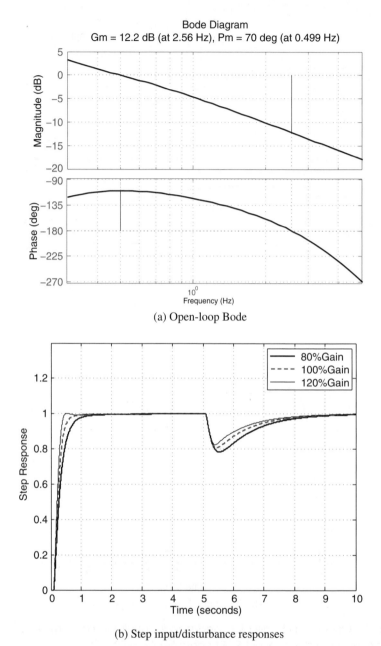

(a) Open-loop Bode

(b) Step input/disturbance responses

Figure 12.12 Open-loop Bode plots and step input/disturbance responses with loop gain variations using designed #2 FOPI controller $C_{fopi2}(s)$ (For a color version of (b), see Plate 19)

In order to show more flexibility and the potential of the FOPI controller, one more FOPI is designed to satisfy the specifications, gain crossover frequency $0.5Hz$ and phase margin $70°$ as follows,

$$C_{fopi2}(s) = 4.0494 + \frac{7.6646}{s^{1.45}}.$$

The open-loop Bode is plotted in Figure 12.12(a), and step input responses and disturbance rejection responses are shown in Figure 12.12(b). It is obvious that, while maintaining the similar step rising time by using $C_{iopi}(s)$, there is almost no step response overshoot using $C_{fopi2}(s)$ with $±20\%$ loop gain variations, which clearly shows the performance advantage of using the properly designed FOPI controller over the optimized IOPI controller and the designed IOPID controller.

12.6 Chapter Summary

This chapter provides a synthesis for the fractional order PI controllers to achieve two pre-specifications, for example, the phase margin and the gain crossover frequency, and the flat phase tuning constraint for the first order plus time systems. This designed FOPI controller is robust to the loop gain variations. The *complete stability region* of the FOPI controller parameters is determined according to a graphical stability criterion. Whereafter, the *relative stability surface* satisfying the pre-specified phase margin is found in the three dimensional parameter space; and the *relative stability curve* guaranteeing two specifications is determined on the *relative stability surface*. Then, by defining a relation function according to the flat phase tuning constraint, the *flat phase stable point* to determine the unique FOPI controller can be located in the three dimensional parameter space. This designed FOPI controller is certainly stable as its parameters are located in the *complete stability region*. This controller can achieve the desired control performance while satisfying two design specifications. This controller is also robust to the loop gain variations following the flat phase constraint. Furthermore, the complete achievable region of the two specifications the (phase margin and the gain crossover frequency) can be obtained for the FOPI controller design. This is an important benefit of this proposed controller synthesis. The detailed design procedures of this FOPI controller synthesis are summarized with an example. The advantage of the FOPI controller is presented from the comparison of the achievable region of the two specifications over the IOPID controller. A simulation illustration is presented to show the performance and benefit of the designed FOPI controller compared with the optimized IOPI controller and the designed integer order PID controller following the similar design scheme for the FOPI, based on the first order plus time delay plants.

Part V
Fractional Order Disturbance Compensations

Part V

13

Fractional Order Disturbance Observer

13.1 Introduction

In practice, a physical motion control system will not be exactly the same as any mathematical model no matter how the model is obtained. In a general sense, one can say that "*All models are wrong but some of them might be useful.*" The disturbance observer regards the difference between the actual output and the output of the nominal model as an equivalent disturbance applied to the nominal model. It estimates the equivalent disturbance and the estimate is utilized as a compensation signal. The disturbance observer (DOB) concept was proposed in [168]. In [217], the framework of disturbance observer theory was refined based on the design of TDOF (two-degree-of-freedom) servo controllers and the factorization approach. Based on an extended pole placement method and a disturbance observer, an accurate motion controller design was proposed in [35]. Recently, DOB was combined with the zero-phase error tracking algorithm (ZPETC) [215] as reported in [75], [108] for digital implementations. It is now common practice to use DOB in many high precision motion control systems, such as, disk drive servo control [54].

Disturbance observers have several attractive features. In the absence of large modeling errors, DOBs allow the independent tuning of disturbance rejection characteristics and the command following characteristics. Furthermore, compared to integral action, disturbances observers allow more flexibilities via selection of the order, the relative degree, and the bandwidth of the low-pass filter known as the disturbance observer filter or the Q-filter. It is well known that by appending disturbance states to a traditional state estimator [79], the disturbance compensation can be handled. However, using the disturbance observer structure allows simple and intuitive tuning of the disturbance observer loop gains independent of the state feedback gains. This explains why DOB is more welcome by the control practitioners.

Fractional Order Motion Controls, First Edition. Ying Luo and YangQuan Chen.
© 2013 John Wiley & Sons, Ltd. Published 2013 by John Wiley & Sons, Ltd.

It has been discovered in a published US patent application (US20010036026) [54] that there is a tradeoff between the phase margin loss and the strength of low frequency vibration suppression when applying DOB. Given the required cutoff frequency of the Q-filter, it turns out that the relative degree of the Q-filter is the major tuning knob for this tradeoff. As a motivation for the fractional order Q-filter, a solution based on integer order Q-filter with variable relative degrees is introduced which is the key contribution of US20010036026 [54]. In this chapter, a fractional order disturbance observer based on the fractional order Q-filter is presented. The fractional order filter is based on "fractional calculus" which is introduced in Chapter 1 [15], [158], [169], [172] [181], [183], [188], [192], [198]. The nice point in this chapter is that the traditional DOB is extended to fractional order DOB (FODOB) with the advantage that the FODOB design will no longer be conservative nor aggressive, that is, given the cutoff frequency of the Q-filter and the desired phase margin, we can uniquely determine the fractional order of the low pass filter [55].

13.2 Disturbance Observer

In the conventional disturbance observer [168], the basic idea is to use a nominal inverse model of the plant to estimate the disturbance. This is illustrated in Figure 13.1 where P_n^{-1} is the inverse of the nominal plant model and Q is usually a low pass filter to restrict the effective bandwidth of the DOB. We remark that this DOB configuration is nothing but another form of loop-shaping to add more attenuation in the lower frequency range at the cost of reduced phase margin and possible amplification of disturbances at other medium and high frequency bands due to the waterbed effect in the sensitivity function. Therefore, it is implicitly implied in DOB that the spectrum of the disturbance d has more low frequency contents than high frequency ones. We argue that, if d is a white noise, little benefit can be gained from using DOB or any other advanced control technique. Note that if the plant has a non-minimum phase zero, P_n^{-1} will have an unstable mode. So, in this case, the DOB shown in Figure 13.1,

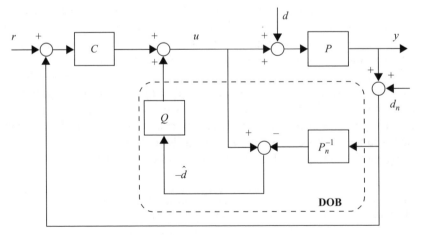

Figure 13.1 The DOB

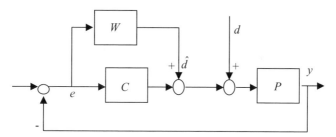

Figure 13.2   Disturbance observer block-diagram: the general form

although physically simple, cannot be directly applied. Some modifications will be necessary. In a more general setting, we can redraw the DOB shown in Figure 13.1 in a form shown in Figure 13.2 where W is a shaping filter. Note that here the disturbance in front of the plant is considered.

13.3 Actual Design Parameters in DOB and their Effects

In practice, the DOB is usually implemented digitally as shown in Figure 13.3.

In Figure 13.3, d' is the "observed" disturbance d; $P_n^{-1}(z^{-1})$ is the stable inverse of P_n, the nominal model of the actual plant P; n_d is the number of pure time delays of the control signal u', the compensated signal of the controller signal u generated by the feedback controller C; Q is a low pass filter with relative degree n_Q and cutoff frequency ω_Q.

There are three key parameters in DOB design as shown in Figure 13.3, namely,

(1) n_d: the number of pure time delays of the control signal u';
(2) n_Q: the relative degree of Q-filter and
(3) ω_Q: the cutoff frequency of Q-filter.

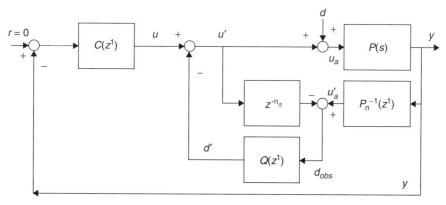

Figure 13.3   Disturbance observer block-diagram: the digital form (the zeroth order holder and the control signal saturator are omitted)

In order to see how the overall system based on the disturbance observer behaves, we examine the error transfer function (ETF) $S(j\omega)$ and the disturbance response transfer function (from d to y) $G_{dy}(j\omega)$ from Figure 13.3. With no DOB,

$$S(j\omega) = \frac{1}{1+PC}, \quad G_{dy}(j\omega) = \frac{P}{1+PC}, \tag{13.1}$$

and with DOB,

$$S(j\omega) = \frac{1}{1+PC+\delta_{PC}}, \quad G_{dy}(j\omega) = \frac{P}{1+PC+\delta_{PC}}, \tag{13.2}$$

where

$$\delta_{PC} = \frac{PP_n^{-1}Q + z^{-n_d}QPC}{1-z^{-n_d}Q} = P\frac{z^{-n_d}Q}{1-z^{-n_d}Q}\left(P_n^{-1}z^{n_d}+C\right). \tag{13.3}$$

Clearly, the disturbance observer cannot be implemented if $Q=1$. Notice that as in many motion control systems, the nominal plant is in the form of $\frac{K}{(\tau s+1)s}$ and P_n^{-1} in this case can be approximated by a finite impulse response (FIR) filter which is always realizable. Therefore, in this case, no constraint is to be put on the relative degree of Q. On the contrary, in the literature, QP_n^{-1} has to be made realizable by letting the relative degree of Q be equal to or greater than that of P_n.

To determine the correct n_d, the major consideration is to minimize the mismatch between the phases of $z^{-n_d}u'$ and u_a as shown in Figure 13.3. It is found that $n_d = 3$ is the best choice for a high TPI (tracks per inch) hard-disk-drive servo system as illustrated by Figure 13.4 [54]. We comment that, for different applications, n_d should be different and n_d should also include the delay effect of the plant P as also pointed out in [108].

With different relative degree of the Q-filter ($n_Q = 1, 2, 3, 4$), the disturbance attenuation performance is achieved differently. For the lowest relative degree ($n_Q = 1$, note again, as pointed out earlier, n_Q cannot be 0), the best disturbance attenuation is achieved. However, this is at the cost of the largest amplification of mid-band frequency contents of both the measurement noises as well as the shock disturbance if any. Therefore, when the disturbance is not presented or is small, $n_Q = 1$ is not a preferred choice. Motivated by this observation, the performance variation with respect to n_Q is illustrated in Figure 13.5.

One may argue that the Bode plots in Figure 13.5 contain the coupled effect from the feedback controller C. It is important to realize that the notion of DOB is to design an add-on compensator for disturbance rejection. Theoretically, however, the joint design of DOB and the re-design of C should be required in a more general loop-shaping framework. In practice, when there is no disturbance, since the feedback controller C should maintain the nominal stability and performance properties, the "add-on" thinking of DOB is preferred.

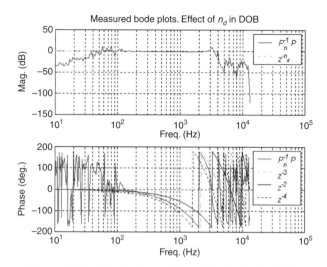

Figure 13.4 Illustration of the effect of n_d in DOB

13.4 Loss of the Phase Margin with DOB

The cut-off frequency of the Q-filter ω_Q is another key design parameter. Too high a ω_Q may result in worse robustness of the overall system. This can be seen from the variation of PM (phase margin) of the overall closed-loop system. The PM is in fact a function of ω_Q as well as n_Q. The basic trend is that the higher the ω_Q, the more PM losses; the larger the n_Q, the less PM losses for a fixed ω_Q. This is demonstrated by a set of measured data shown in Figure 13.6. The dotted plane (approx. $PM(\omega_Q, n_Q) = 37°$) represents the PM of the original system. Therefore, Figure 13.6 can guide us when choosing a right combination of ω_Q and n_Q.

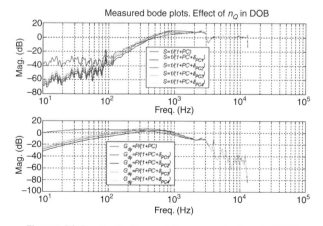

Figure 13.5 Illustration of the effect of n_Q in DOB

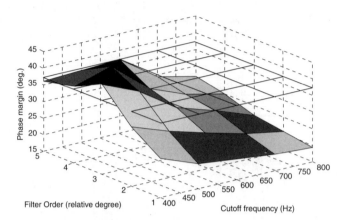

Figure 13.6 Illustration of phase margin (PM) as a function of n_Q and ω_Q in DOB (For a color version of this figure, see Plate 20)

It seems quite hard to achieve a good disturbance attenuation performance without loss of PM. A compromise must be made between the disturbance attenuation performance and the robustness of the original system.

13.5 Solution One: Rule-Based Switched Low Pass Filtering with Varying Relative Degree

This solution is fully documented in [54]. The key motivation is from Figure 13.6–the issue of the loss of GM with DOB. In practice, ω_Q should be pre-determined based on the disturbance attenuation requirement. The only trade off tuning knob will be n_Q. A variable relative degree strategy was used [54] where a switching method is applied based on the amplitude of the output y, the positional error signal (PES), as illustrated in Figure 13.7. A switching policy used in [54] is shown in Figure 13.8 for illustration purposes. As an aside, to avoid the Q-filter initialization problem and the possible large discontinuity in the internal state of Q-filter, all stages of sub-Q-filter in Figure 13.7 should be run at all times. The explanation of Figure 13.8 is straightforward and the deadzone is not always required for all DOB applications.

13.6 The Proposed Solution: Guaranteed Phase Margin Method using Fractional Order Low Pass Filtering

Let us review Figure 13.6 again. In practice, ω_Q is usually specified by the disturbance attenuation requirement. Moreover, the phase margin of the overall compensated system with DOB is also specified. In Figure 13.6, we may find that the required n_Q usually lies between two adjacent integers. For example, from the DOB design, it may turn out that the Q-filter should be of the following form

$$Q(s) = \frac{1}{(\tau s + 1)^{n_Q}}, \quad n_Q = 3.25, \tag{13.4}$$

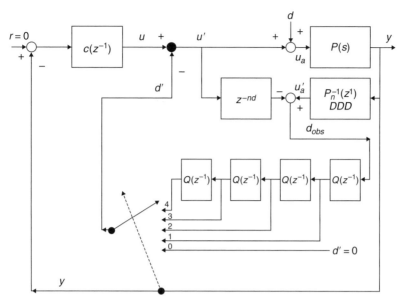

Figure 13.7 Q-filter in DOB with a varying relative degree

which is a fractional order low-pass filter (FO-LPF). When we use a fractional order
Q-filter in DOB, we call it a "fractional order disturbance observer".

13.7 Implementation Issues: Stable Minimum-Phase Frequency Domain Fitting

For the implementation of the fractional order operator, some methods have been
introduced in previous chapters of this book, for example, the impulse response
invariant discretization (IRID) method in time domain as introduced in Chapter 2

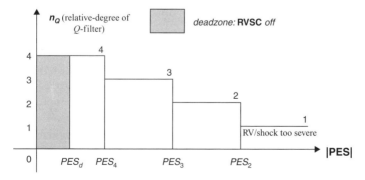

Figure 13.8 A switching policy for the relative degrees of the Q-filter in DOB. (PES:
positional error signal; RVSC: rotational vibration and shock compensator).

and 7 [51], [52] and the Oustaloup Recursive Algorithm in frequency domain [174] as introduced in Chapter 6. In this chapter, a stable minimum-phase frequency domain implementation method for the fractional order operator is introduced.

From the definitions of fractional order derivative, we know that any controller involving a fractional order differentiator or integrator is in fact a filter with an infinite (integer) order, or, we can say that the filter is with an infinite length of memory. In implementing a given FOC, its frequency response is actually exactly known. Clearly, for a given frequency's range of interest, say, $\omega \in [\omega_L, \omega_H]$, a set of frequency response data (magnitude and phase) can be obtained. We can take this set of data as the *measured frequency response* data set and feed it to any frequency-domain system identification software package.

At this point, one may think of the ready MATLAB® function invfreqs or invfreqz in the MATLAB® Signal Processing Toolbox. But this does *not* work for our purpose here mainly due to the bad numerical conditioning in the algorithms used in invfreqs or invfreqz. It is found that the ELiS function in the MATLAB® Frequency Domain Identification Toolbox works fine for our transfer function fitting here. In particular, the stability of the fit transfer function can be guaranteed. Another attractive feature is its professionally designed GUI.[1]

A simple command line example is given by the following script:

```
%Istvan Kollar's stable transfer function fitting
f=logspace(-1,3,200)'; % freq. band of practical interest
Y=1./(j*2*pi*f).^(1/2);% the desired freq. response 1/s^.5
U=ones(size(Y));     % set input to 1 to get I/O data
d=fiddata(Y,U,f);    % build the FIDdata
d.variance=[0,1e-6]; % artificial variance
if ~exist('order'), order=9; end order=yesinput('Order of
model',order,[1,inf]);
%First search for best cost function with stabilization:
disp('First search for best fit among trials...')
[m,finf]=elis(d,'s',order,order,...
    struct('stabilization','r','forceminimumphase','r',...
    'plotdens',-inf,'plot0','off'),...
    struct('displaymessages','off'));
[cfm,itmax]=min(finf.cfv); itmax=itmax-1;
fprintf('Iterate until best fit, itmax=%.0f ...\n',itmax)
figure(1), clf
iterctrl % allow manual iter. on fig. (select 'Finish')
% return model object 'm', order/order, forcing stability
m=elis(d,'s',order,order,...
struct('stabilization','r','forceminimumphase','r',...
'itmax',itmax)); figure(2), plotelpz(m),
zoom on % plot the pole-zero distribution
if ~exist('bi'), bi=2; end,
if order==4, bi=2; end, bi=bi+1;
```

---

[1]Refer to FDIDENT home page at http://elecwww.vub.ac.be/fdident.

```
figure(bi), plot(m), zoom on   % plot the Bode magnitude
xlabel(sprintf('Order: %.0f/%.0f',order,order));drawnow
```

Using get(m), we have

```
Version = 2.2
Date = '02-Dec-2000 15:55:26'
Data = [2x1x200 fiddata]
Algorithm = [1x1 struct]
Variable = 's'
Representation = 'polynomial'
num = [-6.58e-36 -4.56e-31 -4.34e-27 -1.13e-23 -9.4e-21
    -2.46e-18 -1.88e-16 -3.61e-15 -1.44e-14 -8.7537e-15]
denom = [-2.46e-33 -5.33e-29 -2.57e-25 -3.76e-22 -1.76e-19
    -2.53e-17 -9.88e-16 -8.9e-15 -1.48e-14 -2.49e-15]
FreqVect = [200x1 double]
Fscale = 1
Delay = 0
Covariance = [21x21 double]
FitInfo = [1x1 struct]
```

Entering the above fit transfer function coefficients into `CtrlLAB`, the most downloaded package developed by Professor Dingyü Xue for SISO (single input single output) control system analysis and design in MATLAB® Central,[2] we can find, via several mouse clicks, the Bode plot and Nichols chart shown in Figure 13.9 and Figure 13.10 respectively.

Making a good fit with stable poles is sometimes rather difficult. In a high order fitting, there is a good chance that some poles will be driven to the unstable region due to numerical sensitivity problems. The MATLAB® Frequency Domain System Identification Toolbox (Version 3.0 of 22-Nov-00) offers some (artificial) tools to force stable solutions. In the stable fitting script here, the fit transfer function is also restricted to be minimum phase and stable. This is achieved by requesting the reflection/contraction of the unstable zeros and poles.

As a benchmark fitting result, consider the following general filter in lead/lag form [192]

$$C_r(s) = C_0 \left(\frac{1 + s/\omega_b}{1 + s/\omega_h} \right)^r \tag{13.5}$$

where $0 < \omega_b < \omega_h, C_0 > 0$ and $r \in (0, 1)$. Here we give out the fitting result for $C_{0.65}(s)$ [192] with $C_0 = 4280.1$, $\omega_b = 0.5$, $\omega_h = 200$ using the stable frequency fitting method

---

[2]MATLAB® Central URL:http://www.mathworks.com/Matlabcentral/fileexchange/loadFile.do?objectId=18&objectType=file

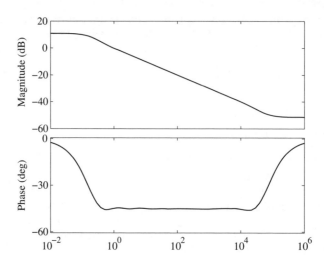

Figure 13.9 Bode plot for order 9/9: stable frequency domain fitting of $1/\sqrt{s}$

introduced in this section. The 4/4 fitting result is that

$$C_{0.65}(s) = 4280.1 \left(\frac{1+2s}{1+0.005s} \right)^{0.65}$$

$$\approx \frac{9.457 \times 10^{-11}s^4 + 1.218 \times 10^{-8}s^3 + 3.07 \times 10^{-7}s^2 + 1.476 \times 10^{-6}s + 9.794 \times 10^{-7}}{4.5 \times 10^{-16}s^4 + 1.161 \times 10^{-13}s^3 + 6.99 \times 10^{-12}s^2 + 9.516 \times 10^{-11}s + 2.14 \times 10^{-10}}$$

with its Bode plot and Nichols chart drawn via CtrlLAB in Figure 13.11 and Figure 13.12, respectively. We can see that Figure 13.11 is quite similar to the characteristic of a frequency-band fractional differentiator.

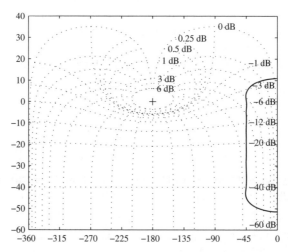

Figure 13.10 Nichols chart for order 9/9: stable frequency domain fitting of $1/\sqrt{s}$

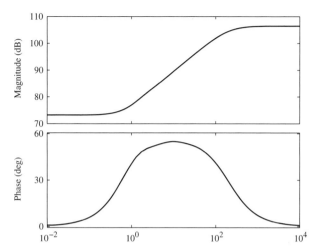

Figure 13.11 Bode plot for order 4/4: stable frequency domain fitting of $C_{0.65}(s)$

As a final remark in this section, the frequency domain fitting idea used in CRONE framework [180, 182] can also be applied in principle. However, since the fractional Q-filter is not in a simple form of $1/s^r$, some additional work needs to be done using the zigzag line fitting in Bode plot for fractional order Q-filter in general form. For the same reason, here the direct or indirect discretization schemes [53] are not used.

13.8 Chapter Summary

In this chapter, we proposed using the fractional order disturbance observer for vibration suppression applications such as hard-disk-drive servo control. The motivation

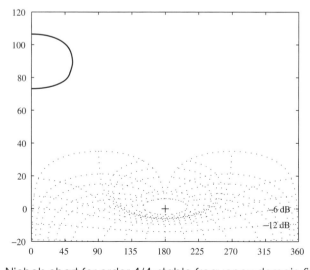

Figure 13.12 Nichols chart for order 4/4: stable frequency domain fitting of $C_{0.65}(s)$

is explained in detail. The major problem is the tradeoff between the phase margin loss and the strength of the low frequency vibration suppression. Given the required cut-off frequency of the Q-filter, it turns out that the relative degree of the Q-filter is the major tuning knob for this tradeoff. To motivate the introduction of the fractional order Q-filter, an existing solution based on integer order Q-filter with a variable relative degree is introduced which is the key contribution of US20010036026 [54]. The fractional order disturbance observer based on the fractional order Q-filter is proposed with the implementation method discussed. The nice point of this chapter is that the traditional DOB has been extended to the fractional order DOB with the advantage that the FODOB design is now no longer conservative or aggressive, that is, given the cutoff frequency and the desired phase margin, we can uniquely determine the fractional order of the low pass filter.

We comment that the experimentation verification of the proposed FODOB seems to be interesting but straightforward in view of the explanation in this chapter and the existing implemented results in [54]. However, during the experimental verification of FODOB, there might be some new issues emerging such as the nonlinearity effects.

14

Fractional Order Adaptive Feed-Forward Cancellation

14.1 Introduction

In practice, the periodic disturbances exist in a variety of electromechanical systems. For example, the repeatable runout (RRO) problem in hard-disk-drive digital servo systems is typically caused by the periodic disturbance which is one of the main contributors to track misregistration [71], [84]; the cogging force in the permanent magnetic motors was defined as the position periodic disturbance in [140] and [5]. The challenges of compensating the periodic disturbances appear in various applications. Numerous control design methods have been developed specifically to eliminate periodic disturbances. Repetitive control [197] and disturbance observer (DOB) based control [194], [243] have demonstrated effective compensation for repeatable disturbances. However, repetitive control tends to amplify nonrepeatable disturbances between the frequencies of the repeatable disturbances while DOB-based control can alter the closed-loop properties of the system.

Another method of designing controllers to cancel periodic disturbances is the internal model principle (IMP) proposed by Francis and Wonham [78] in 1976, which showed that a suitably reduplicated model of the dynamic structure of the disturbance should be included in the feedback loop for perfect disturbance cancellation.

Adaptive feed-forward cancellation (AFC) is also a well-established method to avoid sinusoidal disturbance with a known period but with unknown amplitude and unknown phase [31], which is essentially a special case of more general narrow-band disturbance rejection methods [199]. The adaptive algorithm can be used to estimate the unknown amplitude and unknown phase of the periodic disturbance. So the negative of the disturbance value can be added to the input of the plant, then the periodic disturbance can be simply cancelled out. AFC has been used in hard-disk-drive industry as a standard technique to cancel the once per revolution disturbance

Fractional Order Motion Controls, First Edition. Ying Luo and YangQuan Chen.
© 2013 John Wiley & Sons, Ltd. Published 2013 by John Wiley & Sons, Ltd.

due to the spindle motor runout [197]. AFC technique has also been extended to the case when the period is constant but unknown [30], [209] and to the case when the disturbance is sinusoidal with respect to the state rather than the time [66].

In [31], the authors observed that the AFC algorithm was not only successful in eliminating the first harmonic but also in reducing the third harmonic of the periodic disturbance. The AFC algorithm designed to cancel the first harmonic may be capable of reducing the amplitude of the third harmonic as well. The generation of harmonics in the AFC algorithm was found to be due to the time-variation of the adaptive parameters and was explained using modulation arguments from standard signal and system theory. The results of the analysis originate from the fact that the AFC algorithm is equivalent in some sense to an IMP algorithm.

In this chapter, a fractional order adaptive feed-forward cancellation (FOAFC) scheme is proposed to cancel the periodic disturbance, which offers one more tuning knob, the fractional order, for the performance improvement of the periodic disturbance cancellation according to the interests of the users. The equivalence of the fractional order internal model principle (FOIMP) scheme is derived for FOAFC. Thus, the FOIMP equivalence can be used to analyze the performance of the cancellation for the target periodic disturbance and the suppression for the harmonics and noise, according to the sensitivity function Bode plots of the closed-loop systems. Two FOAFC cases, fractional order $\alpha \in (0, 1)$ and $\alpha \in (1, 2)$, are proposed for the performance analysis, respectively. First, FOAFC with $\alpha \in (0, 1)$ has a narrower slot around the frequency of the target periodic disturbance over the integer order adaptive feed-forward cancellation (IOAFC) in the sensitivity function Bode plots, which means FOAFC with $\alpha \in (0, 1)$ is more selective for the cancellation of the target periodic disturbance with the desired cancellation capability shown as the slot depth in the Bode plots. Meanwhile, the amplitude of FOAFC with $\alpha \in (0, 1)$ is much smaller over that of IOAFC at a higher frequency in the sensitivity function Bode plots, which reveals that the suppression capability of FOAFC with $\alpha \in (0, 1)$ is stronger than that of IOAFC for the high order harmonics of the target periodic disturbance or the high frequency noise. Second, FOAFC with $\alpha \in (1, 2)$ has a deeper slot at the frequency of the target periodic disturbance over IOAFC, and the amplitude of FOAFC with $\alpha \in (1, 2)$ around the frequency of the target periodic disturbance is lower than that of IOAFC, which indicates FOAFC with $\alpha \in (1, 2)$ is not as selective as IOAFC for the cancellation of the target periodic disturbance, but the suppression performance of FOAFC with $\alpha \in (1, 2)$ is better than that of IOAFC for the disturbances or noise around the frequency of the target periodic disturbance. Meanwhile, there is also a disadvantage for FOAFC with $\alpha \in (1, 2)$, the amplitudes of FOAFC with $\alpha \in (1, 2)$ are bigger than that of IOAFC at a range of higher frequency in the sensitivity function Bode plots, which means FOAFC with $\alpha \in (1, 2)$ may amplify the high order harmonics of the target periodic disturbance or the high frequency noise compared with IOAFC. Anyhow, FOAFC with additional tuning knob $\alpha \in (0, 2)$ has advantages and is much more flexible over IOAFC with only $\alpha = 1$ for the cancellation of the target periodic disturbance and the suppression of the harmonics or the noise.

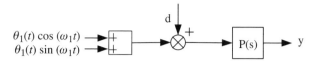

Figure 14.1 Fractional order adaptive feed-forward cancellation

Simulation and experimental results are presented to validate the performance analysis of FOAFC compared with IOAFC [253].

14.2 Fractional Order Adaptive Feed-Forward Cancellation

In this section, a fractional order adaptive feed-forward cancellation scheme is proposed with an extra tuning knob – fractional order. Assuming that the disturbance is set as follows,

$$d(t) = A\sin(\omega_1 t + \phi) = a_1 \cos(\omega_1 t) + b_1 \sin(\omega_1 t). \tag{14.1}$$

As shown in Figure 14.1, the control input of the plant is selected to be,

$$u(t) = \theta_1(t) \cos(\omega_1 t) + \theta_2(t) \sin(\omega_1 t), \tag{14.2}$$

and the plant output can be written as,

$$y(t) = Ł^{-1}[Ł((\theta_1(t) - \theta_1^*) \cos(\omega_1 t) + (\theta_2(t) - \theta_2^*) \sin(\omega_1 t)) P(s)],$$

where, $Ł(\cdot)$ stands for the Laplace transform; θ_1^* and θ_2^* denote the nominal values of $\theta_1(t)$ and $\theta_2(t)$, respectively. So that, when

$$\theta_1^* = -a_1, \theta_2^* = -b_1,$$

and if the parameters $\theta_1(t)$ and $\theta_2(t)$ converge to the nominal values, the disturbance $d(t)$ can be exactly cancelled.

In this study the Caputo definition is adopted for fractional derivative, which allows utilization of initial values of classical integer order derivatives with known physical interpretations [188]. Some definitions and properties can be referred to equations (1.6)–(1.12).

Then, a fractional order adaptive updating law for the adaptive parameters is proposed as follows,

$$_0D_t^\alpha \theta_1(t) = -gy(t) \cos(\omega_1 t), \tag{14.3}$$

$$_0D_t^\alpha \theta_2(t) = -gy(t) \sin(\omega_1 t), \tag{14.4}$$

where the fractional order $\alpha \in (0, 2)$, and $g > 0$ is an arbitrary parameter called the fractional order adaptation gain.

In (3.5) of [31], the traditional integrator was used for the adaptive law, namely, the IOAFC was applied with $\alpha = 1$ in (14.3) and (14.4). The IOAFC algorithm designed to cancel the first harmonic may be capable of reducing the amplitude of the third harmonic as well. The generation of harmonics in the IOAFC algorithm was found to be due to the time-variation of the adaptive parameters and was explained using modulation arguments from standard signals and systems theory in [31].

In order to compare the proposed FOAFC with the IOAFC in [31] fairly, the same plant and disturbance in [31] were used in this FOAFC scheme test as shown below,

$$P(s) = \frac{s + 2}{(s + 1)(s + 3)}, \tag{14.5}$$

$$d(t) = \sin(0.1t) - 0.2\sin(0.3t), \tag{14.6}$$

where all initial conditions are zero, and the adaptive gain is set as $g = 1$. The disturbance has a fundamental component at 0.1 rad/s and a third harmonic at 0.3 rad/s.

Figure 14.2(a) shows the response $y(t)$ of the system without any compensation. The first and third harmonics are clearly visible. Figure 14.2(b) is the response $y(t)$ of the system for $\theta_1(t)$ and $\theta_2(t)$ frozen to $\theta_1^*(t)$ and $\theta_2^*(t)$, respectively. The first harmonic is cancelled exactly and the third harmonic is unchanged. In Figure 14.3, the solid line stands for the response $y(t)$ using the IOAFC, just the same as Figure 5 in [31], not only is the first harmonic eliminated, but the third harmonic is also significantly reduced; and the dashed line shows the response $y(t)$ using the FOAFC with $\alpha = 1.5$ in (14.3) and (14.4). It is obvious that the magnitude of the third harmonic using the FOAFC with $\alpha = 1.5$ is much smaller than that using the IOAFC, which can also be seen clearly from the FFT spectrum with respect to the frequency in Figure 14.4.

The parameter values of $\theta_1(t)$ and $\theta_2(t)$ are shown in Figure 14.5. $\theta_1(t)$ and $\theta_2(t)$ should converge to the nominal values $\theta_1^*(t) = -a_1 = 0$ and $\theta_2^*(t) = -b_1 = -1$, respectively. If the disturbance is only the first harmonic, the parameters indeed converge toward these nominal values. But, because of the third harmonic, there is substantial fluctuation of the parameters in the steady state, and the variations can be recognized as second and fourth harmonics of the fundamental frequency. The solid and dashed lines in Figure 14.5 stand for the parameter values of $\theta_1(t)$ and $\theta_2(t)$ using the IOAFC method; the solid and red lines show the parameter values of $\theta_1(t)$ and $\theta_2(t)$ using the proposed FOAFC scheme with $\alpha = 1.5$. We can see that the fluctuation amplitude of $\theta_1(t)$ and $\theta_2(t)$ using the $\alpha = 1.5$ FOAFC is much bigger than that using the IOAFC.

The time-variation of the parameters $\theta_1(t)$ and $\theta_2(t)$ generates a third harmonic that is not present in the ideal case. This component of the input reduces the amplitude of the third harmonic observed at the output of the plant. This leads to the observation that the adaptive scheme with parameters $\theta_1(t)$ and $\theta_2(t)$ performs better in the presence of higher-order harmonics than the scheme with the fixed, nominal parameters $\theta_1^*(t)$ and $\theta_2^*(t)$. So, with bigger fluctuation amplitude of $\theta_1(t)$ and $\theta_2(t)$, the $\alpha = 1.5$ FOAFC

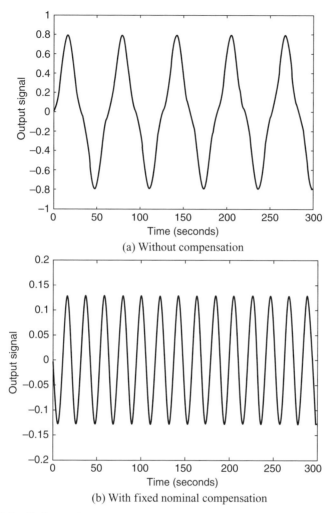

(a) Without compensation

(b) With fixed nominal compensation

Figure 14.2   Output without compensation/with fixed nominal compensation

scheme can reduce the amplitude of the third harmonic more effectively, which can be seen in Figure 14.3 and Figure 14.4.

14.3 Equivalence Between Fractional Order Internal Model Principle and Fractional Order Adaptive Feed-Forward Cancellation

14.3.1 Single-Frequency Disturbance Cancellation

For the fractional order adaptive feed-forward cancellation for a single-frequency disturbance, the input and disturbance signals are shown as (14.1) and (14.2),

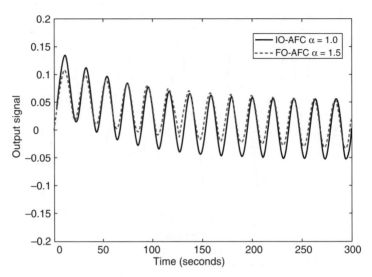

Figure 14.3 Output with IOAFC and $\alpha = 1.5$ FOAFC

respectively. The fractional order adaptive updating law is presented as (14.3) and (14.4).

Since, one has,

$$e^{j\omega_1 t} = \cos \omega_1 t + j \sin \omega_1 t;$$

and,

$$e^{-j\omega_1 t} = \cos \omega_1 t - j \sin \omega_1 t,$$

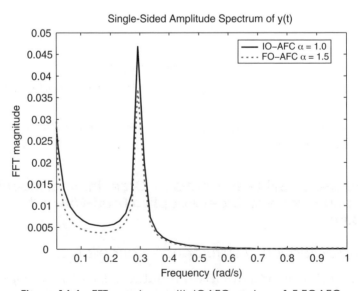

Figure 14.4 FFT spectrum with IOAFC and $\alpha = 1.5$ FOAFC

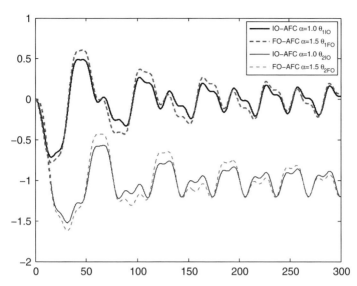

Figure 14.5 Adaptive parameter values of $\theta_1(t)$ and $\theta_2(t)$ with IOAFC and $\alpha = 1.5$ FOAFC

so, from (14.1), we can obtain,

$$u(t) = \theta_1(t)(e^{j\omega_1 t} - j\sin(\omega_1 t))$$
$$+ \theta_2(t)(e^{j\omega_1 t} - \cos(\omega_1 t))(-j), \tag{14.7}$$

$$u(t) = \theta_1(t)(e^{-j\omega_1 t} + j\sin(\omega_1 t))$$
$$+ \theta_2(t)(e^{-j\omega_1 t} - \cos(\omega_1 t))j. \tag{14.8}$$

Then adding (14.7) and (14.8), we can obtain,

$$u(t) = \frac{\theta_1(t)}{2}(e^{j\omega_1 t} + e^{-j\omega_1 t}) + \frac{j\theta_2(t)}{2}(e^{-j\omega_1 t} - e^{j\omega_1 t}). \tag{14.9}$$

In the same way, from (14.3) and (14.4), yield,

$$_0D_t^\alpha \theta_1(t) = -\frac{g}{2}y(t)(e^{j\omega_1 t} + e^{-j\omega_1 t}), \tag{14.10}$$

$$_0D_t^\alpha \theta_2(t) = -\frac{jg}{2}y(t)(e^{-j\omega_1 t} - e^{j\omega_1 t}). \tag{14.11}$$

Denote $U(s)$ and $Y(s)$ as the Laplace transforms of the input $r(t)$ and the output $y(t)$, respectively. Similarly, mark $\Theta_1(s)$ and $\Theta_2(s)$ as the Laplace transforms of $\theta_1(t)$ and $\theta_2(t)$ with initial condition, $\theta_1(t)|_{t=0} = \theta_2(t)|_{t=0} = 0$.

From (14.10) and (14.11) we can derive that [188],

$$\Theta_1(s) = -\frac{g}{2s^\alpha}(Y(s + j\omega_1) + Y(s - j\omega_1)), \tag{14.12}$$

$$\Theta_2(s) = -\frac{jg}{2s^\alpha}(Y(s - j\omega_1) - Y(s + j\omega_1)). \tag{14.13}$$

Then, from (14.9), (14.12) and (14.13) we can obtain,

$$U(s) = \frac{1}{2}(\Theta_1(s - j\omega_1) + \Theta_1(s + j\omega_1))$$

$$+ \frac{j}{2}(\Theta_2(s - j\omega_1) - \Theta_2(s + j\omega_1))$$

$$= -\frac{g}{2}\left(\frac{1}{(s + j\omega_1)^\alpha} + \frac{1}{(s - j\omega_1)^\alpha}\right)Y(s)$$

$$= -gC_{IMP}(s)Y(s), \tag{14.14}$$

where $C_{IMP}(s) = \frac{(s - j\omega_1)^\alpha + (s + j\omega_1)^\alpha}{2(s^2 + \omega_1^2)^\alpha}$.

So, the fractional order adaptive feed-forward cancellation scheme in Figure 14.1 with adaptive rule (14.3) and (14.4) can be equivalent to the fractional order internal model principle scheme in Figure 14.6 with $C(s) = g$. Actually, the controller $C(s)$ is not limited to g. For instance, let Y_c be a signal after a block $F(s)$, for example, a filter, namely, $Y_c(s) = F(s)Y(s)$, and replace $Y(s)$ by $Y_c(s)$ in (14.3) and (14.4), then the FOAFC scheme is equal to the FOIMP scheme with $C(s) = gF(s)$. This modification can also be used to filter the signal u in (14.2) before it is applied to the plant.

14.3.2 Generalization to Multi-Frequency Disturbance Cancellation

This equivalence between FOAFC and FOIMP can easily be extended to the case when multiple frequency components are cancelled. For example, if u is replaced by,

$$u(t) = \theta_1(t)\cos(\omega_1 t) + \theta_2(t)\sin(\omega_1 t)$$

$$+ \theta_3(t)\cos(\omega_2 t) + \theta_4(t)\sin(\omega_2 t), \tag{14.15}$$

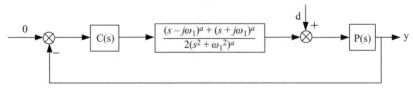

Figure 14.6 Fractional order internal model principle equivalence of the fractional order adaptive feed-forward cancellation

where θ_1, θ_2 are updated by,

$$_0 D_t^\alpha \theta_1(t) = -g_1 y(t) \cos(\omega_1 t),$$

$$_0 D_t^\alpha \theta_2(t) = -g_1 y(t) \sin(\omega_1 t),$$

$$_0 D_t^\alpha \theta_3(t) = -g_2 y(t) \cos(\omega_2 t),$$

$$_0 D_t^\alpha \theta_4(t) = -g_2 y(t) \sin(\omega_2 t),$$

then the FOAFC system is equivalent to a FOIMP control system with,

$$\frac{U(s)}{Y(s)} = -g_1 \frac{(s - j\omega_1)^\alpha + (s + j\omega_1)^\alpha}{2 \left(s^2 + \omega_1^2\right)^\alpha}$$

$$- g_2 \frac{(s - j\omega_2)^\alpha + (s + j\omega_2)^\alpha}{2 \left(s^2 + \omega_2^2\right)^\alpha}.$$

14.4 Frequency-Domain Analysis of the FOAFC Performance for the Periodic Disturbance

As introduced in Section 14.3, the FOAFC scheme proposed can be equivalent to the FOIMP scheme in Figure 14.6 with $C(s) = g$. Thus, the FOAFC performance of the cancellation of the target periodic disturbance and the suppression of the harmonics and noise can be analyzed in the frequency-domain using the FOIMP equivalence.

The sensitivity function of the system with FOIMP in Figure 14.6 is,

$$G_s(s) = \frac{1}{1 + C(s)C_{IMP}(s)P(s)}, \tag{14.16}$$

where $C_{IMP}(s) = \frac{(s - j\omega_1)^\alpha + (s + j\omega_1)^\alpha}{2(s^2 + \omega_1^2)^\alpha}$. Since the FOIMP equivalence is used for the analysis of the FOAFC scheme, the stability/instability boundary of the adaptive system can be predicted, according to the Lyapunov theory [45], the Nyquist criterion [31], or the averaging analysis [29]. It can be seen that the absolute value and the sign of the adaptive gain g are constrained for the stability of the adaptive system [29]. Therefore, for a fair comparison between the integer order AFC and the proposed FOAFC, the same adaptive gain value is used for both schemes, such as, $g = 1$. Then, the disturbance rejection performance using FOAFC can be discussed and compared with that using traditional AFC scheme.

Following the plant (14.5) in Section 14.2 which is from [31], the sensitivity function of $G_s(s)$ can be derived as,

$$G_s(s) = \frac{2 \left(s^2 + \omega_1^2\right)^\alpha (s + 1)(s + 3)}{D(s)}, \tag{14.17}$$

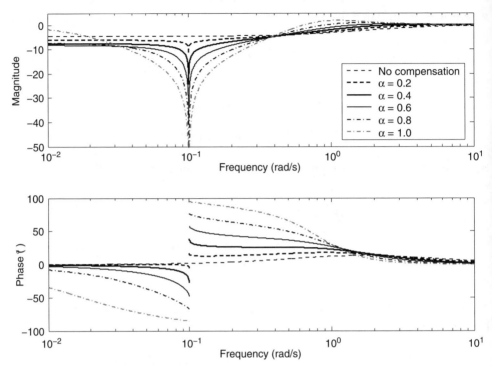

Figure 14.7   Bode plots of the sensitivity function with $\omega_1 = 0.1$ and $\alpha \in (0, 1)$

where

$$D(s) = 2 \left(s^2 + \omega_1^2\right)^\alpha (s + 1)(s + 3)$$

$$+ g(s + 2)[(s + j\omega_1)^\alpha + (s - j\omega_1)^\alpha]. \tag{14.18}$$

Thus, the Bode plots of $G_s(s)$ can be plotted in Figure 14.7 and Figure 14.8 with $\alpha \in (0, 2)$ and $\omega_1 = 0.1 \ rad/s$.

From Figure 14.7, it can be seen that, around the disturbance frequency $\omega_1 = 0.1 \ rad/s$, the magnitudes of the sensitivity function with AFC compensation are much smaller than that without compensation presented by the dashed line. The FOAFC with $\alpha \in (0, 1)$ have a narrower and shallower slot around the frequency $\omega_1 = 0.1 \ rad/s$ over the IOAFC, which means the FOAFC of $\alpha \in (0, 1)$ can also cancel the disturbance at the frequency $\omega_1 = 0.1 \ rad/s$, and the FOAFC with $\alpha \in (0, 1)$ is more selective for the cancellation of the target periodic disturbance with the desired cancellation capability shown as the slot depth in Figure 14.7. Meanwhile, the magnitudes around $\omega_1 = 0.1 \ rad/s$ from $0.02 \ rad/s$ to $0.5 \ rad/s$ with the FOAFC of $\alpha \in (0, 1)$ are bigger than that with the IOAFC where $\alpha = 1$. But after $0.5 \ rad/s$ and before $0.02 \ rad/s$, the magnitude with IOAFC is bigger than that with the FOAFC of $\alpha \in (0, 1)$, which means the FOAFC of $\alpha \in (0, 1)$ have a better effect of suppressing the higher-order harmonic disturbance

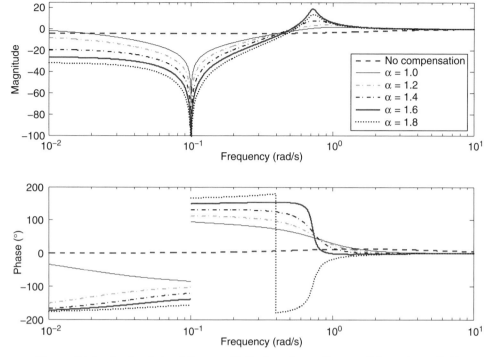

Figure 14.8 Bode plots of the sensitivity function with $\omega_1 = 0.1$ and $\alpha \in (1, 2)$

or high frequency noise after 0.5 rad/s and before 0.02 rad/s; however, from 0.02 rad/s to 0.5 rad/s, the disturbance or noise suppression effect of the FOAFC of $\alpha \in (0, 1)$ is not as good as the IOAFC.

In Figure 14.8, using the FOAFC with $\alpha \in (1, 2)$, we can see that the magnitudes of the sensitivity function with AFC compensation are also much smaller than that without compensation, the FOAFC with $\alpha \in (1, 2)$ have deeper and wider slots around the frequency $\omega_1 = 0.1$ rad/s over the IOAFC, which means the disturbance at the frequency $\omega_1 = 0.1$ rad/s can also be cancelled by the AFC with $\alpha \in [1, 2)$, and the FOAFC with $\alpha \in (1, 2)$ are not as selective as the IOAFC for the cancellation of the target periodic disturbance. At the same time, before 0.6 rad/s, the magnitudes with the FOAFC of $\alpha \in (1, 2)$ are smaller than that with the IOAFC. So, the disturbance or noise before 0.6 rad/s can be suppressed more effectively by the FOAFC of $\alpha \in (1, 2)$ than the IOAFC. However, there is also a trade-off, after the frequency 0.6 rad/s, the disturbance harmonics or noise should be amplified by the FOAFC of $\alpha \in (1, 2)$ compared with the suppression effect by the IOAFC.

To sum up, the FOAFC with the additional tuning knob $\alpha \in (0, 2)$ has many advantages and is much more flexible over the IOAFC with only $\alpha = 1$ for the cancellation of the target periodic disturbance and the suppression of the harmonics or the noise.

14.5 Simulation Illustration

In order to test the frequency-domain analysis results of the disturbance cancellation performance of the FOAFC scheme, the simulation illustration is presented in this section. In the simulation, the system $P(s)$ as (14.5) is also used for the fair comparison, and the disturbance is added as,

$$d(t) = 0.5\sin(0.05t) + \sin(0.1t) - 0.2\sin(0.3t)$$

$$-0.1\sin(0.5t) + N(t), \tag{14.19}$$

which contain the disturbance components in (14.6), where $N(t)$ is the white noise. In the control law (14.2), ω_1 is also chosen as 0.1 rad/s.

For the FOAFC of $\alpha \in (0, 1)$, we choose $\alpha = 0.5$ to test the performance and compare with the IOAFC.

For the repeatability of the work in this chapter, the implementation of the fractional order operator $1/s^\lambda$ is presented in this appendix. $1/s^\lambda$ can be realized by the impulse response invariant discretization (IRID) method [51] in time domain, where a discrete-time finite dimensional (z) transfer function is computed to approximate the continuous irrational transfer function $1/s^\lambda$, s is the Laplace transform variable. The fractional order operator $1/s^{0.5}$ is approximated by a 2nd-order discrete controller using IRID algorithm (sampling period $T_s = 0.01$ sec.) in this chapter:

$$G(z) = \frac{z^2 - 1.463z + 0.471}{0.09354z^2 - 0.08044z + 0.00141}.$$

As and justify shown in Figures 14.9, 14.10 and 14.11. Figure 14.9 is the output comparison of the FOAFC with $\alpha = 0.5$ and the IOAFC, where it is not easy to distinguish the performances of two methods because of the white noise. So, the FFT plots are presented in Figure 14.10 and Figure 14.11. In Figure 14.10, it is obvious that the disturbance at the frequency $\omega_1 = 0.1$ rad/s is almost cancelled completely by not only the IOAFC but also the FOAFC of $\alpha = 0.5$. The disturbance magnitudes at the frequencies $\omega = 0.05$ rad/s, $\omega = 0.3$ rad/s and $\omega = 0.5$ rad/s using the FOAFC of $\alpha = 0.5$ are higher than that using the IOAFC, which corresponds to the Bode plot feature in the frequency range $(0.02, 0.5)$ rad/s in Figure 14.7. At the same time, the disturbance magnitudes at the higher frequency range after 0.5 rad/s using the FOAFC of $\alpha = 0.5$ are lower than that using the IOAFC method, which can be seen in Figure 14.11.

As far as the FOAFC of $\alpha \in (1, 2)$ is considered, we choose $\alpha = 1.5$ to test the performance and compare with the IOAFC, which can be seen in Figures 14.12, 14.13 and 14.14. The output comparison of the FOAFC with $\alpha = 1.5$ and the IOAFC is shown in Figure 14.12, and the FFT plots are presented in Figure 14.13 and Figure 14.14. In Figure 14.13, it can be seen that the disturbance at the frequency $\omega_1 = 0.1$

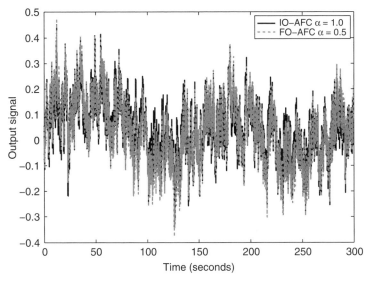

Figure 14.9   Output signals with IOAFC and FOAFC of $\alpha = 0.5$

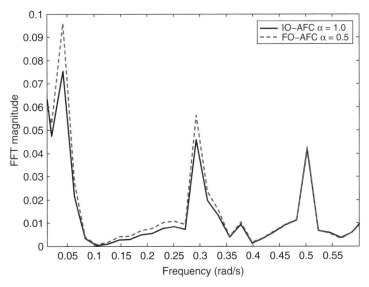

Figure 14.10   FFT spectra of the output with IOAFC and FOAFC of $\alpha = 0.5$ ($\omega \in (0, 0.6)$ rad/s)

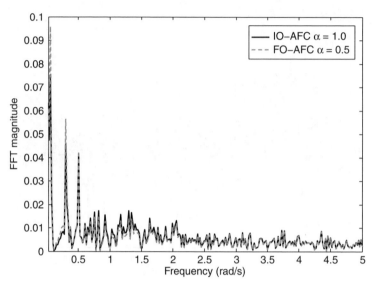

Figure 14.11 FFT spectra of the output with IOAFC and FOAFC of $\alpha = 0.5$ ($\omega \in$ (0, 5) *rad/s*)

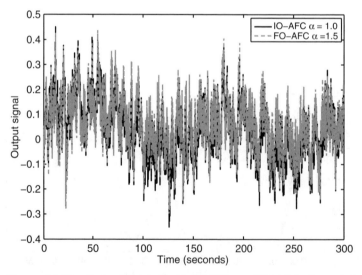

Figure 14.12 Output signals with IOAFC and FOAFC of $\alpha = 1.5$

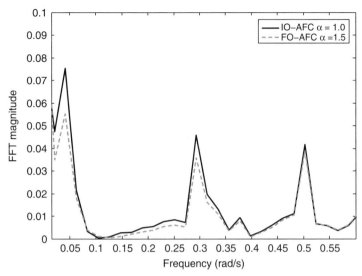

Figure 14.13 FFT spectra of the output with IOAFC and FOAFC of $\alpha = 1.5$ ($\omega \in (0, 0.6)$ *rad/s*)

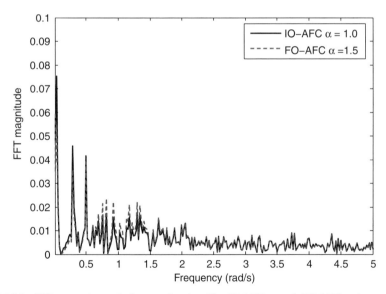

Figure 14.14 FFT spectra of the output with IOAFC and FOAFC of $\alpha = 1.5$ ($\omega \in (0, 5)$ *rad/s*)

is almost cancelled completely by both the IOAFC and the FOAFC of $\alpha = 1.5$. The disturbance magnitudes at the frequencies $\omega = 0.05$, $\omega = 0.3$ and $\omega = 0.5$ using the FOAFC of $\alpha = 1.5$ are lower than that using the IOAFC, which corresponds to the Bode plot feature in the frequency range $\omega < 0.6\ rad/s$ in Figure 14.7. This result is consistent with Figure 14.3 and Figure 14.4 of the simpler disturbance example in Section 14.2. Meanwhile, the disturbance magnitudes at the higher frequency after $0.6\ rad/s$ using the FOAFC of $\alpha = 1.5$ are amplified comparing with that using the IOAFC, which can be seen in Figure 14.14.

14.6 Experiment Validation

The frequency-domain analysis results of disturbance cancellation performance of the FOAFC scheme and the simulation results are validated on the real-time experimental platform with the fractional order $\alpha = 0.5$ in this section.

14.6.1 Introduction to the Experiment Platform

The fractional horsepower dynamometer as a general purpose experiment platform [210] is also applied for the experimental validation in this chapter. The architecture of the dynamometer control system is shown in Figure 3.10. The setup of this experimental platform is introduced in Section 3.6.1. But the DC motor of the dynamometer used in this chapter is different, and the dynamic model has been identified.

Without loss of generality, consider a speed control system modeled by:

$$\dot{v}(t) = u(t) - \frac{B}{J}v - d(t, x). \tag{14.20}$$

where v is the velocity; $d(t, x)$ is the disturbance; B and J are the frictional coefficient and moment of inertia of the dynamometer respectively; and u is the control input. Moreover, the presence of the hysteresis brake allows us to add the designed disturbance $d(t, x)$ to the motor. Thus, the IO/FOAFC controllers can be designed for such a problem and can be tested in the presence of the real disturbance as introduced through the hysteresis brake. In the experiment, the ω_1 in the control law (14.2) is chosen as $0.5\ rad/s$, and the disturbance is added as,

$$d(t) = 0.5\sin(0.2t) + \sin(0.5t) - 0.5\sin(1.5t) + 0.2\sin(2.5t). \tag{14.21}$$

With the open-loop step response as shown in Figure 14.15 for the simple system identification, a DC motor speed control system in the dynamometer real-time platform can be approximately modeled as the first order system below,

$$P_{exp}(s) = G_{DC} = \frac{0.78}{0.34s + 1}.$$

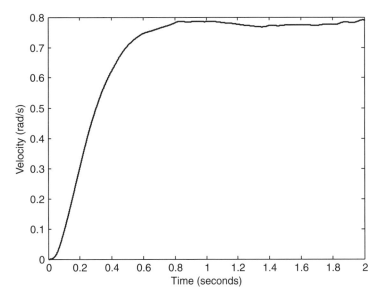

Figure 14.15 Open-loop step response for the system identification of dynamometer

So, the sensitivity function of the experimental system with FOIMP in Figure 14.6 is,

$$G_{s-exp}(s) = \frac{1}{1 + C(s)C_{IMP}(s)P_{exp}(s)}, \tag{14.22}$$

where $C_{IMP}(s) = \frac{(s - j\omega_1)^\alpha + (s + j\omega_1)^\alpha}{2(s^2 + \omega_1^2)^\alpha}$, and $C(s) = g = 1$.

Thus, the Bode plots of $G_{s-exp}(s)$ can be plotted in Figure 14.16 with $\alpha \in (0, 1]$ and $\omega_1 = 0.5 \ rad/s$.

14.6.2 Experiments on the Dynamometer

The proposed method is verified on the real-time dynamometer speed control system. The hysteresis brake force is designed as multi-harmonics disturbance in (14.21).

We choose $\alpha = 0.5$ to test the performance of the FOAFC and compare with the IOAFC, as shown in Figure 14.17 and 14.18. Figure 14.17 is the output comparison of the FOAFC with $\alpha = 0.5$ and the IOAFC. It is obvious that the amplitude of the FOAFC output is smaller than that of IOAFC output. At the same time, the FFT plots are presented in Figure 14.18. It can be seen that the disturbance at the frequency $\omega_1 = 0.5 \ rad/s$ is almost cancelled by not only the IOAFC but also the FOAFC of $\alpha = 0.5$. The disturbance magnitudes at the frequencies $\omega = 0.2 \ rad/s$, $\omega = 1.5 \ rad/s$ and $\omega = 2.5 \ rad/s$ using the FOAFC of $\alpha = 0.5$ are much lower than that using the IOAFC, which corresponds to the Bode plot feature in the frequency range $(0.02, 0.5) \ rad/s$ in Figure 14.16. Meanwhile, the magnitude of the noise in the real-time experiment at the higher frequency range after $0.7 \ rad/s$ using the FOAFC of $\alpha = 0.5$ is almost always lower than that using the IOAFC method.

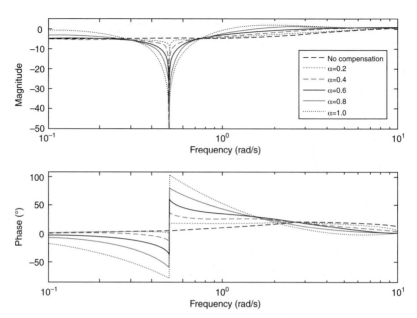

Figure 14.16 Experiment: Bode plots of the sensitivity function with $\omega_1 = 0.5$ and $\alpha \in$ (0, 1) (For a color version of this figure, see Plate 21)

14.7 Chapter Summary

In this chapter, a FOAFC scheme is proposed to cancel the periodic disturbance. This FOAFC offers one more tuning knob, the fractional order, for the performance improvement of the periodic disturbance cancellation according to the interests of the users. The equivalence of the FOIMP scheme is derived for the FOAFC. Thus, the

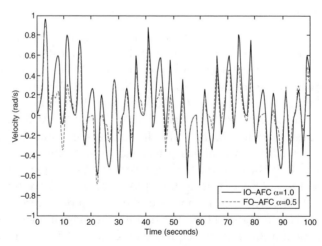

Figure 14.17 Experiment: Output signals with IOAFC and FOAFC of $\alpha = 0.5$

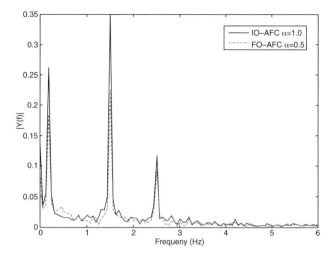

Figure 14.18 Experiment: FFT spectra of the output with IOAFC and FOAFC of $\alpha = 0.5$

FOIMP equivalence can be used to analyze the the the cancellation performance for the target periodic disturbance and the suppression for the harmonics and noise, according to the sensitivity function Bode plots of the closed-loop systems. Two FOAFC cases, fractional order $\alpha \in (0, 1)$ and $\alpha \in (1, 2)$, are proposed for the performance analysis, respectively. The FOAFC with additional tuning knob $\alpha \in (0, 2)$ has many advantages and is much more flexible over the IOAFC with only $\alpha = 1$ for the cancellation of the target periodic disturbance and the suppression of the harmonics or the noise. Simulation and experimental results are presented to validate the performance analysis of FOAFC compared to the IOAFC.

15

Fractional Order Adaptive Compensation for Cogging Effect

15.1 Introduction

In this chapter, a fractional order adaptive compensation (FOAC) method is devised for cogging effect compensation on permanent magnetic synchronous motor (PMSM) position and velocity control system. Cogging effect is a major disadvantage of PMSM, which degrades the servo control performances of applications, especially in a low-speed range [139], [140], [236]. Previous publications [129], [173], [206], [248] focused on how to assume the model structure of the cogging force, and design compensation algorithms. In contrast, what our work does is not to consider the structure of cogging force, but to modify the integer order integral item into a fractional order one. Then, an FOAC scheme is derived from the traditional integer order adaptive control (IOAC) one. In this FOAC scheme, a fractional order adaptive compensator for cogging effect is designed to guarantee the boundedness of all signals. Stability properties have been proven for both systems with the traditional IOAC method and the proposed FOAC scheme. Simulation and experimental results are presented to verify the advantage of the proposed FOAC scheme for cogging effect compensation over the conventional IOAC method [131].

15.2 Fractional Order Adaptive Compensation of Cogging Effect

15.2.1 Cogging Effect Analysis

Cogging effect is generated by the magnetic attraction between the rotors' mounted permanent magnets and the stator [140]. Cogging force is the circumferential

Fractional Order Motion Controls, First Edition. Ying Luo and YangQuan Chen.
© 2013 John Wiley & Sons, Ltd. Published 2013 by John Wiley & Sons, Ltd.

component of attractive force that attempts to maintain the alignment between the stator teeth and the permanent magnets. The cogging force spectrum depends only on the geometry and number of the stator slots. Cogging force harmonics appear at frequencies that are multiples of the $N_{slot-pp} f_s$, where $N_{slot-pp}$ is the number of slots per pole pair and f_s is the electrical frequency of the rotor. Analytical modeling of the cogging force has been changing since its production mechanism involves complex field distribution around state slots [186].

Generally speaking, the cogging effect has been modeled as a Fourier series expansion [95], [111]. However, it has also been modeled as a simple sinusoidal signal such as in (15.1) depending on the displacement signal x,

$$F_{cogging} = A\sin(\omega x + \varphi). \tag{15.1}$$

Some parameter adaptation schemes were proposed to compensate for the cogging effect which was simply identified as the form in (15.1) [112], [206], [241]. However, this model of cogging does not represent high-order items in the Fourier series. So the cogging effect cannot be compensated completely. In this chapter, the cogging force is considered as the general multi-harmonic Fourier expansion form as considered in [140] and [5],

$$a(x) = \sum_{i=1}^{\infty} A_i \sin(\omega_i x + \varphi_i), \tag{15.2}$$

where A_i is the unknown amplitude of the i-th harmonics, ω_i is the known frequency of the position-periodic cogging effect, and φ_i is the phase angle. From physical consideration, $a(x)$ should be bounded as,

$$|a(x)| \leqslant b_0 < +\infty, \tag{15.3}$$

where b_0 is a positive value.

15.2.2 Motivations and Problem Formulation

In this section, a fractional order adaptive compensator for cogging effect compensation is designed. The cogging force can be written as: $-a(\theta)$, where $a(\theta)$ is the function of θ, the angular displacement. In this chapter, to present this idea clearly, without loss of generality, the motion control system is modeled as follows,

$$\dot{\theta}(t) = v(t), \tag{15.4}$$

$$\dot{v}(t) = u - \frac{a(\theta)}{J} - T_{l'} - B_1 v, \tag{15.5}$$

$$u = \frac{1}{J}T_m, \ T_{l'} = \frac{1}{J}T_l, \ B_1 = \frac{B}{J},$$

where θ is the angular position; v is the velocity; u is the control input and $a(\theta)$ is the unknown position-dependent cogging disturbance which is repeating in every pole-pitch; J is the moment of inertia; T_m is the ideal case of the electromagnetic torque generated; and T_l is the load torque applied; B is the viscous friction coefficient.

Remark 15.2.1 *Equations (15.4) and (15.5) can be used to describe the mechanical dynamic system of a PMSM in the synchronous rotating reference frame [140].*

As the cogging effect is a state-dependent signal, it is denoted as $a(x)$ in (15.5) and $a(\theta)$ in (15.2). However, in this chapter, the experienced disturbance information along the state axis is not used in the proposed control law as can be seen in the following statement. Actually, all the signals are indeed in the time domain. Therefore, the cogging force can also be written in the time domain as $a(t)$ in the following statement.
The following notations are defined,

$$e_\theta(t) = \theta_d(t) - \theta(t), \quad e_v(t) = v_d(t) - v(t) = \dot{e}_\theta(t), \tag{15.6}$$

where $\theta_d(t)$ is the reference position, and $v_d(t)$ is the corresponding reference velocity.
The control objective is to track or servo the desired position $\theta_d(t)$ and the corresponding desired velocity $v_d(t)$ with tracking errors as small as possible. In practice, it is reasonable to assume that $\theta_d(t)$, $v_d(t)$ and $\dot{v}_d(t)$ are all bounded signals.
The feedback controller is designed as:

$$u(t) = \dot{v}_d(t) + T_{l'} + \frac{\hat{a}(t)}{J} + \alpha m(t) + \gamma e_v(t), \tag{15.7}$$

with

$$m(t) := \gamma e_\theta(t) + e_v(t), \tag{15.8}$$

where α and γ are positive gains; $\hat{a}(t)$ is an estimated cogging force from a adaptive compensation mechanism to be specified in (15.9); $\dot{v}_d(t)$ is the desired acceleration.
The proposed adaptive law is designed as follows:

$$\hat{a}(t) = z - \mu v, \tag{15.9}$$

where μ is a positive constant; and the following tuning mechanism is designed for z:

$$_0D_t^\nu z(t) = \mu[\dot{v}_d(t) + \alpha m(t) + \gamma e_v(t)] + \frac{e_v(t)}{J}, \tag{15.10}$$

where $\nu \in (0, 1]$, $z(t)|_{t=0} = 0$.
In this study the Caputo definition is adopted for fractional derivative, which allows utilization of initial values of classical integer order derivatives with known physical interpretations [188]. Some definitions and properties can be referred from equations (1.6)–(1.12).

Remark 15.2.2   *With the initial condition, $e_\theta(t)|_{t=0} = 0$ and $e_v(t)|_{t=0} = v_d(0)$, choosing $\nu \in (0, 1)$, the designed control law (15.9) is the proposed fractional order adaptive compensation scheme; on the other hand, if $\nu = 1$, the designed control law (15.9) becomes the integer order adaptive compensation method.*

Remark 15.2.3   *The proposed controller shown in (15.7) and (15.9) actually can handle general bounded disturbances (15.3), not just the bounded cogging force disturbance in the form of (15.2). The fractional order compensation scheme is proposed in this chapter for the real industrial application – cogging effect compensation.*

Now, based on the above discussions, the following stability analysis of the systems with the proposed IOAC and FOAC schemes are presented in the frequency domain. We consider two cases: (1) when $\nu = 1$ for IOAC method of cogging effect compensation; and (2) when $0 < \nu < 1$ for FOAC scheme of cogging effect compensation.

15.2.3 IOAC Stability Analysis

First, with the initial condition, $e_\theta(t)|_{t=0} = 0$ and $e_v(t)|_{t=0} = v_d(0)$, let us consider case (1) Integer order adaptive compensation method with $\nu = 1$.

Theorem 15.2.1   *If the integer order adaptive compensation is used, the equilibrium points e_θ and e_v are bounded as $t \to \infty$.*

Proof: From (15.5), (15.7) and the adaptive law (15.9), one can obtain

$$\dot{v}(t) = (\dot{v}_d(t) + T_{l'} + \frac{1}{J}\hat{a}(t) + \alpha\gamma e_\theta(t) + (\alpha + \gamma)e_v(t))$$

$$- \frac{1}{J}a(t) - T_{l'} - B_1(v_d(t) - e_v(t))$$

$$= \dot{v}_d(t) + \frac{1}{J}(z - \mu v(t)) + \alpha\gamma e_\theta(t) + (\alpha + \gamma)e_v(t)$$

$$- \frac{1}{J}a(t) - B_1(v_d(t) - e_v(t))$$

$$= \dot{v}_d(t) + \frac{1}{J}z + \alpha\gamma e_\theta(t) + (\alpha + \gamma)e_v(t)$$

$$- \frac{1}{J}a(t) - \left(\frac{1}{J}\mu + B_1\right)(v_d(t) - e_v(t))$$

$$= \dot{v}_d(t) + \alpha\gamma e_\theta(t) + \left(\alpha + \gamma + \frac{1}{J}\mu + B_1\right)e_v(t)$$

$$- \frac{1}{J}a(t) - \left(\frac{1}{J}\mu + B_1\right)v_d(t) + \frac{1}{J}z(t), \tag{15.11}$$

and from (15.6), one can obtain

$$\dot{e}_v = \dot{v}_d(t) - \dot{v}(t)$$

$$= -\alpha\gamma e_\theta(t) - \left(\alpha + \gamma + \frac{1}{J}\mu + B_1\right)e_v(t)$$

$$+ \frac{1}{J}a(t) + \left(\frac{1}{J}\mu + B_1\right)v_d(t) - \frac{1}{J}z(t). \tag{15.12}$$

Then from (15.4), one has,

$$\dot{e}_\theta = \dot{\theta}_d(t) - \dot{\theta}(t) = v_d(t) - v(t) = e_v(t). \tag{15.13}$$

Substituting (15.13) into (15.12) yields

$$\ddot{e}_\theta(t) = -\alpha\gamma e_\theta(t) - \left(\alpha + \gamma + \frac{1}{J}\mu + B_1\right)\dot{e}_\theta(t)$$

$$+ \frac{1}{J}a(t) + \left(\frac{1}{J}\mu + B_1\right)v_d(t) - \frac{1}{J}z(t). \tag{15.14}$$

Now, from (15.10), using integer order derivative, namely $v = 1$,

$$\dot{z}(t) = \mu[\dot{v}_d(t) + \alpha m(t) + \gamma e_v(t)] + \frac{e_v(t)}{J}. \tag{15.15}$$

Then, without loss of generality, with the initial condition, $e_\theta(t)|_{t=0} = 0$ and $e_v(t)|_{t=0} = v_d(0)$, using the formula for the Laplace transform of (15.14) and (15.15) leads to

$$s^2 E_\theta(s) + \left(\alpha + \gamma + \frac{1}{J}\mu + B_1\right)s E_\theta(s) + \alpha\gamma E_\theta(s)$$

$$= \frac{1}{J}A(s) + \left(\frac{1}{J}\mu + B_1\right)V_d(s) + v_d(0) - \frac{1}{J}Z(s), \tag{15.16}$$

$$Z(s) = \frac{\mu}{s}(s V_d(s) - v_d(0)) + \frac{\mu}{s}\alpha\gamma E_\theta(s) + \left(\mu(\alpha + \gamma) + \frac{1}{J}\right)E_\theta(s). \tag{15.17}$$

Substituting (15.17) into (15.16), one can obtain

$$E_\theta(s) = \frac{1}{s^2 + a_0 s + b_0 + \frac{1}{s}c_0} F_0(s)$$

$$= \frac{s}{s^3 + a_0 s^2 + b_0 s + c_0} F_0(s), \tag{15.18}$$

where

$$a_0 = \alpha + \gamma + \frac{1}{J}\mu + B_1, \quad b_0 = \alpha\gamma + \frac{1}{J}\mu(\alpha + \gamma) + \frac{1}{J^2},$$

$$c_0 = \frac{1}{J}\mu\alpha\gamma, \quad F_0(s) = \frac{1}{J}A(s) + B_1 V_d(s) + \left(1 + \frac{1}{Js}\mu\right)v_d(0).$$

As α, γ and μ are all positive, so, a_0, b_0 and c_0 are also all positive. By selecting proper α, γ, μ, it is possible to ensure that

$$a_0 b_0 - c_0 = \left(\alpha + \gamma + \frac{1}{J}\mu + B_1\right)\left(\alpha\gamma + \frac{1}{J}\mu\alpha + \frac{1}{J}\mu\gamma + \frac{1}{J^2}\right) - \frac{1}{J}\mu\alpha\gamma$$

$$> \left(\alpha + \frac{1}{J}\mu + B_1\right)\left(\alpha\gamma + \frac{1}{J}\mu\gamma + \frac{1}{J^2}\right)$$

$$> 0. \tag{15.19}$$

From the Routh table technique, it can be concluded that the system (15.18) is stable. Since $a(t)$ and $v_d(t)$ are bounded, from the inverse Laplace transform

$$f_0(t) = L^{-1}[F_0(s)],$$

the input signal $f_0(t)$ in system (15.18) is bounded, and the output signal $e_\theta(t)$ in system (15.18) is also bounded.

Furthermore, from (15.13),

$$E_v(s) = sE_\theta(s) = \frac{s^2}{s^3 + a_0 s^2 + b_0 s + c_0}F_0(s). \tag{15.20}$$

In the same way, the system (15.20) is stable and the error signal $e_\theta(t)$ is also bounded.

So, it can be concluded that the equilibrium points e_θ and e_v, are bounded as $t \to \infty$.   ∎

15.2.4 FOAC Stability Analysis

Now, with the initial condition, $e_\theta(t)|_{t=0} = 0$ and $e_v(t)|_{t=0} = v_d(0)$, let us consider case (2) Fractional order (FO) adaptive compensation scheme with $0 < \nu < 1$. The major result is summarized in Theorem 15.2.2. First of all, the following lemma is needed for the proof of Theorem 15.2.2.

Lemma 15.2.1 *An ordinary input/output LTI system with only integer order derivatives can be written in differential operator polynomial representation*

$$P(\sigma)\xi = Q(\sigma)u, \tag{15.21}$$

$$y = R(\sigma)\xi,$$

where $u \in \mathfrak{R}^{\bar{m}}$ is the control, $\xi \in \mathfrak{R}^{\bar{n}}$ is the partial state, and the $y \in \mathfrak{R}^{\bar{p}}$ is the output; P, Q, and R are polynomial matrices in the variable σ of dimensions $\bar{n} \times \bar{n}$, $\bar{n} \times \bar{m}$, and $\bar{p} \times \bar{n}$, respectively; σ can be seen as the symbol of the usual derivative d^α or s^α, when all initial conditions are zero.

If the triplet (P, Q, R) of polynomial matrices is minimal, one has the following equivalence: the system (15.21) is bounded-input bounded-output iff $\det(P(\sigma)) \neq 0 \, \forall \sigma$, $|arg(\sigma)| < \alpha\frac{\pi}{2}$.

For a proof of Lemma 15.2.1, see [152].

Theorem 15.2.2 *If proper parameters α, γ and μ are chosen to ensure that*

$$|arg(w_i)| > \lambda \frac{\pi}{2},$$

where w_i are the roots of the following equation

$$w^{2pq+p^2} + aw^{pq+p^2} + bw^{pq} + dw^{p^2} + c = 0, \tag{15.22}$$

where $\lambda = 1/pq$, $p/q = v$, p and q are positive integers,

$$a = \alpha + \gamma + \frac{1}{J}\mu + B_1, \quad b = \frac{1}{J}\mu(\alpha + \gamma) + \frac{1}{J^2},$$

$$c = \frac{1}{J}\mu\alpha\gamma, \quad d = \alpha\gamma,$$

the equilibrium points $e_\theta(t)$ and $e_v(t)$ are bounded, as $t \to \infty$.

Proof: From (15.11), (15.12) and (15.13), one can obtain

$$\ddot{e}_\theta(t) = -\alpha\gamma e_\theta(t) - \left(\alpha + \gamma + \frac{1}{J}\mu + B_1\right)\dot{e}_\theta(t)$$

$$+ \frac{1}{J}a(t) + \left(\frac{1}{J}\mu + B_1\right)v_d(t) - \frac{1}{J}z(t). \tag{15.23}$$

According to equations (2.113) and (2.115) in [188]

$$_aD_t^{-p}(_aD_t^p f(t)) = f(t) - \left[_aD_t^{p-1}f(t)\right]_{t=a}\frac{(t-a)^{p-1}}{\Gamma(p)}, \tag{15.24}$$

where $0 < p < 1$,

$$_aD_t^{-p}(_aD_t^q f(t)) = _aD_t^{q-p}f(t) - \sum_{j=1}^{k}\left[_aD_t^{q-j}f(t)\right]_{t=a}\frac{(t-a)^{p-j}}{\Gamma(1+p-j)}, \quad (15.25)$$

where $0 < p$ and $0 < k - 1 < q < k$.

Since

$$_0D_t^v z(t) = \mu[\ddot{v}_d(t) + \alpha m(t) + \gamma e_v(t)] + \frac{e_v(t)}{J}, \quad (15.26)$$

and with the initial conditions, $z(t)|_{t=0} = 0$, $e_\theta(t)|_{t=0} = 0$ and $e_v(t)|_{t=0} = v_d(0)$, one can obtain

$$z(t) = _0D_t^{-v}(_0D_t^v z(t))$$

$$= _0D_t^{-v}\left\{\mu[\ddot{v}_d(t) + \alpha m(t) + \gamma e_v(t)] + \frac{e_v(t)}{J}\right\}$$

$$= _0D_t^{-v}\left\{\mu[\ddot{v}_d(t) + \alpha\gamma e_\theta(t) + (\alpha + \gamma)e_v(t)] + \frac{e_v(t)}{J}\right\}$$

$$= \mu_0D_t^{-v}\ddot{v}_d(t) + \mu\alpha\gamma_0D_t^{-v}e_\theta(t) + \left(\mu(\alpha + \gamma) + \frac{1}{J}\right)_0D_t^{-v}e_v(t)$$

$$= \mu_0D_t^{-v}\ddot{v}_d(t) + \mu\alpha\gamma_0D_t^{-v}e_\theta(t) + \left(\mu(\alpha + \gamma) + \frac{1}{J}\right)_0D_t^{1-v}e_\theta(t),$$

$$= \mu\left(_0D_t^{1-v}v_d(t) - v_d(0)\right) + \mu\alpha\gamma_0D_t^{-v}e_\theta(t)$$

$$+ \left(\mu(\alpha + \gamma) + \frac{1}{J}\right)_0D_t^{1-v}e_\theta(t). \quad (15.27)$$

By substituting (15.27) into (15.23), one has

$$\ddot{e}_\theta(t) + a\dot{e}_\theta(t) + b_0D_t^{1-v}e_\theta(t) + c_0D_t^{-v}e_\theta(t) + de_\theta(t) = f(t), \quad (15.28)$$

where

$$a = \alpha + \gamma + \frac{1}{J}\mu + B_1, \quad b = \frac{1}{J}\mu(\alpha + \gamma) + \frac{1}{J^2},$$

$$c = \frac{1}{J}\mu\alpha\gamma, \quad d = \alpha\gamma,$$

$$f(t) = \frac{1}{J}a(t) + \left(\frac{1}{J}\mu + B_1\right)v_d(t) - \frac{1}{J}\mu\left(_0D_t^{1-v}v_d(t) - v_d(0)\right). \quad (15.29)$$

According to equations (2.242) and (2.248) in [188]

$$L\left\{_0D_t^{-p}f(t);s\right\} = s^{-p}F(s), \tag{15.30}$$

where $0 < p$,

$$L\left\{_0D_t^{p}f(t);s\right\} = s^{p}F(s) - \sum_{k=1}^{n-1}s^{k}\left[_0D_t^{p-k-1}f(t)\right]_{t=0}, \tag{15.31}$$

where $0 \leqslant n - 1 \leqslant p < n$; using formula for the Laplace transform of (15.28) leads to

$$(s^2E_\theta(s) - v_d(0)) + a(s + bs^{1-\nu} + cs^{-\nu} + d)E_\theta(s)$$

$$= F(s)$$

$$= \frac{1}{J}A(s) + \left(\frac{1}{J}\mu + B_1\right)V_d(s) - \frac{\mu s^{1-\nu}}{J}V_d(s) + \frac{\mu}{Js}v_d(0), \tag{15.32}$$

thus

$$E_\theta(s) = \frac{1}{s^2 + as^1 + bs^{1-\nu} + cs^{-\nu} + d}(F(s) + v_d(0))$$

$$= \frac{s^\nu}{s^{2+\nu} + as^{1+\nu} + bs + c + ds^\nu}G(s), \tag{15.33}$$

where

$$G(s)$$

$$= F(s) + v_d(0)$$

$$= \frac{1}{J}A(s) + \left(\frac{1}{J}\mu + B_1\right)V_d(s) - \frac{\mu s^{1-\nu}}{J}V_d(s) + \left(\frac{\mu}{Js} + 1\right)v_d(0). \tag{15.34}$$

Denote that

$$\nu = \frac{p}{q}, \quad s^\nu = s^{\frac{p}{q}},$$

where $\nu = p/q$, p and q are positive integers. So,

$$E_\theta(s) = \frac{s^{\frac{p}{q}}}{s^{2+\frac{p}{q}} + as^{1+\frac{p}{q}} + bs + c + ds^{\frac{p}{q}}}G(s). \tag{15.35}$$

Let $\lambda = \frac{1}{pq}$, $w = s^\lambda$, then

$$E_\theta(w) = \frac{w^{p^2}}{w^{2pq+p^2} + aw^{pq+p^2} + bw^{pq} + dw^{p^2} + c}G(w). \tag{15.36}$$

From Lemma 15.2.1, $\bar{m} = \bar{n} = 1$, since

$$|arg(w)| > \lambda \frac{\pi}{2},$$

therefore, system (15.35) is bound-input bound-output stable. As $a(t)$ and $v_d(t)$ are bounded, from the inverse Laplace transform

$$g(t) = L^{-1}[G(s)],$$

so, the input signal $g(t)$ in system (15.35) is bounded, and the output error signal $e_\theta(t)$ in system (15.35) is also bounded.

Furthermore, from (15.13), one has

$$E_v(w) = w^{pq} E_\theta(w)$$

$$= \frac{w^{p^2+pq}}{w^{2pq+p^2} + aw^{pq+p^2} + bw^{pq} + dw^{p^2} + c} G(w). \qquad (15.37)$$

Similarly, the system (15.37) is stable and the error signal $e_\theta(t)$ is also bounded. In summary, it can be concluded that the equilibrium points e_θ and e_v are bounded as $t \to \infty$. ∎

15.3 Simulation Illustration

In this section, three simulation cases are presented to illustrate the effectiveness of the proposed fractional order adaptive compensation for cogging effect on PMSM servo control system. Figure 15.1 shows the simulation block diagram. Details of this simulation system can be found in [140].

For the simulation tests, the control gains in (15.7) were selected as: $\alpha = 50$, $\gamma = 20$, and $\mu = 3$. The motor parameters are given in Table 15.1 and $T_l = 1[Nm]$. The real cogging force is modeled as the state-periodic sinusoidal signal of multiple harmonics below, as shown in Figure 15.2,

$$F_{cogging} = 2\cos(6\theta) + \cos(12\theta) + 0.5\cos(18\theta).$$

15.3.1 Case-1: IOAC with Constant Reference Speed

In this simulation case, choosing $v = 1$ in adaptive law (15.9), namely, IOAC scheme is used for cogging effect compensation. The cogging effect degrades the performance of PMSM seriously in a low speed range. Normally, the motor speed below 3% of the rated speed can always be treated as in the low speed range. In this simulation model, the rated speed of the PMSM is 2000 rpm, which is given in Table 15.1. So, the reference speed in this simulation is chosen as 5 $rad/s = 47.77$ rpm

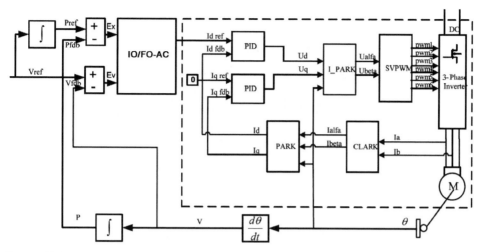

Figure 15.1 Block diagram of PMSM servo simulation system with the IO/FOAC for cogging effect compensation

$\leqslant 60$ rpm. Hence, in this simulation test, the following reference trajectory and velocity signals are used,

$$s_d(t) = 5t \ rad, \tag{15.38}$$

$$v_d(t) = 5 \ rad/s. \tag{15.39}$$

Figure 15.4 shows the estimated cogging force with IOAC. Figures 15.5(a) and 15.5(b) show the position/speed tracking errors with compensation using integer order adaptive compensator. The IOAC method works efficiently compared with the tracking errors without compensation in Figures 15.3(a) and 15.3(b).

15.3.2 Case-2: FOAC with Constant Reference Speed

In this case, for a fair comparison, the reference trajectory (15.38) and velocity (15.39) signals in Case-1 are used. The fractional order adaptive compensator is used for cogging effect compensation. Here, in FOAC law (15.9), choosing $v = 0.5$, and

Table 15.1 PMSM Specifications

Rated power	1.64 Kw	Rated speed	2000 rpm
Rated torque	8 Nm	Stator resistance	2.125 Ω
Stator inductance	11.6 mH	Magnetic flux	0.387 Wb
Number of poles	6	Moment of inertia	0.00289 kgm^2
Friction coefficient	0.0003 Nms		

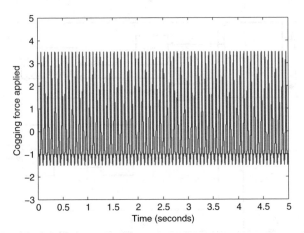

Figure 15.2   Simulation. Cogging force applied

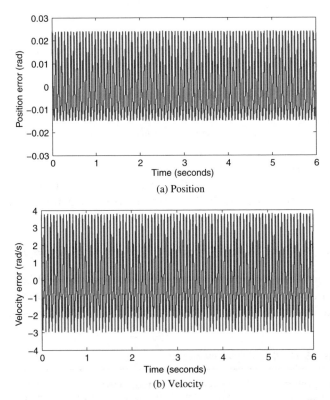

Figure 15.3   Simulation. Constant reference speed tracking errors without compensation

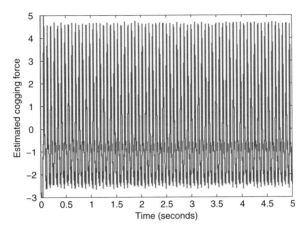

Figure 15.4 Simulation. Estimated cogging force with IOAC ($v = 1$)

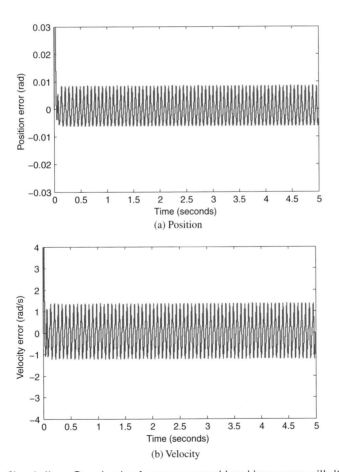

(a) Position

(b) Velocity

Figure 15.5 Simulation. Constant reference speed tracking errors with IOAC ($v = 1$)

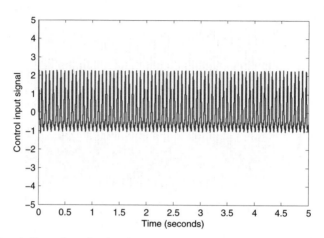

Figure 15.6 Simulation. Constant reference speed tracking control input signal with IOAC ($v = 1$)

$\alpha = 50$, $\gamma = 20$ and $\mu = 3$, substituting the parameter values into (15.2.2), one can obtain five solutions,

$$w_1 = 25.68808 + 55.53037i; \quad w_2 = 25.68808 - 55.53037i;$$

$$w_3 = 0.01285 + 2.32261i; \quad w_4 = 0.01285 - 2.32261i;$$

$$w_5 = -51.40188;$$

and one can obtain

$$|arg(w_1)| = |arg(w_2)| = \frac{67.18^o}{180^o}\pi = 0.3621\pi$$

$$> v\frac{\pi}{2} = 0.5\frac{\pi}{2} = 0.25\pi;$$

$$|arg(w_3)| = |arg(w_4)| = \frac{89.68^o}{180^o}\pi = 0.4982\pi$$

$$> v\frac{\pi}{2} = 0.5\frac{\pi}{2} = 0.25\pi;$$

$$|arg(w_5)| = \pi > v\frac{\pi}{2} = 0.5\frac{\pi}{2} = 0.25\pi.$$

So, the system should be bounded-input bounded-output stable according to the stability analysis presented in Section 15.2.4. For the approximate realization for fractional order derivative, the Oustaloup Recursive Algorithm [183] is used, where the frequency range of practical interest is chosen as [0.001, 1000]. Figure 15.7 shows the estimated cogging force using FOAC, which is closer to the applied cogging force in Figure 15.2 compared with that using IOAC in Figure 15.4. Figures 15.8(a)

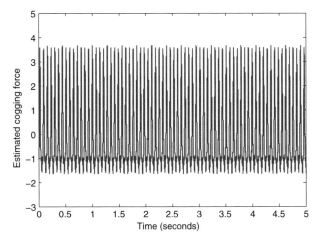

Figure 15.7 Simulation. Estimated cogging force with FOAC ($v = 0.5$)

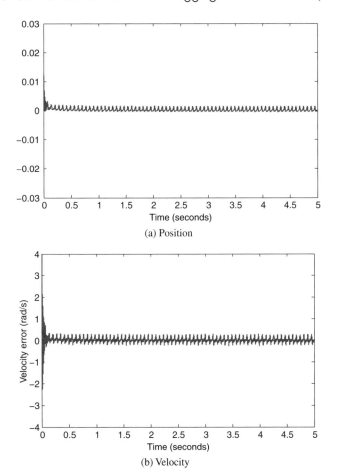

(a) Position

(b) Velocity

Figure 15.8 Simulation. Constant reference speed tracking errors with FOAC ($v = 0.5$)

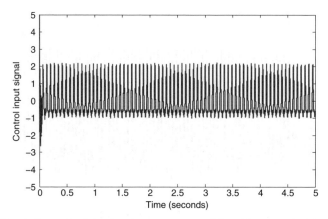

Figure 15.9 Simulation. Constant reference speed tracking control input signal with FOAC ($\nu = 0.5$)

and 15.8(b) show the position/speed tracking errors using FOAC. Comparing with Figures 15.5(a) and 15.5(b), one can clearly see that, the performance using FOAC scheme is much better than that using IOAC for multi-harmonics cogging effect compensation. Furthermore, from Figures 15.6 and 15.9, the magnitude of the control input signal with FOAC is a little bit smaller than that with the integer order one. So, the performance improvement using FOAC does not pay the price of a greater control effort. Instead, the control effort using FOAC is even smaller than that using the integer order one.

15.3.3 Case-3: IO/FOAC with Varying Reference Speed

In this simulation case, the following varying reference trajectory and velocity signals are used:

$$s_d(t) = \int_0^t v_d(\tau)d\tau, \tag{15.40}$$

$$v_d(t) = \begin{cases} 2 \ rad/s & \text{if} \quad js_p \leqslant s < (j+1)s_p, \\ 4 \ rad/s & \text{if} \quad (j+1)s_p \leqslant s < (j+2)s_p, \end{cases} \tag{15.41}$$

where $j = 0, 2, 4, \cdots$; s_p is the known periodicity of the trajectory.

For a fair comparison, both the integer order control law/parameters in Case-1 and fractional order control law/parameters in Case-2 are used in this case.

Figures 15.11(a) and 15.11(b) show the position/speed tracking errors with compensation using IOAC. The IOAC method works efficiently compared with the tracking errors without compensation in Figures 15.10(a) and 15.10(b). Figures 15.13(a) and 15.13(b) show the position/speed tracking errors using FOAC. Compared with Figures 15.11(a) and 15.11(b), one can see that the performance using FOAC scheme is

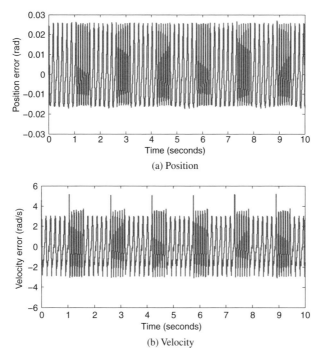

(a) Position

(b) Velocity

Figure 15.10 Simulation. Varying reference speed tracking errors without compensation

much better than that using IOAC for multi-harmonics cogging effect compensation with varying reference speed. From Figures 15.12 and 15.14, with the varying reference speed tracking, the control effort using FOAC is also even smaller than that using the integer order one.

15.4 Experimental Validation

15.4.1 Introduction to the Experimental Platform

The fractional horsepower dynamometer as a general purpose experiment platform [210] is also applied for the experimental validation in this chapter. The architecture of the dynamometer control system is shown in Figure 3.10. The setup of this experimental platform is introduced in Section 3.6.1.

15.4.1.1 Proposed Application

Without loss of generality, consider a servo control system,

$$\dot{x}(t) = v(t), \tag{15.42}$$

$$\dot{v}(t) = u(t) - f(t, x) - \frac{B_d}{J_d}v, \tag{15.43}$$

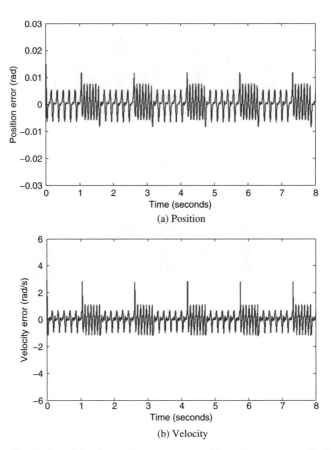

Figure 15.11 Simulation. Varying reference speed tracking errors with IOAC ($\nu = 1$)

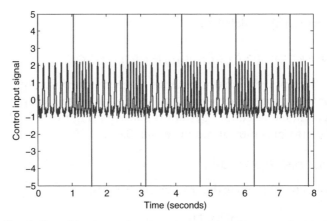

Figure 15.12 Simulation. Varying reference speed tracking control input signal with IOAC ($\nu = 1$)

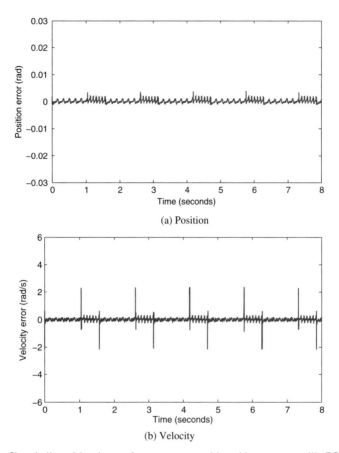

(a) Position

(b) Velocity

Figure 15.13 Simulation. Varying reference speed tracking errors with FOAC ($\nu = 0.5$)

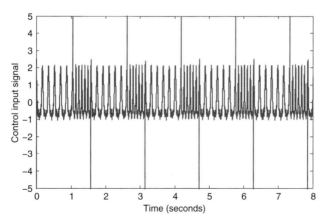

Figure 15.14 Simulation. Varying reference speed tracking control input signal with FOAC ($\nu = 0.5$)

where x is the position or displacement, v is the velocity, $f(t, x)$ is the unknown state-dependent or time-dependent disturbance, B_d and J_d are the frictional coefficient and inertia moment of the DC motor, respectively, and u is the control input. The hysteresis brake allows us to add a time-dependent or state-dependent disturbance to the motor. This lab test-bed can emulate a system similar to the one given by (15.42) and (15.43). The controller can be designed for the disturbances compensation problem, and tested in the presence of the real disturbance as introduced via the hysteresis brake in the dynamometer control system.

15.4.2 Experiments on the Dynamometer

The proposed method is verified on the real-time dynamometer position control system. The hysteresis brake force is designed as the multi-harmonics cogging-like disturbance,

$$f(t, x) = \frac{T_d}{J_d} + \frac{a(t)}{J_d} = \frac{T_d}{J_d} + F_{disturbance}, \tag{15.44}$$

where

$$\frac{T_d}{J_d} = 1,$$

$$F_{disturbance} = 10\cos(x) + 5\cos(2x) + 2.5\cos(3x),$$

where the applied cogging-like disturbance $F_{disturbance}$ is shown in Figure 15.16. When x is replaced by θ, then the system with (15.42) and (15.43) is in the same format with (15.4) and (15.5). As mentioned in [140], after using vector control (control i_d to 0), the nonlinear and coupling characteristics of PMSM can be decoupled. Thus, the torque magnitude control of PMSM only need to control the current in the direction of the q-axis. So, a PMSM can be controlled as easily as a DC motor. Therefore, the developed fractional order adaptive compensator can be validated for cogging effect compensation on this dynamometer platform. Figure 15.15 shows the block diagram of dynamometer control system where the block in dashed line is emulated by the dynamometer. The control gains in (15.7) are selected as: $\alpha = 5$, $\gamma = 10$ and $\mu = 0.5$.

As in the simulation part, three experimental cases are presented similarly in this section.

In the experimental Case-1e and Case-2e, the following reference trajectory and velocity signals are also used as in the simulation,

$$s_d(t) = 5t \ rad,$$

$$v_d(t) = 5 \ rad/s.$$

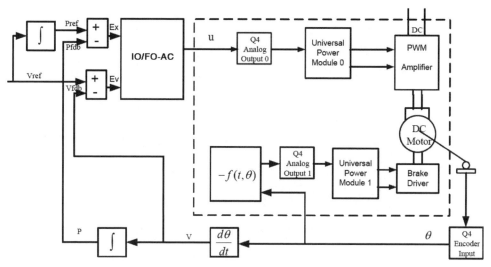

Figure 15.15 Block diagram of the cogging-like disturbance compensation using IO/FOAC in the dynamometer control system

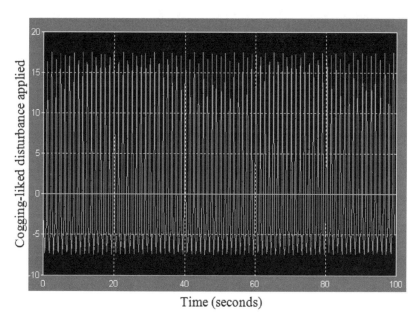

Figure 15.16 Experiment. Cogging-like disturbance applied

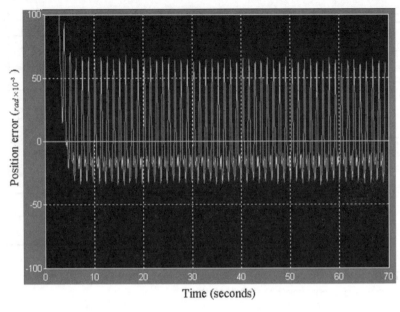

(a) Position

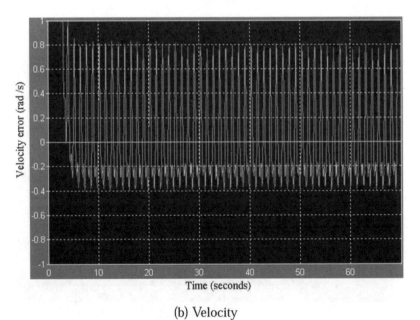

(b) Velocity

Figure 15.17 Experiment. Constant reference speed tracking errors without compensation

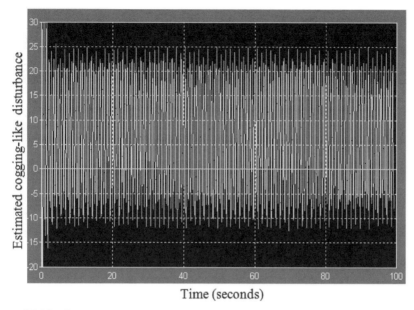

Time (seconds)

Figure 15.18 Experiment. Estimated cogging-like disturbance with IOAC ($v = 1$)

15.4.2.1 Case-1e: IOAC for Cogging-Like Disturbance Compensation with Constant Reference Speed

In this experiment case, choosing $v = 1$ in the adaptive law (15.9), the integer order adaptive compensator is implemented to obtain cogging-like effect compensation. The estimated cogging-like disturbance is shown in Figure 15.18 using IOAC.

Figures 15.19(a) and 15.19(b) show the position/speed tracking errors using the IOAC method. The IOAC method works efficiently compared with the tracking errors without compensation in Figures 15.17(a) and 15.17(b).

15.4.2.2 Case-2e: FOAC for Cogging-Like Disturbance Compensation with Constant Reference Speed

In this case, the fractional order adaptive compensator is applied for cogging-like effect compensation. In the adaptive law (15.9), the parameters are chosen as $v = 0.5$, $\alpha = 5$, $\gamma = 10$ and $\mu = 0.5$. By substituting the parameter values into (15.2.2), one can obtain five solutions,

$$w_1 = 24.18965 + 44.08336i; \ w_2 = 24.18965 - 44.08336i;$$
$$w_3 = -0.00015 + 0.26592i; \ w_4 = -0.00015 - 0.26592i;$$
$$w_5 = -48.37900;$$

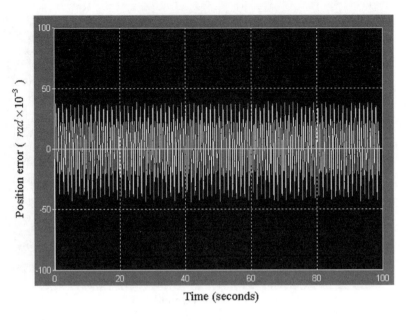

(a) Position

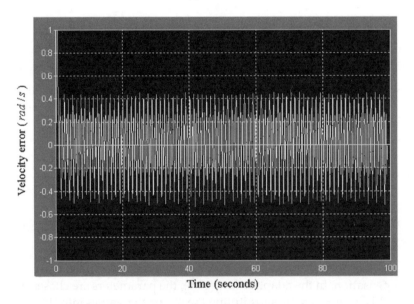

(b) Velocity

Figure 15.19   Experiment. Constant reference speed tracking errors with IOAC ($v = 1$)

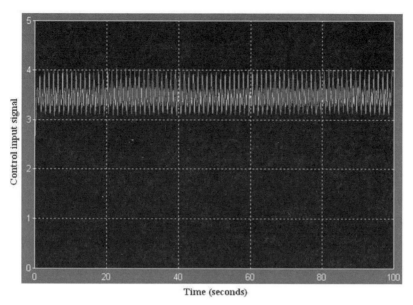

Figure 15.20   Experiment. Constant reference speed tracking control input signal with IOAC ($v = 1$)

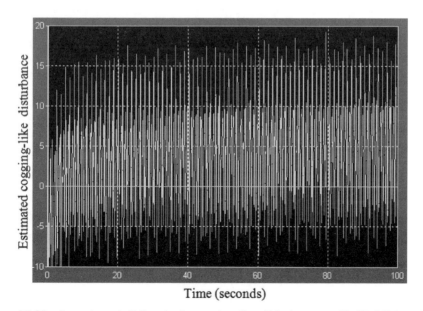

Figure 15.21   Experiment. Estimated cogging-like disturbance with FOAC ($v = 0.5$)

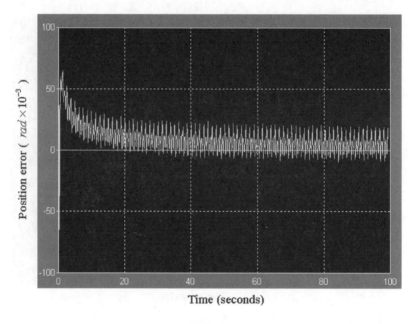

(a) Position

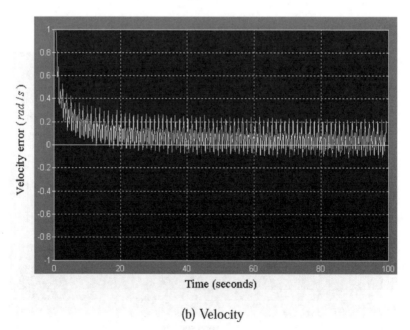

(b) Velocity

Figure 15.22 Experiment. Constant reference speed tracking errors with FOAC ($\nu = 0.5$)

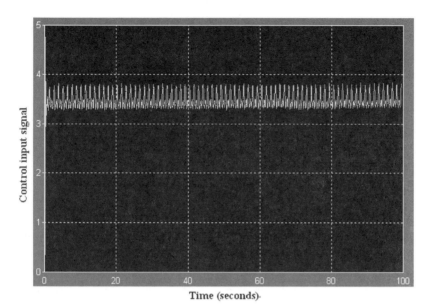

Figure 15.23 Experiment. Constant reference speed tracking control input signal with FOAC ($v = 0.5$)

so, one can obtain,

$$|arg(w_1)| = |arg(w_2)| = \frac{65.175^o}{180^o}\pi = 0.3402\pi$$

$$> v\frac{\pi}{2} = 0.5\frac{\pi}{2} = 0.25\pi;$$

$$|arg(w_3)| = |arg(w_4)| = \pi - \frac{89.968^o}{180^o}\pi = 0.5002\pi$$

$$> v\frac{\pi}{2} = 0.5\frac{\pi}{2} = 0.25\pi;$$

$$|arg(w_5)| = \pi > v\frac{\pi}{2} = 0.5\frac{\pi}{2} = 0.25\pi.$$

Therefore, the system is bounded-input bounded-output stable according to the stability analysis in Section 15.2.4. The same Oustaloup Recursive Algorithm [183] is used for the approximate realization of the fractional order derivative, and the frequency range of practical interest is [0.01, 100] since the bandwidth of the dynamometer is less than 10 Hz. The estimated cogging-like disturbance is shown in Figure 15.21 using FOAC, which is closer to the applied cogging-like disturbance in Figure 15.16 compared with that using IOAC in Figure 15.18. Figures 15.22(a) and 15.22(b) show

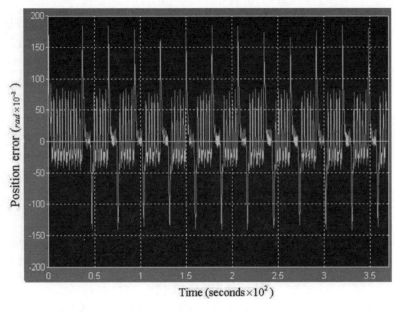

(a) Position

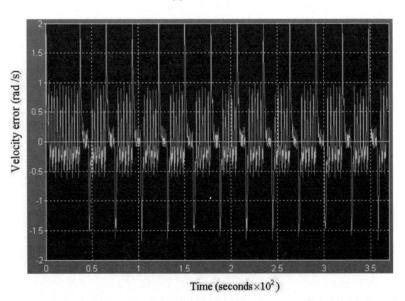

(b) Velocity

Figure 15.24 Experiment. Varying reference speed tracking errors without compensation

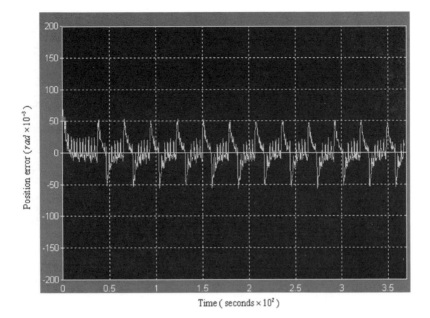

(a) Position

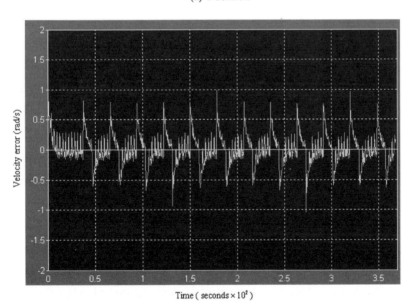

(b) Velocity

Figure 15.25   Experiment. Varying reference speed tracking errors with IOAC ($v = 1$)

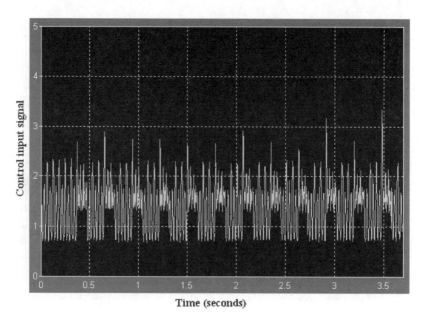

Time (seconds)

Figure 15.26 Experiment. Varying reference speed tracking control input signal with IOAC ($v = 1$)

the position/speed tracking errors using the FOAC method. Compared with Figures 15.19(a) and 15.19(b), the performance using the FOAC scheme is much better than that using the IOAC method for cogging-like disturbance compensation in real-time experimental platform. Moreover, from Figures 15.20 and 15.23, it can be seen that the magnitude of the control input signal using FOAC is obviously smaller than that using the integer order one. Therefore, the performance improvement using the fractional order control does not pay the price of the greater control effort. In fact, the control effort using FOAC is even smaller than that using the integer order one.

15.4.2.3 Case-3e: IO/FOAC for Cogging-Like Disturbance Compensation with Varying Reference Speed

In this case, the varying reference trajectory (15.40) and velocity (15.41) signals in the simulation are also used.

In order to have a fair comparison, the integer order control law and parameters in Case-1e and the fractional order control law and parameters in Case-2e are also used.

Figures 15.25(a) and 15.25(b) show the position/speed tracking errors with compensation using IOAC. The IOAC method also works efficiently compared with the

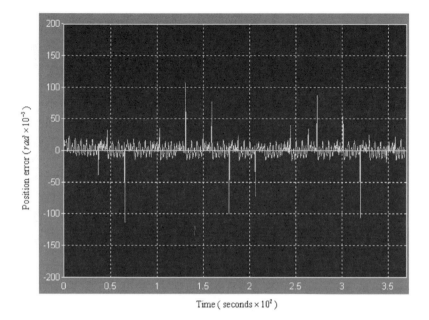

(a) Position

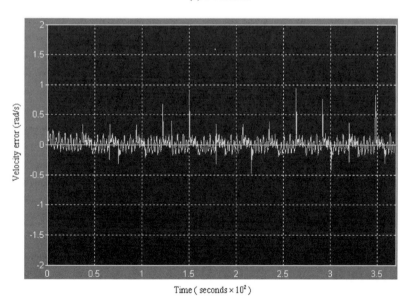

(b) Velocity

Figure 15.27   Experiment. Varying reference speed tracking errors with FOAC ($\nu = 0.5$)

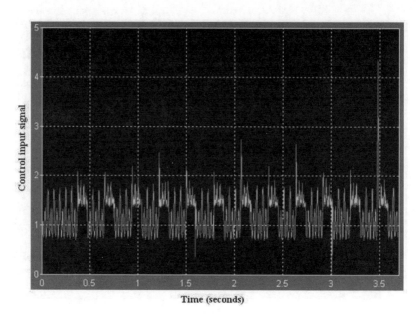

Figure 15.28 Experiment. Varying reference speed tracking control input signal with FOAC ($\nu = 0.5$)

tracking errors without compensation in Figures 15.24(a) and 15.24(b). Figures 15.27(a) and 15.27(b) show the position/speed tracking errors using FOAC. Compared with Figures 15.25(a) and 15.25(b), the performance using FOAC scheme is much better than that using IOAC for the multi-harmonics cogging effect compensation with varying reference speed. Furthermore, from Figures 15.26 and 15.28, it can also be seen that the control effort using the fractional order adaptive compensator is also much smaller than that using the integer order one.

Remark 15.4.1 *In the course of simulation and experiment, it is found that changing the parameter μ does not improve the performance of the position and velocity tracking effectively in the PMSM position servo simulation system as well as in the dynamometer real-time control experiments.*

15.5 Chapter Summary

In this chapter, a fractional order adaptive compensation method is proposed to compensate the cogging effect in the PMSM position and the velocity control system. A fractional order adaptive compensator is designed for the cogging effect and guarantee the boundedness of all signals. Stability properties have been proven for both systems with the traditional integer order adaptive compensation and the proposed fractional order adaptive compensation. Simulation illustration and experimental validation are presented to show the advantage of the proposed FOAC scheme for cogging effect compensation. This new fractional order scheme performs better than the traditional integer order method. The position and speed tracking steady-state errors using FOAC are much smaller than those using IOAC. It also can be observed that the control effort using FOAC is even smaller than that using IOAC. Although the suggested FOAC scheme is developed for the cogging force compensation, this method can also be used to compensate for other kinds of uncertain bounded disturbances.

16

Fractional Order Periodic Adaptive Learning Compensation

16.1 Introduction

State-dependent periodic disturbance (SDPD) exists in many motion control engineering problems. For instance, the cogging effect discussed in Chapter 15 with the permanent magnet (PM) motor is a typical SDPD [95], [100], [132], [133], [142], which degrades the servo control performance, especially in low-speed applications. In [70], the friction force is demonstrated as a state-periodic parasitic effect and in [6], the friction and eccentricity in the wheeled mobile robots are shown to be SDPD. There also are some state-periodic external disturbances in rotary systems [66], [88]. Since the SDPD is widespread in practice, the suppression of this kind of disturbance has been paid more and more attention in recent years. Cogging effect compensation techniques have been proposed in some literatures for example, [100] and [95]. In [206], [236], the SDPD can be well compensated by using the so-called learning control method. Taking advantage of the state-dependent periodicity, adaptive learning control is applied to compensate the cogging and Coulomb friction in [5] and [206]. In the previous work [5], an adaptive learning controller for compensating the cogging effect and coulomb friction in permanent-magnet linear motors is designed.

In this chapter, we devised a fractional order periodic adaptive learning compensation (FO-PALC) method for the SDPD compensation. In the first trajectory period of the proposed FO-PALC scheme, the fractional order adaptive compensator as shown in Chapter 15 is designed to guarantee the boundedness of the system state, input and output signals. From the second repetitive trajectory period and onward, information previously stored for one period along the state axis is used in the current

Fractional Order Motion Controls, First Edition. Ying Luo and YangQuan Chen.
© 2013 John Wiley & Sons, Ltd. Published 2013 by John Wiley & Sons, Ltd.

adaptation law. Asymptotical stability proof of the system with the proposed FO-PALC is presented in the frequency domain. Both simulation illustration and real-time experimental validation are presented to show the benefits of using fractional calculus in periodic adaptive learning compensation for the general SDPD [142].

16.2 Fractional Order Periodic Adaptive Learning Compensation for State-Dependent Periodic Disturbances

16.2.1 The General Form of State-Dependent Periodic Disturbances

This chapter is mainly concerned with the general state-dependent periodic disturbance. This kind of disturbance could be any type of periodic function depending on a state variable x which usually represents the linear displacement or rotational angle. In the Fourier series, the general SDPD can be expressed by

$$F_{disturbance} = \sum_{i=1}^{\infty} A_i \sin(\omega_i x + \varphi_i),$$

where A_i is the amplitude, ω_i is the periodic disturbance frequency, and φ_i is the phase angle. This general form can well represent the SDPD in practice, for example, the state-dependent friction in [4] and the cogging effect of the permanent magnet synchronous motor in [95], [131].

16.2.2 Problem Formulation

In this section, a fractional order periodic adaptive learning compensator for SDPD is designed. The SDPD can be written as: $-a(\theta)$, where a is the function of state variable θ. In this chapter, to present our ideas clearly, without loss of generality, the motion control system is modeled as follows

$$\dot{\theta}(t) = v(t), \tag{16.1}$$

$$\dot{v}(t) = u(t) - \frac{a(\theta)}{J} - T_{l'} - B_1 v(t), \tag{16.2}$$

where θ is the state (displacement); v is the velocity; u is the control input signal; J is a constant which can be the moment of inertia of the motor when the rotary motion system is considered; and $a(\theta)$ is the unknown SDPD, actually, $a(\theta)$ should be bounded in practice, for example,

$$|a(\theta)| \leqslant b_0 < \infty, \quad \forall \theta, \tag{16.3}$$

where, b_0 is a positive real number. Moreover, based on the same physical reason, the change of profile shape in $a(\theta)$ can also be regarded as bounded, that is,

$$|\partial a(\theta)/\partial\theta| \leqslant b_a < \infty \tag{16.4}$$

where b_a is a positive real number.

Definition 16.2.1 *The total passed trajectory is given as:*

$$s(t) = \int_0^t \left|\frac{d\theta}{d\tau}\right| d\tau = \int_0^t |v(\tau)| d\tau,$$

where θ is the angle position, and v is the velocity. Physically, $s(t)$ is the total passed trajectory, hence it has the following property:

$$s(t_1) \geqslant s(t_2), \quad \text{if} \quad t_1 \geqslant t_2.$$

The position corresponding to $s(t)$ is denoted as $\theta(s)$ and the SDPD corresponding to $s(t)$ is denoted as $a(s)$. In our definition, since $s(t)$ is the summation of absolute position increasing along the time axis, just like t, $s(t)$ is a monotonously growing signal, so we have

$$a(\theta(s)) = a(s) = a(t). \tag{16.5}$$

The time instant for one past trajectory from the time instant t is denoted as τ_{k-1}, and its corresponding cycle is completed in $P_k(t)$ amount of time.
  Let us define

$$e_a(s) = a(s) - \hat{a}(s), \tag{16.6}$$

where $\hat{a}$ is an estimated state-dependent periodic disturbance, and $\hat{a}(s) = \hat{a}(t)$ (note: t is the current time corresponding to the current total passed trajectory s). Here, let us change (16.6) into the time-domain such as

$$e_a(s) = a(s) - \hat{a}(s) = a(t) - \hat{a}(t) = e_a(t). \tag{16.7}$$

In the same way, the following relationships are true

$$v_d(s) = v_d(t), \quad v(s) = v(t),$$

and the following notations are also defined

$$e_\theta(t) = \theta_d(t) - \theta(t), \quad e_v(t) = v_d(t) - v(t). \tag{16.8}$$

The control objective is to track or servo the given desired position $\theta_d(t)$ and the corresponding desired velocity $v_d(t)$ with tracking errors as small as possible. In practice, it is reasonable to assume that $\theta_d(t)$ and $v_d(t)$ are both bounded signals.

The feedback controller is designed as

$$u(t) = \dot{v}_d(t) + T_{l'} + \frac{\hat{a}(t)}{J} + \alpha m(t) + \gamma e_v(t), \tag{16.9}$$

with

$$m(t) := \gamma e_\theta(t) + e_v(t), \tag{16.10}$$

where α and γ are positive gains; $\dot{v}_d(t)$ is the desired acceleration.

Our fractional order adaptive control law in the first trajectory period and the periodic adaptive learning control law after the first trajectory period are designed as follows

$$\hat{a}(t) = \begin{cases} \hat{a}_1(t) + \frac{K}{J}m_1(t) & \text{if } s \geqslant s_p \\ z - \mu v & \text{if } s < s_p \end{cases} \tag{16.11}$$

where z is designed following the fractional order tuning mechanism as below

$$_0D_t^\nu z(t) = \mu[\dot{v}_d(t) + \alpha m(t) + \gamma e_v(t)] + \frac{e_v(t)}{J}, \tag{16.12}$$

$$\nu \in (0, 1], \quad z(t)|_{t=0} = 0;$$

$$\hat{a}_1(t) =: \hat{a}(t - P_k), \quad m_1(t) =: m(t - P_k),$$

where K is a positive periodic adaptation gain and μ is also a positive design parameter. As mentioned, the time instant for one past trajectory from current time instant t is denoted as τ_{k-1}, and P_k denotes the amount of time to complete the corresponding cycle.

In this study the Caputo definition is adopted for the fractional derivative, which allows utilization of the initial values of the classical integer order derivatives with known physical interpretations [188]. Equations (1.6)–(1.12) can be referred to for the definitions and properties.

16.2.3 Stability Analysis

Now, based on the above discussions, the following stability analysis of the proposed fractional order adaptive compensation method in the first period and the PALC scheme after the first period is presented.

Consider two cases: (1) when $0 \leqslant t < P_1(0 \leqslant s < s_p)$ and (2) when $t \geqslant P_1(s \geqslant s_p)$. The key idea is that, for case (1), it is required to show the finite time boundedness of the system state, input and output signals; for Case-2, it is necessary to show the asymptotic stability of equilibrium points.

First, let us consider the case (1) when $t < P_1(s < s_p)$.

If choosing $v = 1$, the control law $\hat{a} = z - \mu v$ in (16.11) is the integer order adaptive control scheme. We have the following theorem:

Theorem 1.2.1 *If $|\partial a(\theta)/\partial \theta| < b_a$ (bound of changes in $a(\theta)$), and $\alpha + \gamma > \frac{J(b_a + \mu B_1)^2}{4\mu} - B_1$, the equilibrium points $e_\theta(t)$ and $e_v(t)$ are bounded, when $t < P_1(s < s_p)$.*

Proof: For a proof of this Theorem 16.2.1, the proof of Theorem 15.2.1 can be referred to.   ∎

If choosing $0 < v < 1$, the control law $\hat{a} = z - \mu v$ in (16.11) is the fractional order adaptive compensation scheme. Theorem 15.2.2 can be referred to for the stability proof of this case.

Now, let us investigate Case-2 when $t \geqslant P_1(s \geqslant s_p)$. We have the following stability result:

Theorem 16.2.2 *When $t \geqslant P_1(s \geqslant s_p)$, the control law (16.9) and the periodic adaptation law (16.11) guarantee that $e_\theta(t)$, $e_v(t)$ and $e_a(t)$ all approach 0 as $t \to \infty(s \to \infty)$.*

Proof:

From now on, denoting $t' = t - P_1$. So, when $t = P_1$, $t' = 0$.   ∎

From (16.2) and (16.9),

$$\dot{v}(t') = \dot{v}_d(t') + T_{l'} + \frac{1}{J}\hat{a}(t') + (\alpha + \gamma)e_v(t') + \alpha \gamma e_\theta(t')$$

$$- \frac{1}{J}a(t') - T_{l'} - B_1(v_d(t') - e_v(t'))$$

$$= \dot{v}_d(t') + \frac{1}{J}\hat{a}(t') + \alpha \gamma e_\theta(t') + (\alpha + \gamma)e_v(t')$$

$$- \frac{1}{J}a(t') - B_1(v_d(t') - e_v(t')), \tag{16.13}$$

and from (16.1),

$$\dot{e}_v = \dot{v}_d(t') - \dot{v}(t') = -\alpha \gamma e_\theta(t') - (\alpha + \gamma + B_1)e_v(t')$$

$$+ \frac{1}{J}a(t') - \frac{1}{J}\hat{a}(t') + B_1 v_d(t'). \tag{16.14}$$

Then, from (16.15),

$$\dot{e}_\theta = \dot{\theta}_d(t') - \dot{\theta}(t') = v_d(t') - v(t') = e_v(t'), \tag{16.15}$$

substituting (16.15) into (16.14) yields

$$\ddot{e}_\theta(t') = -\alpha\gamma e_\theta(t')$$
$$-(\alpha + \gamma + B_1)\dot{e}_\theta(t') + \frac{1}{J}a(t') - \frac{1}{J}\hat{a}(t') + B_1 v_d(t'). \qquad (16.16)$$

So, we have

$$\ddot{e}_\theta(t') + (\alpha + \gamma + B_1)\dot{e}_\theta(t') + \alpha\gamma e_\theta(t')$$
$$= \frac{1}{J}a(t') - \frac{1}{J}\hat{a}(t') + B_1 v_d(t'). \qquad (16.17)$$

From Theorem 16.2.1, when $t = P_1$, $e_\theta(t')$ and $e_v(t')$ are bounded and denoting

$$e_\theta(t')|_{t=P_1} = e_\theta(t')|_{t'=0} \le b_\theta, \qquad (16.18)$$
$$\dot{e}_\theta(t')|_{t=P_1} = e_v(t')|_{t=P_1} = e_v(t')|_{t'=0} \le b_v, \qquad (16.19)$$
$$\hat{a}(t')|_{t=P_1} = \hat{a}(t')|_{t'=0} \le b_{\hat{a}}. \qquad (16.20)$$

Performing the Laplace transform of (16.17) with operator variable s' 1 (s' is used to avoid confusion) leads to

$$s'^2 E_\theta(s') + (\alpha + \gamma + B_1)s' E_\theta(s') + \alpha\gamma E_\theta(s')$$
$$= F(s') - \frac{1}{J}\hat{A}(s') + e_v(t')|_{t'=0} + (s' + \alpha + \gamma + B_1)e_\theta(t')|_{t'=0}, \qquad (16.21)$$

where

$$F(s') = L\{f(t')\} = L\left\{\frac{1}{J}a(t') + B_1 v_d(t')\right\}, \qquad (16.22)$$
$$\hat{A}(s') = L\{\hat{a}(t')\}, \qquad (16.23)$$

From our adaptation law (16.11) as $s \geqslant s_p$, and

$$m_1(t) = \gamma e_\theta(t - P_k) + e_v(t - P_k), \qquad (16.24)$$
$$e_\theta(t)|_{t=0} = \theta_d(t)|_{t=0} - \theta(t)|_{t=0} = 0, \qquad (16.25)$$

we can obtain the Laplace transforms

$$L\{e_\theta(t - P_k)\} = e^{-P_k s'} E_\theta(s'), \qquad (16.26)$$
$$L\{e_v(t - P_k)\} = L\{\dot{e}_\theta(t - P_k)\}$$
$$= (s' E_\theta(s') - e_\theta(t)|_{t=0})e^{-P_k s'}$$
$$= s' E_\theta(s')e^{-P_k s'}, \qquad (16.27)$$

then we have

$$\hat{A}(s') = L\{\hat{a}(t')\} = e^{-P_k s'}\hat{A}(s') + \frac{K}{J}e^{-P_k s'}(\gamma E_\theta(s') + s' E_\theta(s')),$$ (16.28)

we have

$$\hat{A}(s') = \frac{\frac{K}{J}(s' + \gamma)e^{-P_k s'}}{1 - e^{-P_k s'}}E_\theta(s').$$ (16.29)

Then from (16.21) and (16.29), we can obtain

$$G(s') = \frac{E_\theta(s')}{H(s')} = \cfrac{1}{s'^2 + as' + b + \cfrac{\frac{K}{J^2}(s'+\gamma)e^{-P_k s'}}{1 - e^{-P_k s'}}}$$

$$= \cfrac{\frac{1}{s'^2 + as' + b}(1 - e^{-P_k s'})}{1 - [1 - \frac{\frac{K}{J^2}(s'+\gamma)}{s'^2 + as' + b}]e^{-P_k s'}},$$ (16.30)

where

$$H(s') = F(s') + e_v(t')|_{t'=0} + (s' + \alpha + \gamma + B_1)e_\theta(t')|_{t'=0},$$ (16.31)
$$a = \alpha + \gamma + B_1, \quad b = \alpha\gamma.$$

Equation (16.30) can be treated as the transfer function of the system in Figure 16.1. Once the equivalence has been shown, in order to guarantee the stability of the system (16.30), one must establish under which conditions each block in Figure 16.1 is stable. The first block is $\frac{1}{s'^2 + as' + b}$, which should be stable if all its characteristic roots have negative real parts. The second block is nothing more than a time delay, so it is always stable. Finally, the third block can be described as a positive-feedback closed-loop system with the term $(1 - \frac{\frac{K}{J^2}(s'+\gamma)}{s'^2 + as' + b})e^{-P_k s'}$ in the feedback path.

According to the Small Gain Theorem [109], the sufficient condition for the stability of the system (16.30) can be split into two conditions.

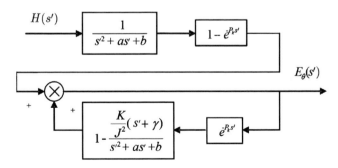

Figure 16.1 Alternative block diagram for the closed-loop control system (16.30)

First stability condition

$$\Re(\lambda) < 0, \quad \forall \lambda \in \left\{ \lambda_i | \lambda_i^2 + a\lambda_i + b = 0 \right\}. \tag{16.32}$$

Second stability condition

$$\left| 1 - \frac{\frac{K}{J^2}(s' + \gamma)}{s'^2 + as' + b} \right|_{s'=j\omega} < 1, \quad \forall \omega. \tag{16.33}$$

These two stability conditions are easy to satisfy when designing the proposed learning control systems. There is a choice of K, γ, a and b which can satisfy the second stability condition (16.33). For example, when $\omega = 0$, $|1 - \frac{K\gamma/J^2}{b}| < 1$. From Theorem 16.2.1, Theorem 15.2.2, and equations (16.3) (16.22), the input signal $f(t')$ in system (16.30) is bounded. So, the system $E(s') = H(s')G(s')$ is input-output stable according to the Small Gain Theorem. In order to establish $e_\theta(t') = 0$ as $t \to \infty$, let us apply the Final Value Theorem with the condition that $H(0) \neq \infty$ from (16.3.1), and the stable condition of system $E(s')$ proved above,

$$\lim_{t \to \infty} e_\theta(t') = \lim_{s' \to 0} s' H(s')G(s') = 0, \tag{16.34}$$

where we used the obvious fact that $G(0) = 0$ in (16.30).
    Finally, from (16.15), we have

$$E_v(s') = s' E_\theta(s') - e_\theta(t')|_{t'=0} = s' H(s')G(s') - e_\theta(t')|_{t'=0}. \tag{16.35}$$

So, the system $E_v(s')$ is also input-output stable. With this stability condition and $|e_\theta(t')|_{t'=0}| < +\infty$, we can conclude from (16.35) that the error signal $e_v(t')$ also approaches 0 as $t \to \infty$ since

$$e_v(\infty) = \lim_{s' \to 0} s' E_v(s') = 0.$$

16.3 Simulation Illustration

In this section, we present two simulation tests to demonstrate the effectiveness of the proposed PALC for the cogging effect on the PMSM position servo control system model, and Figure 16.2 shows the simulation block diagram.

- Case-1: Integer order periodic adaptive learning compensation for multi-harmonics cogging effect.
- Case-2: Fractional order periodic adaptive learning compensation for the same cogging effect as in Case-1.

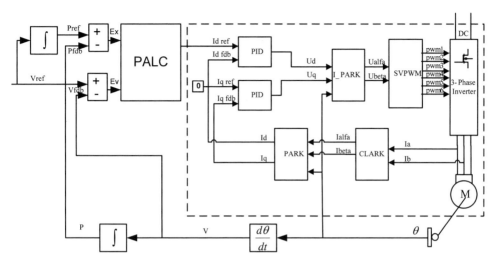

Figure 16.2 Block diagram of the cogging PALC in the PMSM position servo system model

According to the experience from our previous work [5], [132], for our simulation tests, the control gains in (16.9) are selected as: $\alpha = 80$, $\gamma = 20$ and $\mu = 3$. The periodic adaptation gain K is selected as 0.005. For the fairness issue, both in the integer order PALC control test and in the proposed fractional order PALC control test, all the same setting parameters above are used. The only changed parameter between the integer order periodic adaptive learning compensation (IO-PALC) and FO-PALC is just the order ν. The motor parameters are given in Table 15.1 with $T_l = 1[Nm]$. The actual cogging force is modeled as the state-period sinusoidal signal of multiple harmonics as below:

$$F_{cogging} = 2\cos(6\theta) + \cos(12\theta) + 0.5\cos(18\theta).$$

16.3.1 Case-1: Integer Order PALC

In this case simulation test, choosing $\nu = 1$ in adaptive law (16.11), namely, the integer order periodic adaptive learning compensation scheme is used for cogging effect compensation. As the cogging effect degrades the performance of PMSM seriously in a low speed range, and normally, the motor speed below 3% of the rated speed always can be treated as in the low speed range, in our simulation system model, the rated speed of the PMSM is 2000 rpm, which is given in Table 15.1. So we choose the reference speed in this simulation as $5\,rad/s = 47.77\,\text{rpm} \leqslant 60\,\text{rpm}$. So, for this simulation test case, the following reference trajectory and velocity signals are used:

$$s_d(t') = 5t\ rad, \tag{16.36}$$

$$v_d(t') = 5\ rad/s. \tag{16.37}$$

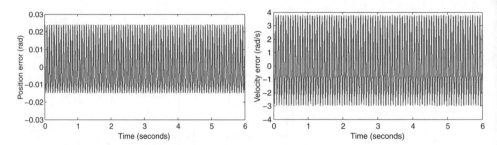

Figure 16.3   Simulation. Tracking errors without any compensation

Figures 16.4(a) and 16.4(b) show the position/speed tracking errors using integer order periodic adaptive learning compensation. We can observe that, as time increases, the positive/speed tracking errors become smaller and smaller. The integer order periodic adaptive learning compensation method works efficiently compared with the tracking errors without compensation in Figures 16.3(a) and 16.3(b).

16.3.2 Case-2: Fractional Order PALC

In this case, we use fractional order periodic adaptive learning compensation for the cogging effect, thus, in adaptive law (16.11), we choose $v = 0.5$. The reference trajectory (16.36) and velocity (16.37) signals in Case-1 are also used.

Figures 16.5(a) and 16.5(b) show the position/speed tracking errors using the fractional order periodic adaptive learning compensation. Compared with Figures 16.4(a) and 16.4(b), we can clearly see that the performance using the fractional order periodic adaptive learning compensation method is much better than that when using the integer order periodic adaptive learning compensation for the multi-harmonics cogging effect.

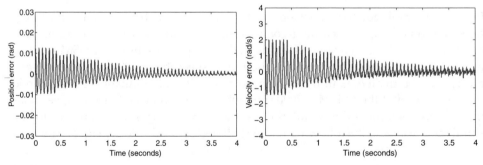

Figure 16.4   Simulation. Tracking errors with integer order periodic adaptive learning compensation

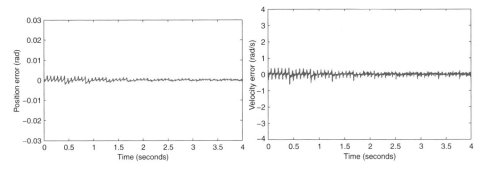

Figure 16.5 Simulation. Tracking errors with fractional order periodic adaptive learning compensation

16.4 Experimental Validation

16.4.1 Introduction to the Experiment Platform

The fractional horsepower dynamometer as a general purpose experiment platform [210] is also applied for the experimental validation in this chapter. The architecture of the dynamometer control system is shown in Figure 3.10. The setup of this experimental platform is introduced in Section 3.6.1.

Without loss of generality, consider a servo control system modeled by:

$$\dot{x}(t) = v(t), \tag{16.38}$$

$$\dot{v}(t) = u(t) - f(t, x) - \frac{B_d}{J_d}v. \tag{16.39}$$

where x is the position or displacement, v is the velocity, $f(t, x)$ is the unknown disturbance, which may be state-dependent or time-dependent, B_d and J_d are the frictional coefficient and moment of inertia of the DC motor respectively, and u is the control input. The system under consideration, that is the dynamometer position control system, has a transfer function $1.52/s(1.01s + 1)$. Moreover, the hysteresis brake allows us to add any time-dependent or state-dependent disturbance (load) to the motor. These factors combined can emulate a system similar to the one given by (16.38) and (16.39). A controller can be designed for such a problem and can be tested in the presence of the real disturbance as introduced through the hysteresis brake.

16.4.2 Experiments on the Dynamometer

The proposed method is validated on the real-time dynamometer position control system. The hysteresis brake force is designed as the load and the state-dependent periodic disturbance

$$f(t, x) = \frac{T_d}{J_d} + \frac{a(t)}{J_d} = \frac{T_d}{J_d} + F_{disturbance}, \tag{16.40}$$

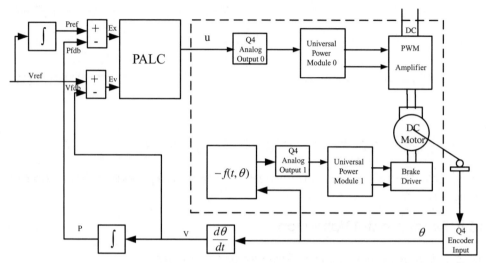

Figure 16.6 Block diagram of PALC for SDPD in the dynamometer position control system

where

$$\frac{T_d}{J_d} = 1, \quad F_{disturbance} = 10\cos(x) + 5\cos(2x) + 2.5\cos(3x).$$

When x is replaced by θ, the system (16.38) (16.39) is in the same format as the system (16.1) (16.2). So we can verify the proposed FO-PALC for SDPD on this dynamometer platform. Figure 16.6 shows the block diagram of the position and velocity control system where again the block in dashed line is emulated by the dynamometer with the cogging effect emulated by the hysteresis brake. According to the experience from our previous work [132], the control gains in (16.9) and (16.11) are selected as, $\alpha = 20$, $\gamma = 10$ and $\mu = 0.5$; the periodic adaptation gain K is selected as 0.01 in our experimental tests. For fair comparisons, both in the IO-PALC and the FO-PALC control experiments, the same setting parameters above are used. The only changed parameter between IO-PALC and FO-PALC is just the order ν.

As a demonstration, two experimental cases are implemented:

- Case-1e: Integer order periodic adaptive learning compensation for SDPD;
- Case-2e: Fractional order periodic adaptive learning compensation for SDPD.

In both experimental cases, the following reference trajectory and velocity signals are used:

$$s_d(t) = 5t \, rad,$$
$$v_d(t) = 5 \, rad/s.$$

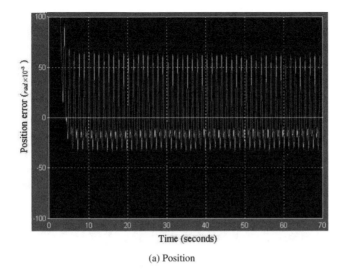

(a) Position

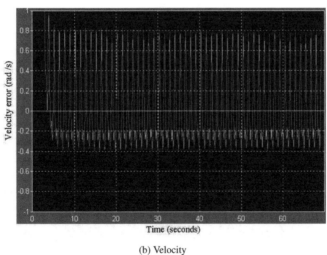

(b) Velocity

Figure 16.7 Tracking errors without compensation

16.4.2.1 Case-1e IO-PALC

For this case, let $\nu = 1$ in adaptive law (16.11) as the integer order PALC for SDPD. Figures 16.8(a) and 16.8(b) show the position and speed tracking errors the using the IO-PALC method, respectively.

16.4.2.2 Case-2e FO-PALC

In this case, if choosing $\nu = 0.5$ in adaptive law (16.11), then the fractional order PALC is used for SDPD. Figures 16.9(a) and 16.9(b) show the position and speed tracking errors using the FO-PALC method, respectively.

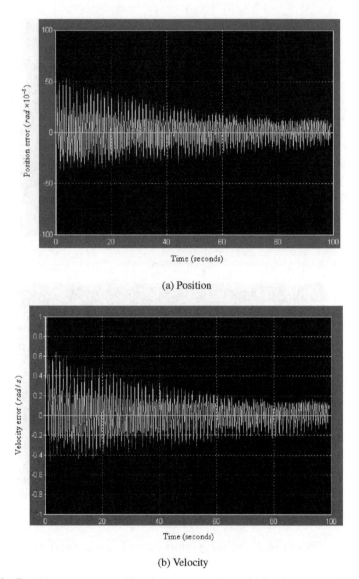

(a) Position

(b) Velocity

Figure 16.8 Tracking errors with integer order periodic adaptive learning compensation

We can observe that IO-PALC works efficiently comparing the tracking errors in Figures 16.8(a) and 16.8(b) with that in Figures 16.7(a) and 16.7(b) where no compensation approach is implemented in the tracking control. Meanwhile, comparing with Figures 16.8(a) and 16.8(b), it can be seen clearly the performance in Figures 16.9(a) and 16.9(b) when using FO-PALC is much better than that when using IO-PALC for state-dependent periodic disturbance in real-time position and velocity servo system.

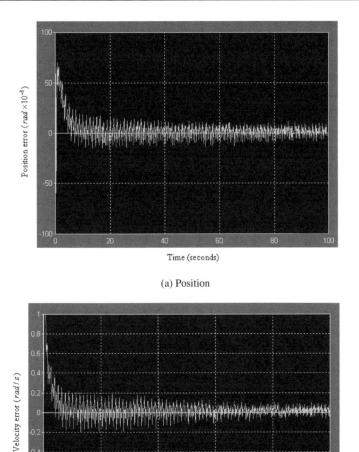

(a) Position

(b) Velocity

Figure 16.9 Tracking errors with fractional order periodic adaptive learning compensation

16.5 Chapter Summary

In this chapter, a fractional order periodic adaptive learning compensation method is proposed for state-dependent periodic disturbances in position and velocity servo systems. In our FO-PALC scheme, in the first trajectory period, a fractional order adaptive compensator is designed to guarantee the boundedness of the system state, input and output signals. From the second repetitive trajectory period and onward, information previously stored for one period along the state axis is used in the current adaptation law. Asymptotical stability is proven for the proposed FO-PALC. Simulation illustration and experimental validation are presented to show the benefits of using fractional calculus in periodic adaptive learning compensation for the general form of state-dependent periodic disturbances.

Part VI
Effects of Fractional Order Controls on Nonlinearities

Part VI

17

Fractional Order PID Control of a DC-Motor with Elastic Shaft

17.1 Introduction

In this chapter, we focus on using a fractional order PID (FOPID) controller for an integer order (IO) plant– "DC-Motor with elastic shaft" [98] to show the effect of fractional order control on nonlinearities, such as, Coulomb friction, backlash, deadzone, and so on [238].

Intuitively, with non-integer order controllers for integer order plants, there is a better flexibility in adjusting the gain and phase characteristics than using IO controllers. This flexibility makes fractional order (FO) control a powerful tool in designing a robust control system with less controller parameters to tune as introduced in Parts II to IV of this book. One key point is that by using few tuning knobs, the FO controller achieves a similar robustness to that achieved when using very high-order IO controllers. Since the tradeoff between the stability and other control specifications always exists, introducing fractional order control makes it more straightforward to achieve a better tradeoff.

17.2 The Benchmark Position System

The benchmark position servomechanism system used for the study in this chapter is a "DC-motor with elastic shaft" [98]. To make this chapter self-contained, in this section, we restate the whole model used in our simulation, as depicted in Figure 17.1 where we can see the benchmark system consists of a DC motor, a gearbox, an elastic shaft and a load.

Fractional Order Motion Controls, First Edition. Ying Luo and YangQuan Chen.
© 2013 John Wiley & Sons, Ltd. Published 2013 by John Wiley & Sons, Ltd.

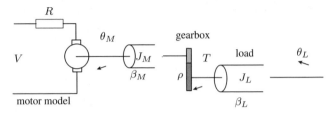

Figure 17.1 The benchmark position servomechanism model

The benchmark system in Figure 17.1 can be represented by the following differential equations:

$$\dot{\omega}_L = -\frac{k_\theta}{J_L}\left(\theta_L - \frac{\theta_M}{\rho}\right) - \frac{\beta_L}{J_L}\omega_L \tag{17.1}$$

$$\dot{\omega}_M = \frac{k_T}{J_M}\left(\frac{V - k_T\omega_M}{R}\right) - \frac{\beta_M\omega_M}{J_M} + \frac{k_\theta}{\rho J_M}\left(\theta_L - \frac{\theta_M}{\rho}\right) \tag{17.2}$$

where V is the applied voltage, T is the torque acting on the load, $\omega_L = \dot{\theta}_L$ is the load's angular velocity, $\omega_M = \dot{\theta}_M$ is the motor shaft's angular velocity, k_θ and k_T the torsional rigidity and motor constant, J_M and J_L the motor and nominal load inertia, β_M and β_L the motor viscous friction coefficient and load viscous friction coefficient, ρ the gear ratio and R the armature resistance.

Defining the state variables as $x_p = [\theta_L\ \omega_L\ \theta_M\ \omega_M]^T$, the above model can be converted to an LTI state-space form:

$$\dot{x}_p = \begin{bmatrix} 0 & 1 & 0 & 0 \\ -\dfrac{k_\theta}{J_L} & -\dfrac{\beta_L}{J_L} & \dfrac{k_\theta}{\rho J_L} & 0 \\ 0 & 0 & 0 & 1 \\ \dfrac{k_\theta}{\rho J_M} & 0 & -\dfrac{k_\theta}{\rho^2 J_M} & -\dfrac{\beta_M + k_T^2/R}{J_M} \end{bmatrix} x_p + \begin{bmatrix} 0 \\ 0 \\ 0 \\ \dfrac{k_T}{R J_M} \end{bmatrix} V, \tag{17.3}$$

$$\theta_L = [1\ \ 0\ \ 0\ \ 0]x_p, \tag{17.4}$$

$$T = \left[k_\theta\ \ 0 - \frac{k_\theta}{\rho}\ \ 0\right]x_p. \tag{17.5}$$

The only measurement available for feedback is θ_L. The load's angular position must be set at a desired value by adjusting the applied voltage V. The elastic shaft has a finite shear strength, so the torque T must stay within specified limits. From an input/output viewpoint, the plant has a single input V, which is manipulated by

Table 17.1 The system parameters of the servomechanism

Symbol	Value (SI units)	Symbol	Value (SI units)
k_θ	1280.2	ρ	20
k_T	10	β_M	0.1
J_M	0.5	β_L	25
J_L	$50 J_M$	R	20

the controller. It has two outputs, one measured feedback to the controller θ_L and one unmeasured T.

The parameters of the experimental position servomechanism system are shown in Table 17.1.

The designed controller must set the load's angular position θ_L at a given value. The elastic shaft has a finite shear strength, so the torque, T, must stay within specified limits $|T| \leqslant 78.5 \text{Nm}$. Also, the applied voltage must stay within the range $|V| \leqslant 220$ V [98].

17.3 A Modified Approximate Realization Method

Here, a different approximate realization method for fractional derivative is introduced in the frequency range of interest $[\omega_b, \omega_h]$. Our proposed method here gives a better approximation than Oustaloup's fractional operator approximation method in both low frequency and high frequency parts.

Let

$$s^\alpha \approx \left(\frac{1 + \dfrac{s}{\frac{d}{b}\omega_b}}{1 + \dfrac{s}{\frac{b}{d}\omega_h}} \right)^\alpha, \tag{17.6}$$

where $0 < \alpha < 1, s = j\omega, b > 0, d > 0$. Thus

$$s^\alpha \approx \left(\frac{bs}{d\omega_b} \right)^\alpha \left(1 + \frac{-ds^2 + d}{ds^2 + bs\omega_h} \right)^\alpha. \tag{17.7}$$

Then, within $[\omega_b, \omega_h]$, using Taylor series expansion

$$s^\alpha \approx \frac{(d\omega_b)^\alpha b^{-\alpha}}{\left(1 + \alpha p(s) + \frac{\alpha(\alpha-1)}{2} p^2(s) + \cdots \right)} \left(\frac{1 + \dfrac{s}{\frac{d}{b}\omega_b}}{1 + \dfrac{s}{\frac{b}{d}\omega_h}} \right)^\alpha, \tag{17.8}$$

where

$$p(s) = \frac{-ds^2 + d}{ds^2 + bs\omega_h}.$$ (17.9)

Truncating the Taylor series to the first order term, then

$$s^\alpha \approx \left(\frac{d\omega_b}{b}\right)^\alpha \left(\frac{ds^2 + bs\omega_h}{d(1-\alpha)s^2 + bs\omega_h + d\alpha}\right) \left(\frac{1 + \frac{s}{\frac{d}{b}\omega_b}}{1 + \frac{s}{\frac{b}{d}\omega_h}}\right)^\alpha.$$ (17.10)

Note that (17.10) is stable if all the poles are on the left half s-plane. It is easy to observe that the expression (17.10) has three poles. One is $-b\omega_h/d$, which is negative because $\omega_h, b, d > 0$. The other two are roots of

$$d(1-\alpha)s^2 + as\omega_h + d\alpha = 0,$$ (17.11)

whose real parts are negative as $0 < \alpha < 1$.

Based on the well-known zig-zag line approximation idea in the Bode plot, let

$$\left(\frac{1 + \frac{s}{\frac{d}{b}\omega_b}}{1 + \frac{s}{\frac{b}{d}\omega_h}}\right)^\alpha = \lim_{N\to\infty} \prod_{k=-N}^{N} \frac{1 + s/\omega_k'}{1 + s/\omega_k},$$ (17.12)

where $-\omega_k'$ and $-\omega_k$ are zero and pole of range k

$$\omega_k' = \left(\frac{d\omega_b}{b}\right)^{\frac{\alpha-2k}{2N+1}}, \quad \omega_k = \left(\frac{b\omega_h}{d}\right)^{\frac{\alpha+2k}{2N+1}}.$$ (17.13)

Hence

$$s^\alpha \approx K\left(\frac{ds^2 + bs\omega_h}{d(1-\alpha)s^2 + bs\omega_h + d\alpha}\right) \prod_{k=-N}^{N} \frac{s + \omega_k'}{s + \omega_k},$$ (17.14)

where

$$K = \left(\frac{d\omega_b}{b}\right)^\alpha \prod_{k=-N}^{N} \frac{\omega_k}{\omega_k'}.$$ (17.15)

In this chapter, we used $b = 10$ and $d = 9$.

The procedures for the modified approximation can be briefly summarized in the following:

(1) Given the frequency range $[\omega_b, \omega_h]$ and N.
(2) Based on the fractional order α, calculate ω'_k and ω_k according to (17.13).
(3) Compute K from (17.15).
(4) Obtain the approximate rational transfer function from (17.14) to replace s^α.

17.4 Comparative Simulations

17.4.1 Best IOPID versus Best FOPID

Simulations of the position servomechanism controlled by the normal integer order PID controller and the fractional order PID controller are carried out based on the parameters setting in Table 17.1 with maximum output torque limitation ± 78.5 Nm. We used a constrained optimization routine to search for the best controller parameters. Two optimization criteria are used. One is the integral-time-absolute-error (ITAE) and the other is the integral of squared error (ISE), where the constraint is $|T| < 78.5$Nm. The reference signal is the unit step function.

For the optimally searched IOPID using the ITAE,

$$G_{c1}(s) = 41.94 + \frac{21.13}{s} - 8.26s. \tag{17.16}$$

For the optimally searched IOPID using the ISE,

$$G_{c2}(s) = 110.09 + \frac{10.65}{s} + 30.97s. \tag{17.17}$$

Figure 17.2 shows the responses to the unit step of the angular position controlled by two integer order PID controllers $G_{c1}(s)$ and $G_{c2}(s)$, respectively, with the Bode plots of the open-loop controlled system shown in the same figure.

Now let us look at the best FOPID controllers. As the first attempt, let us first fix $\lambda = 0.5$ and $\mu = 0.6$. Doing the numerical search, we obtain the best ITAE of 2.22 and the corresponding fractional order PID controller is

$$G_{c3}(s) = 135.12 + \frac{0.01}{s^{0.7}} - 31.6s^{0.6}, \tag{17.18}$$

while the best ISE is 0.87 and the corresponding fractional order controller is

$$G_c(s) = 61.57 + \frac{91.95}{s^{0.5}} + 2.33s^{0.6}. \tag{17.19}$$

The step responses are compared in Figure 17.3 with corresponding Bode plots.

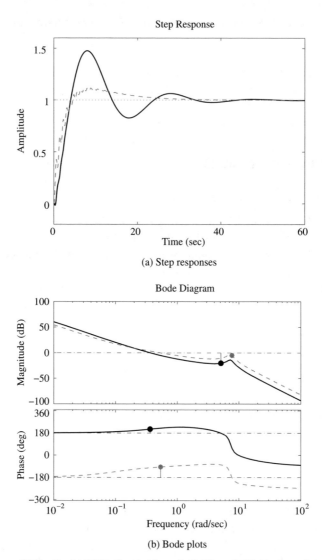

(a) Step responses

(b) Bode plots

Figure 17.2 Best IOPID Controllers. Solid line: ITAE; Broken line: ISE

The observation is clear. The best FOPID performs better than the best IOPID. This is not surprising but this may not be fair since the FOPID has two more extra parameters in the optimal search.

17.4.2 How to Decide λ and μ?

In the last section, we got a flavor that FOPID performs better in side by side comparison. We simply fixed $\lambda = 0.5$ and $\mu = 0.6$. But in reality, how to decide on these two

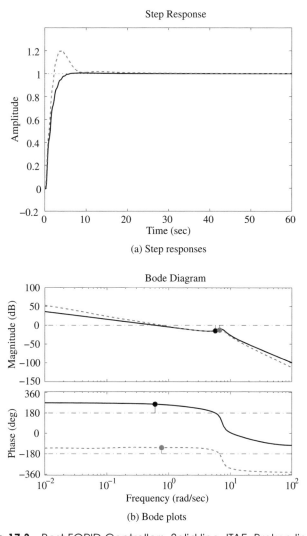

Figure 17.3   Best FOPID Controllers. Solid line: ITAE; Broken line: ISE

magic orders? In this chapter, we only show a brutal force search result to partially justify why we fixed $\lambda = 0.5$ and $\mu = 0.6$.

Here, we build two tables of optimal ITAE and ISE, respectively, with respect to λ and μ which are enumerated from 0.5 to 1.5 with step of 0.1. In other words, we did $2 \times 11 \times 11$ optimal searches. These two tables are illustrated in Figure 17.4.

Note that, in this investigation, we used the approximate order $N = 4$. It is unfortunate that no definite relationship can be established in Figure 17.4. However, in general, we can qualitatively tell that, the integer case $\lambda = 1$ and $\mu = 1$ is *not* optimal. In other words, the optimal case most like corresponds to the non-integer case.

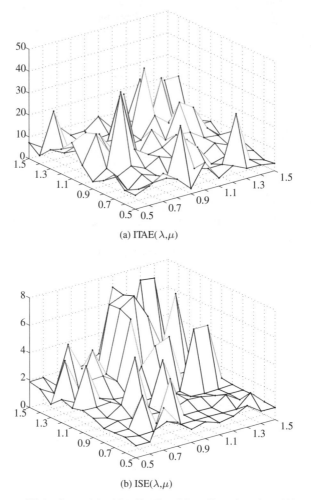

(a) ITAE(λ,μ)

(b) ISE(λ,μ)

Figure 17.4 Searching for the best fractional orders ($N = 4$)

Moreover, we can tell that, in this benchmark system, we prefer the low order integral and the lower order derivative.

17.4.3 Which N is Good Enough?

In implementing the FOPID, we need to decide what the finite order is for the finite order approximation. In our case, we need to decide the N. Let us first repeat Figure 17.4 using $N = 6$.

We conclude that the difference between Figure 17.4 and Figure 17.5 is very small. To illustrate, let us run two examples with an emphasis on the effect of a different approximation order N. Figure 17.6 suggests that the changes in N will not contribute to the differences in this benchmark problem.

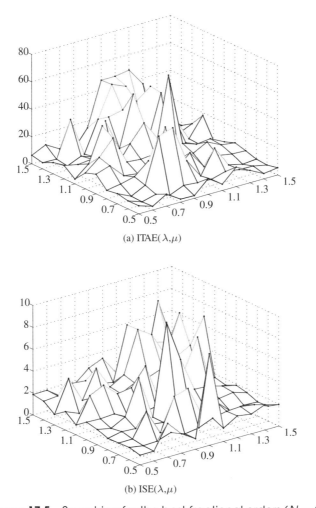

(a) ITAE(λ,μ)

(b) ISE(λ,μ)

Figure 17.5 Searching for the best fractional orders ($N = 6$)

17.4.4 Robustness against Load Variations

In the last subsection, we saw the robustness with respect to N. Here, the position ser-vomechanism control system is controlled by the best fractional order PID controller (17.18) and the best IOPID controller (17.17) when the load changes $\pm50\%$. The results are summarized in Figure 17.7. The robustness against load variations is clearly seen from Figure 17.7.

17.4.5 FOPI Controllers

It will be interesting to check if we can see the case that "the best FOPI works better than the best IOPI". We repeat what we performed for the PID controllers and summarize briefly in the following.

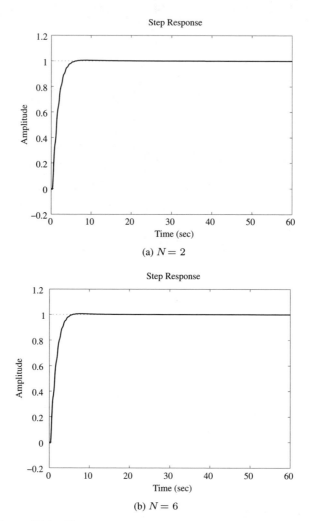

Figure 17.6 Step responses comparisons with different Ns

Under the ITAE criterion, the following optimal IOPI controller is sought:

$$G_c(s) = 107.35 + \frac{0.14}{s},$$ (17.20)

and by the ISE criterion,

$$G_c(s) = 106.82 + \frac{3.36}{s}.$$ (17.21)

Figure 17.8 shows the step responses and the Bode plots.

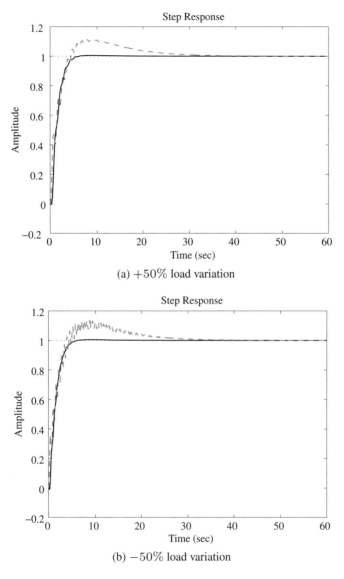

(a) $+50\%$ load variation

(b) -50% load variation

Figure 17.7 Step responses comparison with $N = 4$ (Dotted: Best PID; Solid line: Best FOPID)

Similar to Figure 17.4, we can draw Figure 17.9 which provides the basis for λ selection. From Figure 17.9, if we use the ITAE, we should choose $\lambda = 0.05$. Then the optimal fractional order PI controller is

$$G_c(s) = 39.82 + \frac{72.3}{s^{0.05}}. \tag{17.22}$$

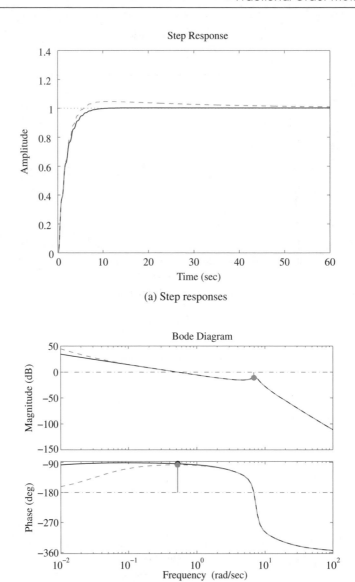

(a) Step responses

(b) Bode plots

Figure 17.8 Best IOPI Controllers. Solid line: ITAE; Broken line: ISE

However, if we use the ISE, we should choose $\lambda = 0.2$. After a search, the optimal fractional order PI controller becomes

$$G_c(s) = -48.38 + \frac{198.26}{s^{0.2}}. \tag{17.23}$$

Figure 17.10 summarizes the comparison of step responses and Bode plots.

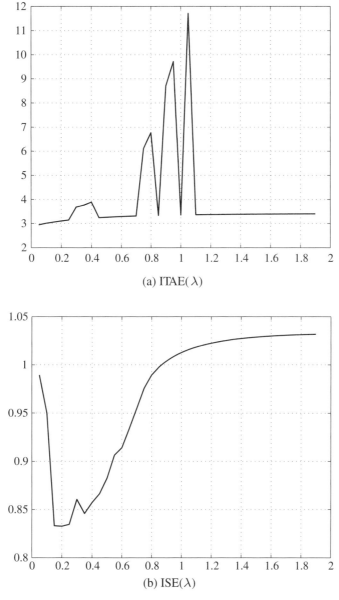

(a) ITAE(λ)

(b) ISE(λ)

Figure 17.9 Searching for the best fractional order ($N = 4$)

The observation is again clear. The best FOPI performs better than the best IOPI. Again this is not surprising but this may not be fair since FOPI has an extra parameter in optimal search.

When the load increases $\pm 50\%$, the corresponding results are summarized in Figure 17.11. Again the robustness against the load variations can be observed.

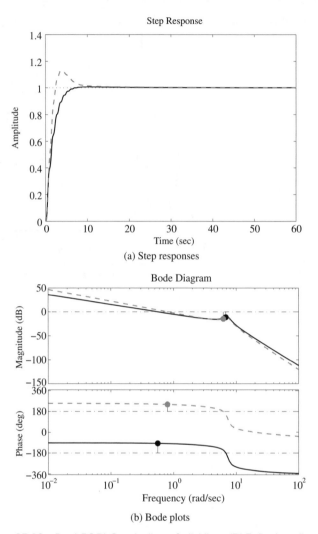

(a) Step responses

(b) Bode plots

Figure 17.10   Best FOPI Controllers. Solid line: ITAE; Broken line: ISE

17.4.6 Robustness to Mechanical Nonlinearities

Using the square wave as the reference input signal (period $T = 40$ sec.) and adding Coulomb friction 0.1, the output responses of the controlled angular position are shown in Figure 17.12. Similarly, we simulated the nonlinear case with backlash, with the deadband width of 0.5. The symmetrical square wave position tracking responses are compared in Figure 17.13. Finally, we checked the case of the deadzone with its parameter set as ±0.5. In this case, the responses are compared in Figure 17.14. So, we can observe that the best FOPID controllers perform much better than their IO counterparts.

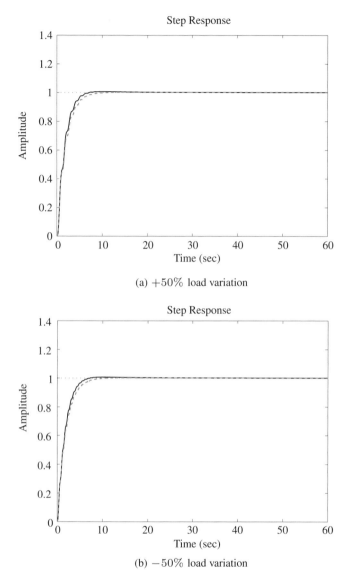

(a) $+50\%$ load variation

(b) -50% load variation

Figure 17.11 Step responses comparison with $N = 4$ Dashed line: Best IOPI; Solid line: Best FOPI

17.4.7 Robustness to Elasticity Parameter Change

Finally, we are interested in checking the robustness with respect to the changes of the elasticity parameter k_θ in Table 17.1. When k_θ varies $\pm 50\%$, the corresponding results are summarized in Figures 17.15 and 17.16, respectively. We can again observe that the best FO controllers perform better than the best IO controllers.

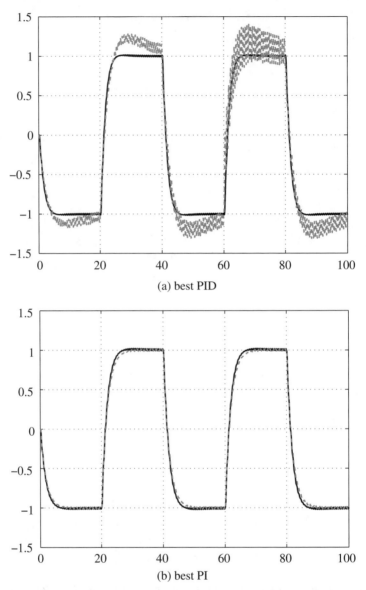

(a) best PID

(b) best PI

Figure 17.12 Responses comparison with Coulomb friction Dashed line: Best IO Controllers; Solid line: Best FO Controllers

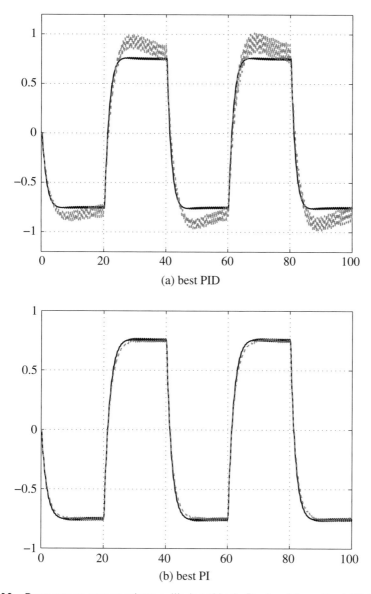

(a) best PID

(b) best PI

Figure 17.13 Responses comparison with backlash Dashed line: Best IO Controller; Solid line: Best FO Controller

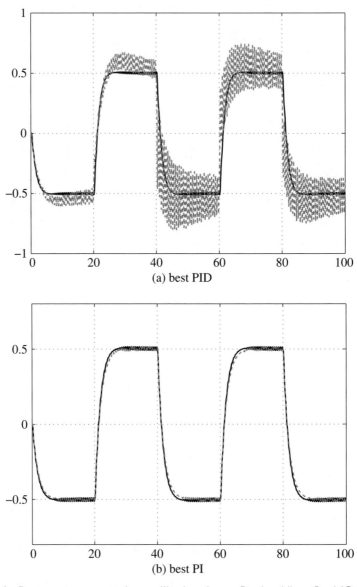

(a) best PID

(b) best PI

Figure 17.14   Responses comparison with deadzone Dashed line: Best IO Controllers; Solid line: Best FO Controllers

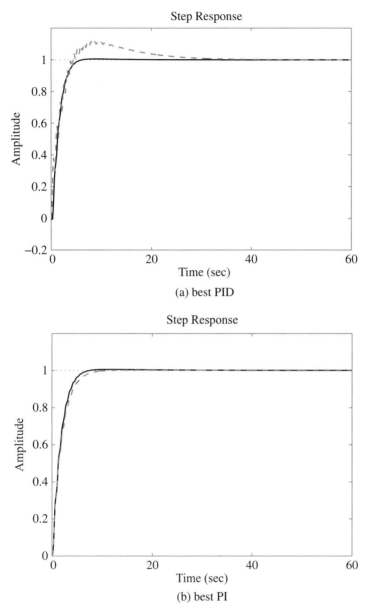

(a) best PID

(b) best PI

Figure 17.15   Responses comparison when k_θ increases 50% Dashed line: Best IO Controllers; Solid line: Best FO Controllers

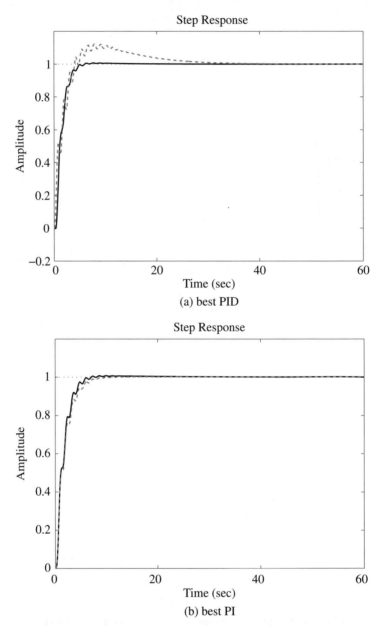

Figure 17.16 Responses comparison when k_θ decreases 50% Dashed line: Best IO Controllers; Solid line: Best FO Controllers

17.5 Chapter Summary

In this chapter, a fractional order PID controller is examined on a benchmark position servomechanism control system considering the actuator saturation and the shaft torsional flexibility. For actual implementation, we introduced a modified approximation method to realize the designed fractional order PID controller. Using numerical optimization, numerous simulation comparisons presented in this chapter indicate that the fractional order PID controller, if properly designed and implemented, will outperform the conventional integer order PID controller. It can be seen that, for the position system with nonlinearities, the best FOPID works better than the IOPID, and the effect of the fractional order control on nonlinearities is obvious.

13.5 Chapter Summary

18

Fractional Order Ultra Low-Speed Position Control

18.1 Introduction

Friction is the force resisting the relative lateral motion of solid surfaces, fluid layers, or material elements in contact. This common nonlinear phenomenon has an impact in all regimes of operation on mechanisms, and produces an undesirable behavior in control systems, such as tracking errors and limit cycles [123]. Especially in the high-precision position control systems, the performance is inherently affected by the frictional effect. Compensation for friction and attenuation of its effect have been addressed in many papers over the years [9], [67], [123], [171]. Meanwhile, the describing function (DF) method is widely used as a common tool for nonlinear system analysis [73], [154], [164], [211], [216]. In [171], using the DF method, different approaches to limit cycle prediction in control systems with friction are discussed based on a simple stick-slip motion example. The existing limit cycle cannot be predicted using the DF when setting only the friction as the nonlinear part. However, the DF of the combining part of the plant with the friction model as the nonlinear part, can capture the behavior of the friction at zero velocity. Then the limit cycles can be predicted. This DF depends on three parameters, the amplitude, frequency and offset, compared with the normal DF depending on two parameters, the amplitude and frequency.

In Chapter 6, a systematic design method for a fractional order proportional and derivative (FOPD) controller is proposed for a class of typical second-order position control plants. The tuned FOPD controller can ensure that the given gain crossover frequency and phase margin are fulfilled, and the phase derivative w.r.t. the frequency is zero, i.e., phase Bode plot is flat, around the given gain crossover frequency. So that, the closed-loop system is robust to the system gain variations. Simulation and experimental results show that, at normal speed, the dynamic performance and robustness with the designed FOPD controller outperform that with the Integral of Time and Absolute Error (ITAE) optimized, traditional, integer order proportional integral (IOPI) controller. Furthermore, we found that, for ultra low-speed position tracking with a severe friction effect in Chapter 6 [119], the tracking performance using the designed FOPD controller is much better than that using the optimized IOPI controller. However, the system, considering the friction effect which affects the control performance severely in ultra low-speed tracking, is nonlinear and complicated, the reason for the performance advantage using the designed FOPD for the ultra low-speed tracking is not clear. Based on this favorable experimental phenomenon, theoretical analysis is needed for a clear understanding.

In this chapter, to explain the observed advantage of the designed FOPD controller over the optimized IOPI controller, Bode plots analysis is applied using the described function method with ultra low-speed position tracking where the nonlinear friction effect is severe. For the performance test and comparison, extended experimental tests are demonstrated to verify the theoretical explanation [141].

18.2 Ultra Low-Speed Position Tracking using Designed FOPD and Optimized IOPI

In this section, the main idea of the FOPD controller design and the performance comparison with the ITAE optimized IOPI controller in Chapter 6 are discussed briefly. The experimental comparisons of using the designed FOPD and the optimized IOPI are presented for the ultra low-speed position tracking with a severe friction effect.

18.2.1 FOPD Design for the Position Tracking without Considering the Friction Effect

A systematic design method for the FOPD controller is proposed for a class of typical second-order plants without considering the friction effect in Chapter 6. The key points of this FOPD controller systematic design scheme are that the designed FOPD controller can be ensured to fulfill the given gain crossover frequency and phase margin, and furthermore the phase derivative with respect to the frequency is zero at the gain crossover frequency. Thus, the closed-loop system is robust to the loop gain variations.

For the FOPD controller design, the class of second-order plants $P(s)$ is described by (6.2). The transfer function of the FOPD controller is shown in (6.1). The three specifications in Section 2.3 of Chapter 6 are applied to design the fractional order controller.

With these specifications, the gain crossover frequency is set as $\omega_c = 10\,rad/s$, and the desired phase margin is set as $\phi_m = 70°$. Moreover, the robustness to gain variations is required. According to the experimental model identified in Chapter 6, we can obtain the parameters of the FOPD controller as $\mu = 0.844$, $K_d = 0.368$ and $K_p = 13.860$. Meanwhile, the parameters of the ITAE optimized IOPI controller are designed as $K_p = 2.6531$ and $K_i = 1.1662$.

Simulation and experimental results in Chapter 6 show that, at normal speed, the dynamic performance and robustness of the position ramp response with the designed FOPD controller outperform that with the ITAE optimized IOPI controller.

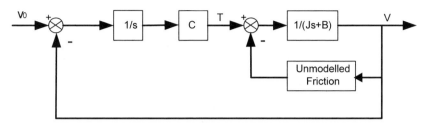

Figure 18.1 Position tracking control equivalent diagram with constant speed reference and friction

18.2.2 Ultra Low-Speed Position Tracking Performance with Designed FOPD and Optimized IOPI

In the experimental demonstration of Chapter 6, the nonlinear unmodeled friction is not considered. In the same experimental plant, and using the same designed FOPD and optimized IOPI controllers in Chapter 6, the closed-loop system can be presented in Figure 18.1, considering both the nonlinear unmodeled friction effect and viscous friction, with C as the designed FOPD or the optimized IOPI. In this system, the position ramp reference is generated by the integration of a speed reference, so the system's input and output can be treated as the speed signals with an integrator in the closed-loop system.

When the speed reference used for the tracking is reduced to a very small value such as $0.05\ rad/s$, then the friction effect is severe and cannot be ignored. So, Figure 18.1 should be used to describe the closed-loop experimental system. As demonstrated in Chapter 6, the position control system is identified as $P(s) = 1.52/s(0.4s + 1)$. The ITAE optimized IOPI controller is designed as $C_{IOPI} = 2.5631(1 + 1.1662/s)$ according to [68]. From [68], the ITAE optimization method can be used to obtain the optimum PID controllers for many kinds of motion control system responses. For the ramp response of the second-order position control system, the optimized IOPI controller is presented with the system's parameters in [68], which is applied to the baseline of the FOPD controller design and comparison in Chapter 6 and this chapter. For the fairness issue of the controller design and comparison, the integer order PD controller may be designed properly for the comparison since the fractional order PD is designed. As mentioned in Chapter 6, "that means the specifications (i) and (ii) cannot be satisfied simultaneously for the traditional integer order PD controller", and the traditional PD controller is not properly designed for the system $P(s)$ in (6.2). The designed FOPD is implemented and compared with the optimized IOPI following the ITAE method in [68].

The designed fractional order PD controller is $C_{FOPD} = 13.86(1 + 0.368s^{0.844})$. Figures 18.2(a), 18.2(b), 18.3(a) and 18.3(b) show the speed and position outputs of the position tracking with constant ultra low-speed reference, it is obvious that the tracking performance using the designed FOPD in Figure 18.2(b) is much better than that using the optimized IOPI in Figure 18.2(a).

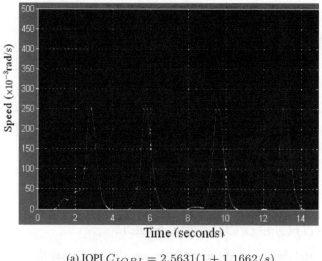

(a) IOPI $C_{IOPI} = 2.5631(1 + 1.1662/s)$

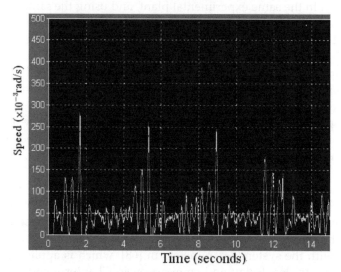

(b) FOPD $C_{FOPD} = 13.86(1 + 0.368s^{0.844})$

Figure 18.2 Speed outputs of position tracking with constant speed reference using optimized IOPI and designed FOPD

18.3 Static and Dynamic Models of Friction and Describing Functions for Friction Models

In this section, the different friction models and two uncoupling methods of the linear part and nonlinear part are presented.

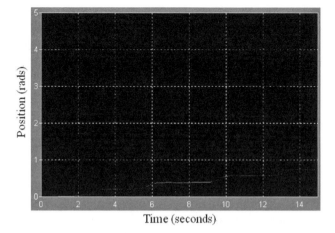

(a) IOPI $C_{IOPI} = 2.5631(1 + 1.1662/s)$

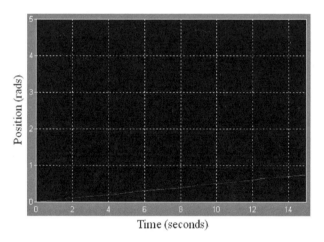

(b) FOPD $C_{FOPD} = 13.86(1 + 0.368s^{0.844})$

Figure 18.3   Position outputs of position tracking with constant speed reference using optimized IOPI and designed FOPD

18.3.1 Static and Dynamic Models of Friction

In general, the friction models are described by a discontinuous relation between the relative velocity in between the contacted surfaces and the resulting friction force. The friction models can be briefly divided into the traditional static models and the dynamic models. The static models are expressed by the static equations and the combinations of the coulomb friction, viscous friction, and so on [8]. The friction can be also be modeled by the dynamic models proposed in the past few decades with the differential equations [63], [67]. However, many friction models are defined without considering zero velocity. With this particular velocity, the friction depends on the applied force. As presented in [67], the proposed LuGre dynamic friction model combines the stiction behavior, that is, the Dahl effect, with arbitrary steady state friction characteristics which can include the Stribeck effect and the zero velocity friction. The typical LuGre model useful for various control tasks, is given below [67],

$$\dot{z} = v - \frac{|v|}{g(v)}z, \tag{18.1}$$

$$F = \sigma_0 z + \sigma_1 \dot{z} + \sigma_2 v, \tag{18.2}$$

$$\sigma_0 g(v) = F_C + (F_S - F_C)e^{-(v/v_s)^2}, \tag{18.3}$$

where the average deflection of the bristles is denoted by z; v is the relative velocity between the two surfaces; the function g is positive and depends on many factors such as material properties, lubrication and temperature; σ_0 is the stiffness, and σ_1, σ_2 are damping coefficients; F_C is the Coulomb friction level; F_S is the level of the stiction force; and v_s is the Stribeck velocity. The values of those parameters in (18.1) are presented in Table 18.1.

18.3.2 Describing Functions for Friction Models and Two Uncoupling Methods of Linear and Nonlinear Parts

The experimental platform for the position tracking with consideration of the friction effect is shown in Figure 18.1. Using the describing function method, the transfer function of the nonlinear block is defined by the relationship between the output response $y(t)$ and the frequency ω. In the output response, only the first harmonic $y_1(t)$ with the same frequency as that in the input signal is considered,

$$y_1(t) = a\cos\omega t + b\sin(\omega t) = c\sin(\omega t + \varphi), \tag{18.4}$$

where

$$a = \frac{2}{\pi}\int_0^{\pi} y(t)\cos(\omega t)\mathrm{d}(\omega t), \quad b = \frac{2}{\pi}\int_0^{\pi} y(t)\sin(\omega t)\mathrm{d}(\omega t),$$

$$c = \sqrt{a^2 + b^2}, \quad \varphi = \arctan(a/b). \tag{18.5}$$

Table 18.1 Parameter values in the LuGre model

σ_0	10^5	[N/m]
σ_1	$\sqrt{10^5}$	[Ns/m]
σ_2	0.4	[Ns/m]
F_C	1	[N]
F_S	1.5	[N]
v_s	0.001	[m/s]

The describing function of the nonlinear block can be expressed by the gain and phase shift between the first harmonic of the output and the sinusoid input $A\sin \omega t$, as below,

$$N_I(A, \omega) = \frac{c}{A}e^{j\varphi}. \tag{18.6}$$

In order to use the describing function method to analyze the system performance, we need to find the approximated closed-loop system with uncoupling linear and nonlinear parts as in Figure 18.4, from the nonlinear position tracking system with friction effect in Figure 18.1. There are two methods for uncoupling the linear and nonlinear parts as shown in Figure 18.5(a) and Figure 18.5(b). The straightforward way is to treat only the friction as the nonlinear part, and the other items as the linear part in Figure 18.5(a). Following the analysis in [171], with the simple example of the stick-slip motion, the velocity is the input and the friction is the output for the describing function analysis. Then the nonlinear part will not be affected by the friction force with exact zero velocity. The intricate behavior of the friction at zero velocity unfortunately is ignored. Meanwhile, the Nyquist curves for the linear and nonlinear parts are plotted in [171]. These two curves have no intersection where the limit cycle frequency can be obtained. So the analysis with the uncoupling method in Figure 18.5(a) does not predict any limit cycle for all the average velocity v_0, frequency ω and amplitude A.

For the other uncoupling method in Figure 18.5(b), the force T is the input and the velocity V is the output of the nonlinear part which includes not only the friction but also the system's dynamics G. In this case, the friction force has the possibility of counteracting the applied force and keeping the velocity as zero. During one period of the sinusoid input, sticking may occur. Therefore, the essential characteristics of the friction can be captured in the describing function of the nonlinear part in Figure 18.5(b). This time, a mean value of the force T has to be included as one of the input parameters for the nonlinear block [171]. So, this method in Figure 18.5(b) is chosen to uncouple the linear and nonlinear parts of our experimental nonlinear system. The describing function will hence depend on three parameterss: the amplitude A, the frequency ω and the mean force T_0. The output will be an oscillation with a mean value different from zero. The describing function in this case should be denoted as below,

$$N(T_0, A, \omega) = [|N|e^{\phi}, v_a], \tag{18.7}$$

where the $|N|$ and ϕ are the gain and phase shift of the describing function, v_a is the average value of the output velocity of the nonlinear part.

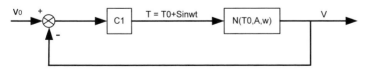

Figure 18.4 Approximation closed-loop system with linear and nonlinear parts

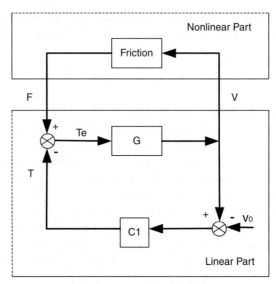

(a) Nonlinear part with only friction

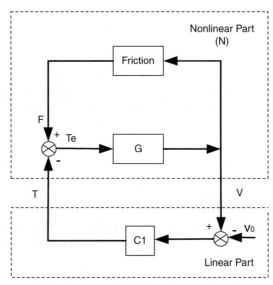

(b) Nonlinear part with friction and system dynamics

Figure 18.5 Two methods of uncoupling linear and nonlinear parts

Using the uncoupling method in Figure 18.5(b) with the designed FOPD and the optimized IOPI controller in Chapter 6, the output signals of the nonlinear part can be seen in Figures 18.2 and 18.3, which show the speed and position outputs of the tracking with ultra low-speed reference in the experiment introduced in Section 18.2. It can be seen that both the outputs with IOPI and FOPD have limit cycles.

18.4 Simulation Analysis with IOPI and FOPD Controllers using Describing Function

In order to reveal the potential advantage of the designed FOPD controller over the optimized IOPI controller in Chapter 6, for the nonlinear ultra low-speed position tracking system with friction effect, the Bode plot analysis with the describing function is applied in this section.

First, the open-loop Bode plots for the position tracking systems in Chapter 6 are drawn in Figure 18.6 with the optimized IOPI and designed FOPD respectively, without considering the friction effect. It can be seen that, for the frequency around the gain crossover frequency (12 rad/s), the phase of the open-loop system is flat, and the amplitude with FOPD is bigger than that with IOPI. Meanwhile, the phase delay with FOPD is much smaller than that with IOPI, which means that the relative stability of the system is significantly improved by the designed FOPD controller compared with the optimized IOPI controller.

Second, for the block diagram as shown in Figure 18.5(b), the mean forces T_{0IOPI} and T_{0FOPD} of the input signals T of the nonlinear block with IOPI and FOPD are measured as 0.577 and 0.622, respectively. So the describing function of the nonlinear part can be calculated following the method introduced in Section 18.3.2. Using the optimized IOPI with $T_{0IOPI} = 0.577\ Nm$, the 3D/2D Bode plots of the amplitudes and phases with respect to ω and A are drawn in Figure 18.7 and Figure 18.8, respectively; using the designed FOPD with $T_{0FOPD} = 0.622\ Nm$, the 3D/2D Bode plots are drawn in Figure 18.9 and Figure 18.10, respectively. From Figure 18.7 and Figure 18.9, and their enlarged parts of the 2D Bode plots in Figure 18.11(a) and Figure 18.11(b), it is obvious that the

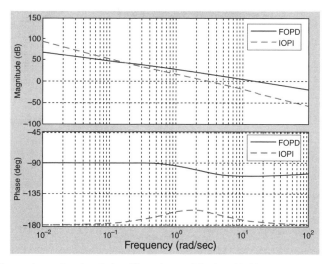

Figure 18.6 Open-loop Bode plots without considering friction using the optimized IOPI and designed FOPD

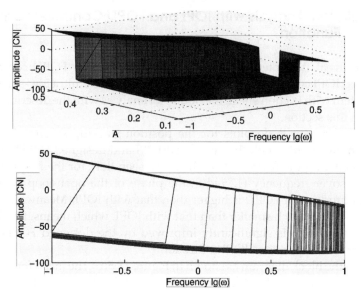

Figure 18.7 3D/2D Bode plot of the amplitude with respect to ω and A using IOPI with $T_{0IOPI} = 0.577$ Nm

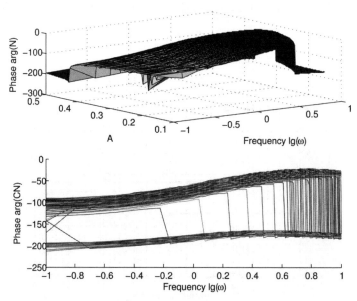

Figure 18.8 3D/2D Bode plot of the phase with respect to ω and A using IOPI with $T_{0IOPI} = 0.577$ Nm

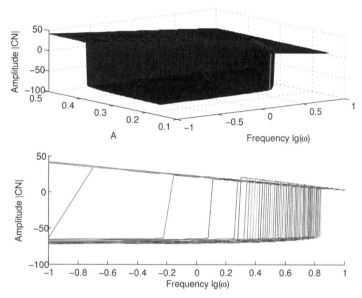

Figure 18.9 3D/2D Bode plot of the amplitude with respect to ω and A using FOPD with $T_{0FOPD} = 0.622 \ Nm$

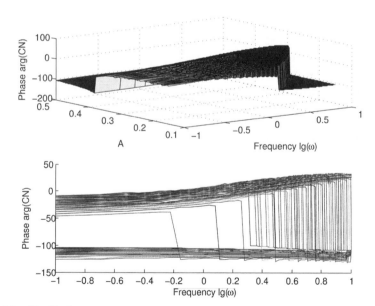

Figure 18.10 3D/2D Bode plot of the phase with respect to ω and A using FOPD with $T_{0FOPD} = 0.622 \ Nm$

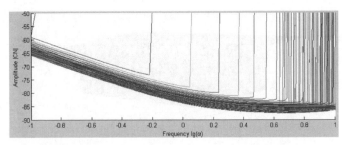

(a) Enlarged 2D Bode plot of the amplitude w. r. t. ω using IOPI with $T_{0IOPI} = 0.577\,Nm$

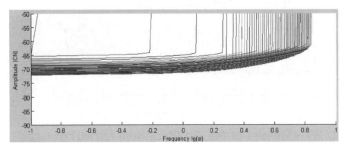

(b) Enlarged 2D Bode plot of the amplitude w. r. t. ω using FOPD with $T_{0FOPD} = 0.622\,Nm$

Figure 18.11   Enlarged 2D Bode plot of the amplitude with respect to ω (For a color version of this figure, see Plate 22)

amplitude with the designed FOPD is greater than that with the optimized IOPI, in ultra low-speed position tracking with the limit cycles as shown in Figure 18.2(a) and Figure 18.2(b). So the tracking performance with the designed FOPD should be better than that with the optimized IOPI. At the same time, comparing Figure 18.8 and Figure 18.10, the phase delay with the designed FOPD is much smaller than that with the optimized IOPI, which means the tracking system with the designed FOPD controller is more stable than that with the optimized IOPI.

18.5 Extended Experimental Demonstration

In this section, an extended experiments for the varying ultra low-speed position tracking is presented to validate the theoretical analysis in this chapter. This experimental validation is performed on the same experimental platform – the dynamometer position control system as in Chapter 6.

The ramp position tracking reference, with the slope of $\pm 0.05\,rad/s$ which is the varying ultra low-speed, is shown in Figure 18.12. The optimum integer order PI controller $C_{IOPI} = 2.5631(1 + 1.1662/s)$ and the designed fractional order PD controller $C_{FOPD} = 13.86(1 + 0.368s^{0.844})$ in Chapter 6 are implemented and tested. It can be seen that the position tracking performance with the designed FOPD in

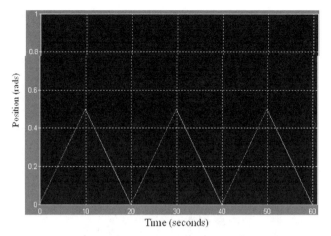

Figure 18.12 Varying low speed (± 0.05 *rad/s*) position reference for tracking

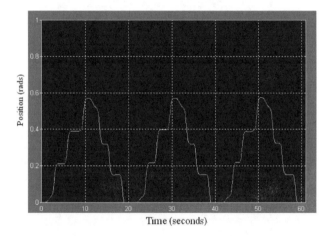

(a) IOPI $C_{IOPI} = 2.5631(1 + 1.1662/s)$

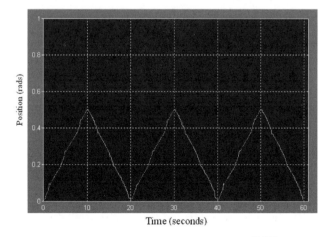

(b) FOPD $C_{FOPD} = 13.86(1 + 0.368s^{0.844})$

Figure 18.13 Position tracking outputs with varying low speed reference using IOPI/FOPD

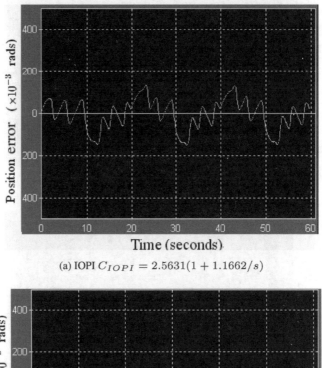

(a) IOPI $C_{IOPI} = 2.5631(1 + 1.1662/s)$

(b) FOPD $C_{FOPD} = 13.86(1 + 0.368s^{0.844})$

Figure 18.14 Position tracking errors with varying low speed reference using IOPI/FOPD

Figure 18.13(b) is much better than that with the optimized IOPI in Figure 18.13(a). It is easy to check the position tracking errors using these two controllers in Figure 18.14(a) and Figure 18.14(b). The maximum position error with FOPD is 20×10^{-3} *rads* which is much smaller than the maximum position error 160×10^{-3} *rads* with IOPI. The performance advantage of the designed FOPD over the optimized IOPI can also be supported by the speed output comparison in Figure 18.15(a) and Figure 18.15(b).

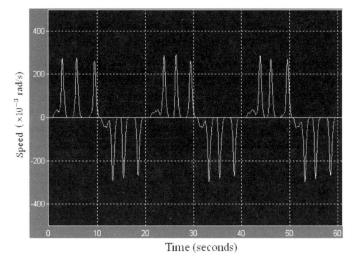

(a) IOPI $C_{IOPI} = 2.5631(1 + 1.1662/s)$

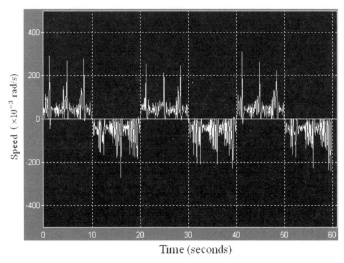

(b) FOPD $C_{FOPD} = 13.86(1 + 0.368s^{0.844})$

Figure 18.15 Speed output with varying low speed position tracking reference using IOPI/FOPD

18.6 Chapter Summary

In the previous work in Chapter 6, the simulation and experimental results demonstrate that, with a *normal* speed reference, the dynamic performance and robustness of the control system tracking using the designed FOPD controller are better than that using the ITAE optimized IOPI controller. Furthermore, it has been shown that, for the tracking control with an ultra low-speed reference, the designed FOPD out-performs the optimized IOPI, too. However, the system considering friction effect which affects the system significantly in ultra low-speed tracking is nonlinear and complicated, the reason for the performance advantage using the designed FOPD for the ultra low-speed tracking was not clear. In this chapter, using the describing function analysis method and the Bode plots analysis, the observed advantage of the designed FOPD controller over the optimized IOPI controller in Chapter 6 is explained clearly for the nonlinear ultra low-speed position tracking system with friction effect. This theoretical analysis is consistently demonstrated by some further experimental results.

19

Optimized Fractional Order Conditional Integrator

19.1 Introduction

The integrator is widely used to find the zero steady errors to constant signals in system control. However, the 90 degrees phase lag at all frequencies is the cost, which means a loss in relative stability, for the benefit of zero steady state errors. The reset or Clegg conditional integrator (CCI) was proposed for the first time in [58] to reduce this phase lag while retaining the integrator's desirable magnitude frequency response, to provide more flexibilities in linear controllers design. The potential advantages of using CCI to meet some stringent design specifications have been presented in [25], [46], [96], [124], [245].

In [125] and [229], an intelligent conditional integrator (ICI) was presented, where the integrator "holds" the output value from the zero-crossing point of the input signal derivative to the zero-crossing point of the input signal, and resets the output value at the zero-crossing point of the input signal as in CCI. The phase delay of ICI was reduced further compared with CCI. In order to achieve better control performance, we can proportionally tune the "holding" output value of ICI with a parameter k in between the zero-crossing point of the input signal derivative and the zero-crossing point of the input signal, which can be treated as a modified intelligent conditional integrator (MICI).

On the other hand, from the very beginning of using the fractional calculus in control [150], [178], [222], the fractional integrator has been considered as an alternative approach for control purposes [90], [119], [132], [134], [136]. So the fractional integrator can also be used in the conditional integrator as the fractional order conditional integrator (FOCI). The potentially improved control performance may be achievable with the extra tunable parameter – fractional order α of FOCI. Meanwhile, the describing function is widely applied for the analysis of the nonlinear systems

Fractional Order Motion Controls, First Edition. Ying Luo and YangQuan Chen.
© 2013 John Wiley & Sons, Ltd. Published 2013 by John Wiley & Sons, Ltd.

control [72], [133], [154], [211]. Obviously, the conditional integrator which is a non-linear block can be analyzed using the describing function method.

In this chapter, an optimized fractional order conditional integrator (OFOCI) is proposed. By tuning the fractional order α and the parameter k for the "holding" value between the zero-crossing point of the input signal derivative and the zero-crossing point of the input signal following the analytical optimality design specifications, this presented OFOCI can achieve the optimized performance not achievable in integer order conditional integrators. The optimality specifications for FOCI are proposed to satisfy the desired phase delay in the reachable range and to minimize the concerned magnitude ratio, which can be calculated from certain high order harmonic magnitude or the weighted magnitude of several high order harmonics divided by the fundamental wave magnitude. The numerical solution of calculating the optimized parameters α and k of the OFOCI is introduced, the phase delays and magnitude ratios of four OFOCIs designed according to four different given phase delays are compared with the integer order conditional integrators (IOCIs). Simulation results with the FFT spectra are also presented to illustrate the theoretical analysis [252].

19.2 Clegg Conditional Integrator

The Clegg conditional integrator (CCI) represents an attempt to synthesize a nonlinear circuit possessing the amplitude-frequency characteristic of a linear integrator while avoiding the 90 degrees phase lag associated with the linear transfer function. A functional diagram of the Clegg conditional integrator, which switches on input zero crossings, was illustrated in Figure 1 of [82]. Basically, the operation consists of the input being gated through one of two integrators (the input of the other is simultaneously reset) in accordance with zero-crossing detector (ZCD) commands. An analog implementation of this integrator is given in [245], including four diodes, four RC networks, and two operational amplifiers.

In the interval $0 < \varphi < \pi$, $\varphi = \omega t_1$; for the input $x(t) = A\sin(\omega t)$, the output is

$$y_C(\varphi) = \int_0^{\varphi/\omega} x(t)dt = \frac{1}{\omega}\int_0^{\varphi} A\sin(\omega t)d(\omega t)$$
$$= \frac{A}{\omega}(1 - \cos(\varphi)). \tag{19.1}$$

Input and output waveforms are shown in Figure 19.1 where $A = 1$. The describing function (DF) is given by

$$N_C(A, \omega) = \frac{2j}{\pi A}\int_0^{\pi} \frac{A}{\omega}(1 - \cos\varphi)e^{-j\varphi}d\varphi = \frac{4}{\pi\omega}\left(1 - j\frac{\pi}{4}\right), \tag{19.2}$$

$$|N_C(A, \omega)| = \frac{1}{\omega}\sqrt{1 + \left(\frac{4}{\pi}\right)^2} = \frac{1.62}{\omega}, \tag{19.3}$$

$$\arg N_C(A, \omega) = -\arctan\frac{\pi}{4} \simeq 0.212\pi(rad.) = 38.15°. \tag{19.4}$$

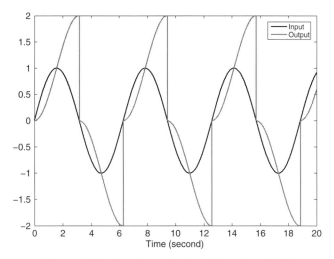

Figure 19.1 Input and output signals of CCI

So the DF of CCI is independent of the input amplitude, has a constant phase of $-38.15°$, and a magnitude slope of $-20db/dec$. A disadvantage of the CCI is that it may induce oscillations [102].

19.3 Intelligent Conditional Integrator

An intelligent conditional integrator (ICI) was proposed in 1990 [229]. The input and output signals of ICI are denoted as $u(t)$ and $y_I(t)$, respectively. As the input is the sinusoidal signal $u(t) = A\sin(\omega t)$, the output waveform can be described as the piecewise analytic expression below,

$$
y_I(t) = \begin{cases}
\int_0^t u(\tau)d\tau = \frac{A}{\omega}(1 - \cos(\omega t)), 0 \leqslant \omega t < \frac{\pi}{2}; \\
\frac{A}{\omega}, \frac{\pi}{2} \leqslant \omega t < \pi; \\
0, \omega t = \pi; \\
\int_\pi^t u(\tau)d\tau = -\frac{A}{\omega}(1 + \cos(\omega t)), \pi \leqslant \omega t < \frac{3\pi}{2}; \\
-\frac{A}{\omega}, \frac{3\pi}{2} \leqslant \omega t < 2\pi; \\
0, \omega t = 2\pi.
\end{cases} \tag{19.5}
$$

The input and output signals of ICI are shown in Figure 19.2 with $A = 1$. It can be seen that, ICI "holds" the output value from the zero-crossing point of the input signal derivative to the zero-crossing point of the input signal, and resets the output value at

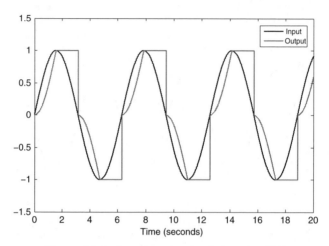

Figure 19.2 Input and output signals of ICI

the zero-crossing point of the input signal as in CCI. Obviously, it is a non-sinusoidal periodic function in (19.10), and the coefficients of its fundamental wave are,

$$
\begin{aligned}
a_{1I} &= \frac{2}{\pi} \int_0^{\pi} y_I(t) \cos(\omega t) \mathrm{d}(\omega t) \\
&= \frac{2}{\pi} \left[\int_0^{\frac{\pi}{2}} \frac{A}{\omega}(1 - \cos(\omega t)) \cos(\omega t) \mathrm{d}(\omega t) + \int_{\frac{\pi}{2}}^{\pi} \frac{A}{\omega} \cos(\omega t) \mathrm{d}(\omega t) \right] \\
&= -\frac{0.5A}{\omega},
\end{aligned}
\tag{19.6}
$$

$$
\begin{aligned}
b_{1I} &= \frac{2}{\pi} \int_0^{\pi} y_I(t) \sin(\omega t) \mathrm{d}(\omega t) \\
&= \frac{2}{\pi} \left[\int_0^{\frac{\pi}{2}} \frac{A}{\omega}(1 - \cos(\omega t)) \sin(\omega t) \mathrm{d}(\omega t) + \int_{\frac{\pi}{2}}^{\pi} \frac{A}{\omega} \sin(\omega t) \mathrm{d}(\omega t) \right] \\
&= \frac{0.955A}{\omega}.
\end{aligned}
\tag{19.7}
$$

So the fundamental wave of $y_I(t)$ can be obtained,

$$
\begin{aligned}
y_{1I}(t) &= -\frac{0.5A}{\omega} \cos \omega t + \frac{0.955A}{\omega} \sin(\omega t) \\
&= c_{1I} \sin(\omega t + \varphi_{1I}),
\end{aligned}
\tag{19.8}
$$

where

$$
c_{1I} = \sqrt{a_{1I}^2 + b_{1I}^2} = \frac{1.08A}{\omega},
$$

$$
\varphi_{1I} = \arctan \left(\frac{a_{1I}}{b_{1I}} \right) \simeq 0.153\pi \,(rad.) = -27.6^{\circ}.
$$

The describing function of ICI is given by,

$$N_I(A, \omega) = \frac{c_{1I}e^{j\varphi_{1I}}}{Ae^{j0}} = \frac{1.08}{\omega}e^{-j0.153\pi}. \tag{19.9}$$

Thus, we can see that the phase delay of ICI is only $27.6°$. Compared with CCI, the phase delay is decreased by $10.5°$.

19.4 The Optimized Fractional Order Conditional Integrator

In this section, an optimized fractional order conditional integrator (OFOCI) is proposed. The key feature of this OFOCI is to tune the fractional order α and the parameter k for the "holding" output value simultaneously following the optimality design specifications, and to achieve the optimized performance. The proposed optimality specifications are designed to satisfy the given phase delay in the reachable range, and to minimize the concerned harmonics magnitude ratio, which can be calculated from a certain high order harmonic magnitude or the weighted magnitude of several high order harmonics divided by the fundamental wave magnitude. The numerical method of calculating the optimized parameters of OFOCI is also introduced.

19.4.1 Fractional Order Conditional Integrator

First, a fractional order conditional integrator (FOCI) is proposed. In this study the Caputo definition is adopted for the fractional derivative (see Chapter 1). The FOCI input is $u(t)$, and the FOCI output is $y_F(t)$. This input is a sinusoidal signal $u(t) = A\sin(\omega t)$. With the initial condition $u(0) = 0$, the fractional order integral of the input signal is [188],

$$_{-\infty}D_t^{-\alpha}A\sin(\omega t) = \frac{A}{\omega^\alpha}\sin\left(\omega t - \frac{\alpha\pi}{2}\right), \tag{19.10}$$

where $\alpha \in (0, 2)$ is the fractional order of the fractional order integral. Then, we can obtain:

$$_0D_t^{-\alpha}A\sin(\omega t) = \frac{A}{\omega^\alpha}\left[\sin\left(\omega t - \frac{\alpha\pi}{2}\right) + \sin\left(\frac{\alpha\pi}{2}\right)\right], \tag{19.11}$$

$$_\pi D_t^{-\alpha}A\sin(\omega t) = \frac{A}{\omega^\alpha}\left[\sin\left(\omega t - \frac{\alpha\pi}{2}\right) - \sin\left(\frac{\alpha\pi}{2}\right)\right]. \tag{19.12}$$

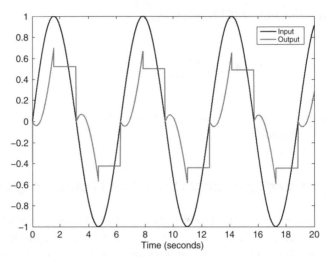

Figure 19.3 Input and output signals of the FOCI with $A = 1$, $\alpha = 1.209$ and $k = 0.753$

The output waveform of FOCI can be described as the piecewise analytical expression below,

$$
y_F(t) = \begin{cases}
{}_0D_t^\alpha u(t) = \frac{A}{\omega^\alpha}[\sin(\omega t - \frac{\alpha\pi}{2}) + \sin(\frac{\alpha\pi}{2})], \ 0 \leqslant \omega t < \frac{\pi}{2}; \\
\qquad\qquad\qquad\qquad\qquad\qquad k\frac{A}{\omega}, \ \frac{\pi}{2} \leqslant \omega t < \pi; \\
\qquad\qquad\qquad\qquad\qquad\qquad\qquad\qquad 0, \ \omega t = \pi; \\
{}_\pi D_t^\alpha u(t) = \frac{A}{\omega^\alpha}[\sin(\omega t - \frac{\alpha\pi}{2}) - \sin(\frac{\alpha\pi}{2})], \ \pi \leqslant \omega t < \frac{3\pi}{2}; \\
\qquad\qquad\qquad\qquad\qquad\qquad -k\frac{A}{\omega}, \ \frac{3\pi}{2} \leqslant \omega t < 2\pi; \\
\qquad\qquad\qquad\qquad\qquad\qquad\qquad\qquad 0, \ \omega t = 2\pi;
\end{cases}
$$

where k is the "holding" parameter for the output value between the zero-crossing point of the input signal derivative and the zero-crossing point of the input signal. As an example, the input and output signals of the FOCI are shown in Figure 19.3 with the parameters $A = 1$, $\alpha = 1.209$ and $k = 0.753$. The output signal is a non-sinusoidal periodic function, the coefficients of its fundamental wave can be calculated as follows,

$$
\begin{aligned}
a_{1F} &= \frac{2}{\pi} \int_0^\pi y_F(t) \cos(\omega t) \mathrm{d}(\omega t) \\
&= \frac{2}{\pi} \left[\int_0^{\frac{\pi}{2}} \frac{A}{\omega^\alpha} \left(\sin\left(\omega t - \frac{\alpha\pi}{2}\right) + \sin\left(\frac{\alpha\pi}{2}\right) \right) \cos(\omega t) \mathrm{d}(\omega t) \right. \\
&\quad \left. + \int_{\frac{\pi}{2}}^\pi \frac{A}{\omega^\alpha} \left(\sin\left(\omega t - \frac{\alpha\pi}{2}\right) - \sin\left(\frac{\alpha\pi}{2}\right) \right) \cos(\omega t) \mathrm{d}(\omega t) \right] \\
&= \frac{2A}{\pi\omega^\alpha} \left[\left(1 - \frac{\pi}{4} - k\right) \sin\left(\frac{\alpha\pi}{2}\right) + \left(\frac{1}{2} - k\right) \cos\left(\frac{\alpha\pi}{2}\right) \right],
\end{aligned}
\qquad (19.13)
$$

similarly, we have,

$$
b_{1F} = \frac{2}{\pi} \int_0^\pi y_F(t) \sin(\omega t) d(\omega t)
$$

$$
= \frac{2A}{\pi \omega^\alpha} \left[\left(\frac{1}{2} + k \right) \sin \left(\frac{\alpha \pi}{2} \right) + \left(\frac{\pi}{4} + k \right) \cos \left(\frac{\alpha \pi}{2} \right) \right]. \tag{19.14}
$$

Thus, the fundamental wave of $y_F(t)$ can be described as,

$$
y_{1F}(t) = a_{1F} \cos(\omega t) + b_{1F} \sin(\omega t)
$$

$$
= c_{1F} \sin(\omega t + \varphi_{1F}), \tag{19.15}
$$

where

$$
c_{1F} = \sqrt{a_{1F}^2 + b_{1F}^2}, \quad \varphi_{1F} = \arctan \left(\frac{a_{1F}}{b_{1F}} \right).
$$

In the same way, we can calculate the coefficients of the third-order harmonic wave,

$$
a_{3F} = \frac{2}{\pi} \int_0^\pi y_F(t) \cos(3\omega t) d(\omega t)
$$

$$
= \frac{2A}{\pi \omega^\alpha} \left[\left(\frac{k}{3} - \frac{1}{3} \right) \sin \left(\frac{\alpha \pi}{2} \right) + \left(\frac{k}{3} - \frac{1}{2} \right) \cos \left(\frac{\alpha \pi}{2} \right) \right], \tag{19.16}
$$

$$
b_{3F} = \frac{2}{\pi} \int_0^\pi y_F(t) \sin(3\omega t) d(\omega t)
$$

$$
= \frac{2A}{\pi \omega^\alpha} \left[\left(\frac{k}{3} - \frac{1}{6} \right) \sin \left(\frac{\alpha \pi}{2} \right) + \frac{k}{3} \cos \left(\frac{\alpha \pi}{2} \right) \right]. \tag{19.17}
$$

Then, we can obtain the third-order harmonic wave of $y_F(t)$,

$$
y_{3F}(t) = a_{3F} \cos(3\omega t) + b_{3F} \sin(3\omega t)
$$

$$
= c_{3F} \sin(\omega t + \varphi_{3F}), \tag{19.18}
$$

where

$$
c_{3F} = \sqrt{a_{3F}^2 + b_{3F}^2}, \quad \varphi_{3F} = \arctan \left(\frac{a_{3F}}{b_{3F}} \right).
$$

19.4.2 Optimality Criteria

In order to find the optimized fractional order conditional integrator, two specifications of the conditional integrator are proposed in this chapter:

(i) A cost function $J_{xF}(\alpha, k)$ for the optimization of FOCI represents the concerned high order harmonic component ratio of FOCI based on the fundamental wave,

$$J_{xF}(\alpha, k) = \frac{c_{xF}}{c_{1F}}, \tag{19.19}$$

where, c_{1F} is the magnitude of the fundamental wave, and c_{xF} is the magnitude of certain high order harmonic wave or even the weighed value of several high order harmonic waves.

This cost function is to be minimized with respect to the parameter k, that is,

$$\frac{\partial J_{xF}(\alpha, k)}{\partial k} = 0. \tag{19.20}$$

(ii) The phase delay of FOCI achieves the designed value in the reachable range.

19.4.3 Optimization of FOCI

In order to compare with different conditional integrators fairly, we used the same specification (i) that the third-order harmonic component ratio of FOCI based on the fundamental wave is the minimum with respect to the parameter k, to design four different optimized fractional order conditional integrators. Based on the magnitude of the fundamental wave, the cost function (19.10) of the third-order harmonic component is presented below,

$$J_{3F}(\alpha, k) = \frac{c_{3F}}{c_{1F}}. \tag{19.21}$$

The minimum third-order harmonic component ratio of FOCI based on the fundamental wave with respect to the parameter k, can be achieved by solving the minimization problem of the cost function $J_{3F}(\alpha, k)$, thus, we can obtain a condition in terms of k and α,

$$\frac{\partial J_{3F}(\alpha, k)}{\partial k} = 0. \tag{19.22}$$

Meanwhile, from specification (ii), as the designed phase delay is φ^*, we can obtain the other condition in terms of k and α,

$$\varphi_{1F} = \arctan\left(\frac{a_{1F}}{b_{1F}}\right) = \varphi^*. \tag{19.23}$$

From (19.22) and (19.23), the optimized solution α^* and k^* of α and k can be solved. Thus the OFOCI is achieved to satisfy the two specifications proposed in Section 19.4.2.

Four different OFOCIs are designed following four desired phase delays. In order to compare with CCI fairly, we designed the $OFOCI_1$ following the same phase delay with that in Section 19.2 as $\varphi^* = -38.1°$. From (19.22) and (19.23), it is obvious that the optimal values of α and k can be solved analytically in theory. However, because of the complexity of these two equations, it is hard to find the analytical solutions. Fortunately, the graphical numerical method in MATLAB® can be used to find the optimized solutions $\alpha^* = 1.273$ and $k^* = 0.788$ according to the intersection of the two curves drawn from the two equations (19.22) and (19.23) in Figure 19.4(a).

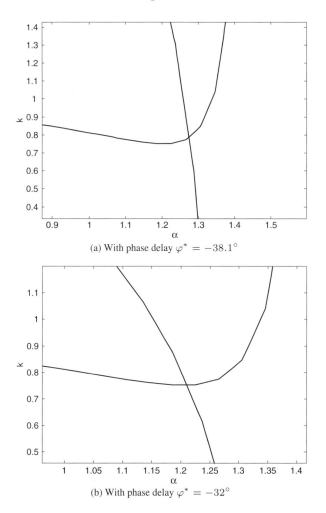

(a) With phase delay $\varphi^* = -38.1°$

(b) With phase delay $\varphi^* = -32°$

Figure 19.4 Optimized solutions with phase delays $-38.1°$ and $-32°$ according to the intersection of two curves from (19.22) and (19.23)

Table 19.1 Performance comparison with the third-order harmonic component ratio minimum specification

Type	$N(j\omega)$	φ_1	$\frac{c_3}{c_1}$ (%)	$\frac{c_5}{c_1}$ (%)	$\frac{c_7}{c_1}$ (%)	$\frac{c_9}{c_1}$ (%)	$\sum$ (%)
CCI	$\dfrac{1.62}{\omega}e^{-j0.212\pi}$	$-38.1°$	26.2	15.8	11.2	8.7	61.9
ICI	$\dfrac{1.08}{\omega}e^{-j0.153\pi}$	$-27.6°$	9.84	13.78	7.03	7.22	37.87
$OFOCI_1$	$\dfrac{0.42}{\omega^{1.273}}e^{-j0.212\pi}$	$-38.1°$	6.23	17.35	5.24	8.51	37.33
$OFOCI_2$	$\dfrac{0.52}{\omega^{1.209}}e^{-j0.178\pi}$	$-32.0°$	0.31	16.29	5.77	8.14	30.51
$OFOCI_3$	$\dfrac{0.67}{\omega^{1.126}}e^{-j0.153\pi}$	$-27.6°$	4.38	15.04	6.25	7.65	33.32
$OFOCI_4$	$\dfrac{1.33}{\omega^{0.530}}e^{-j0.122\pi}$	$-22.0°$	12.48	12.73	7.10	6.75	39.06

So the describing function $N_{1O}(j\omega)$ of $OFOCI_1$ can be derived as $\frac{0.42}{\omega^{1.273}}e^{-j0.212\pi}$ with the designed phase delay $\varphi_{1O} = \varphi^* = -38.1°$. Meanwhile, the third-order, fifth-order, seventh-order and ninth-order harmonic component ratios based on the fundamental wave y_{1O} are also presented in Table 19.1.

In the same way, we also designed $OFOCI_2$ specifying the phase delay as $\varphi^* = -32°$, the optimized solution $\alpha^* = 1.209$ and $k^* = 0.753$ can be obtained from the intersection in Figure 19.4(b). Similarly, we can find the optimized solution of $OFOCI_3$ as $\alpha^* = 1.126$ and $k^* = 0.767$ satisfying the same phase delay $\varphi^* = -27.6°$ with that of the ICI in Section 3. Finally, the $OFOCI_4$ with phase delay $\varphi^* = -22°$ is designed, and the optimized solution $\alpha^* = 0.530$ and $k^* = 0.966$. For comparing clearly, all the describing functions and high order harmonic component ratios based on the fundamental waves of the $OFOCI_2$, $OFOCI_3$ and $OFOCI_4$ are presented in Table 19.1.

From Table 19.1, comparing CCI with $OFOCI_1$, it can be seen that, with the same phase delay, the third-order harmonic component ratio based on the fundamental wave of $OFOCI_1$ is much smaller than that of CCI. At the same time, except for the fifth-order harmonic component ratio which is a little bit higher, the seventh-order and ninth-order harmonic component ratios of $OFOCI_1$ are both smaller than that of CCI. Focusing on ICI and $OFOCI_3$, we can see that, also with the same phase delay, there is not a large difference between these two integrators in the fifth-order, seventh-order and ninth-order harmonic component ratios, but the third-order harmonic component ratio of $OFOCI_3$ is even less than a half of that of ICI. Then, let us see $OFOCI_2$, the phase delay of $OFOCI_2$ is between that of CCI and ICI. Comparing the high order harmonic component ratios of $OFOCI_2$ with those of CCI and ICI, the fifth-order harmonic component ratio of $OFOCI_2$ is a little bit bigger, the seventh-order harmonic component ratio of $OFOCI_2$ is smaller, and the ninth-order harmonic component ratio of $OFOCI_2$ is almost the same as those of CCI and ICI. However, the most important is that the third-order harmonic component ratio of $OFOCI_2$ is much smaller, even close to zero, namely, almost without this third-order harmonic component. Exactly, this third-order harmonic component ratio of $OFOCI_2$ based on the fundamental wave is the smallest one we can obtain following our proposed OFOCI optimality specifications, which is going to be explained in Figure 19.7. Finally, the feature of $OFOCI_4$ is also shown in Table 19.1. It can be seen that, following the proposed optimization method, $OFOCI_4$ with a much smaller phase delay $-22°$ can also be obtained, but there is a trade-off, that is the third-order harmonic component ratio is bigger than that of other conditional integrators except CCI.

In order to perform a more comprehensive comparison, four optimized FOCIs are designed following another optimality specification (i) which minimizes the weighed high order harmonic component ratio of the third-order, fifth-order, seventh-order and ninth-order harmonics with the same weighting coefficients. All the phases and high order harmonic component ratios of the conditional integrators are presented in Table 19.2.

Using the proposed optimization method of FOCI with the minimum third-order harmonic component ratio, and the numerical method of the parameters calculation above, the reachable range of the phase delay is in between ($20°$ and $55°$), and beyond this phase delay range, the intersection cannot be found in the figures for the optimized parameters. In order to reveal the relationships between the designed phase delay φ_1^*, the optimized parameters k^*, the optimized fractional order α^* and the third-order harmonic magnitude ratio J_3 based on that of the fundamental wave, the curves α^* with respect to φ_1^*, k^* with respect to φ_1^*, and J_3 with respect to φ_1^* are plotted in Figures 19.5, 19.6 and 19.7, respectively. From Figure 19.7, we can see that the minimum third-order harmonic component ratio J_3 following the proposed optimality specifications is 0.31% when the phase delay equals $-32°$.

Table 19.2 Performance comparison with the sum component ratio minimum specification of four harmonics

Type	$N(j\omega)$	φ_1	$\dfrac{c_3}{c_1}$ (%)	$\dfrac{c_5}{c_1}$ (%)	$\dfrac{c_7}{c_1}$ (%)	$\dfrac{c_9}{c_1}$ (%)	$\sum$ (%)
CCI	$\dfrac{1.62}{\omega}e^{-j0.212\pi}$	$-38.1°$	26.2	15.8	11.2	8.7	61.9
ICI	$\dfrac{1.08}{\omega}e^{-j0.153\pi}$	$-27.6°$	9.84	13.78	7.03	7.22	37.87
$OFOCI_5$	$\dfrac{0.47}{\omega^{1.267}}e^{-j0.212\pi}$	$-38.1°$	6.33	16.64	5.64	8.25	36.86
$OFOCI_6$	$\dfrac{0.58}{\omega^{1.196}}e^{-j0.178\pi}$	$-32.0°$	2.52	15.66	5.99	7.89	32.06
$OFOCI_7$	$\dfrac{0.75}{\omega^{1.103}}e^{-j0.153\pi}$	$-27.6°$	5.52	14.61	6.39	7.49	34.01
$OFOCI_8$	$\dfrac{1.22}{\omega^{0.812}}e^{-j0.122\pi}$	$-22.0°$	11.73	12.90	6.99	6.81	38.45

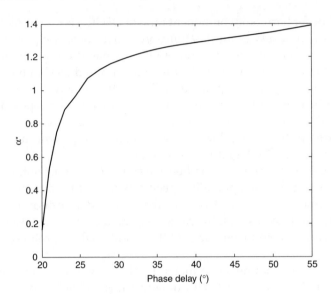

Figure 19.5 α^* with respect to φ_1^*

19.5 Simulation Illustration

In this section, the theoretical analysis and comparison between CCI, ICI and four designed OFOCIs in Section 19.4 are validated by the simulation results with the FFT spectra.

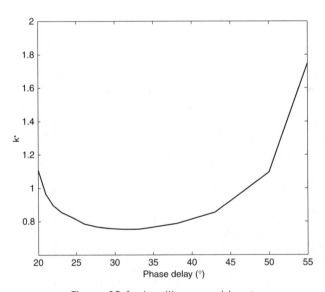

Figure 19.6 k^* with respect to φ_1^*

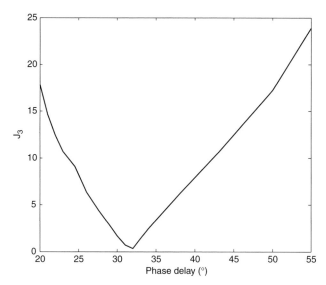

Figure 19.7 J_3 with respect to φ_1^*

In this simulation, the fractional order operator s^λ for OFOCIs is implemented by the impulse response invariant discretization (IRID) method in time domain [51], where a discrete-time finite dimensional (z) transfer function is computed to approximate the continuous irrational transfer function s^λ, where $\lambda \in (-2, 0)$ and s is the Laplace transform variable.

From Table 19.1, it can be seen that not only the phase delay but also the high order harmonic component ratios of *ICI* are better than those of *CCI*. So, the simulation comparison of *ICI* and *OFOCI$_3$* with the same phase delay is performed to test and illustrate the analysis results in Section 19.4.

In Figure 19.8, using the same input signal $u(t) = \sin(t)$, the dashed line shows the output of *ICI*, and the solid line stands for the output of *OFOCI$_3$* with $\alpha^* = 1.126$ and $k^* = 0.767$. In order to verify the high order harmonic component ratios of these two methods, the FFT spectra are presented in Figure 19.9, where the darker line is the FFT spectrum of *ICI* output signal in Figure 19.8, and the pale gray line shows the FFT spectrum of *OFOCI$_3$* output in Figure 19.8, correspondingly. It is obvious that the third-order harmonic component ratio of *OFOCI$_3$* is much smaller than that of *ICI*, the magnitudes from this figure can be used to calculate the values of the high order harmonic component ratios in Table 19.1, which can be illustrated by the simulation results.

19.6 Chapter Summary

In this chapter, a class of optimized fractional order conditional integrator is proposed. The key feature of this OFOCI is to tune the fractional order α and the "holding"

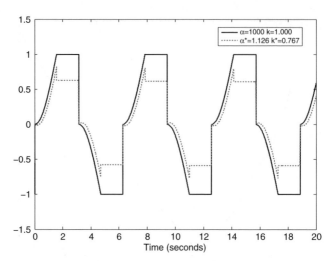

Figure 19.8 Output comparison of the *ICI* and the *OFOCI*₃ (For a color version of this figure, see Plate 23)

parameter k following the optimality criteria to achieve the optimized performance not achievable in integer order conditional integrators. The proposed optimality criteria are designed to satisfy the given phase delay in the reachable range and to minimize the concerned magnitude ratio, which can be calculated from certain high order harmonic magnitude or the weighted magnitude of several high order harmonics divided by the fundamental wave magnitude. The graphical numerical method of calculating the two optimized parameters of OFOCI is introduced, the phase delays and magnitude ratios of four OFOCIs designed according to four different given

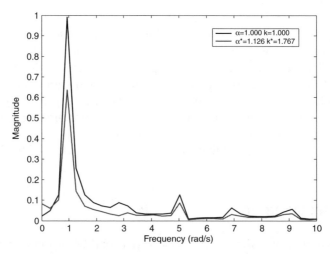

Figure 19.9 FFT spectrum comparison of the *ICI* and the *OFOCI*₃

phase delays are compared with the integer order conditional integrators. Simulation results with the FFT spectra comparison are also presented to illustrate the theoretical analysis. This proposed optimized fractional order conditional integrator can be typically applied in many nonlinear control systems requiring integration action, to achieve less phase delay than the traditional integer order integrator. Consequently, the harmonics of the nonlinear system response can be optimized to achieve the desirable control performance.

Part VII
Fractional Order Motion Control Applications

Part VII

20

Lateral Directional Fractional Order Control of a Small Fixed-Wing UAV

20.1 Introduction

In recent years, small fixed-wing unmanned aerial vehicles (UAVs) have been attracting more and more interest. With short wingspan and light weight, small fixed-wing UAVs can be built easily and operated by only one or two people with hand-carrying and hand-launching [41]. There has been an increasing demand of UAVs in both military and civilian applications. In military use, small fixed-wing UAVs can be used in the battlefield for target acquisition, tracking, damage assessment, and even weapon payloads delivery [76]. Developing the capabilities of small fixed-wing UAVs has also led to increased civilian usage, such as environmental monitoring, border patrol, remote sensing, agriculture mapping, etc. Especially at low altitudes with less interference of clouds [40], small fixed-wing UAVs can even provide cheaper, more prompt and more accurate information than manned aircrafts or satellites. Furthermore, with the developments of micro-electromechanical systems and wireless communication techniques, groups or even swarms of small fixed-wing UAVs can be used to perform more challenging missions such as real-time large-scale surveillance, mapping, and cooperative remote sensing in the near future [156].

In [234], [235], two fourth-order autoregressive with exogenous input (ARX) models are identified to present the attitude characteristics of the longitudinal (pitch) and the lateral (roll) control channels of a small fixed-wing UAV prototype. However, their results are constrained by the technology of about eight or nine years ago. It is relatively hard to estimate the attitude information using only accelerometers. Recently, more and more accurate electronic instruments can be used to build the

Fractional Order Motion Controls, First Edition. Ying Luo and YangQuan Chen.
© 2013 John Wiley & Sons, Ltd. Published 2013 by John Wiley & Sons, Ltd.

UAV control systems, for example, the commercial off-the-shelf inertial measurement units (IMU) [213]. In fact, small fixed-wing UAVs are designed to perform tasks at a low altitude, so they can observe the ground objects closely. This low altitude flight is easily affected by various disturbances from the environment and the UAV is prone to crash. Therefore, a robust flight controller in the autopilot system is indispensable for small fixed-wing UAVs to perform the desired tasks successfully [43].

In this chapter, in order to improve the flight control performance and robustness of small fixed-wing UAVs, a fractional order $(PI)^\lambda$ controller is proposed and designed for the lateral directional (roll-channel) flight control. The roll-channel control system of a small fixed-wing airplane is approximately decoupled and identified as a first order plus time delay (FOPTD) system. An integer order PI controller is designed based on this identified roll-channel control model following the traditional modified Ziegler-Nichols tuning rule. According to three design pre-specifications, the integer order PID, fractional order PI^λ, and $(PI)^\lambda$ controllers are also designed for the roll-channel flight control. These four designed controllers are compared side by side in simulation and the results are validated by real flight tests. It can been seen that the designed fractional order controllers outperform the designed integer order PID (IOPID) and modified Ziegler-Nichols PI (MZNPI) controllers. Furthermore, the $(PI)^\lambda$ controller can achieve even better performance and robustness with respect to gain variations than the designed PI^λ controller [135].

20.2 Flight Control System of Small Fixed-Wing UAVs

20.2.1 Dynamics of Small Fixed-Wing UAVs

The small fixed-wing UAV dynamics can be modeled using system states [16], [36]: longitude (p_x), latitude (p_y) and altitude (p_z) for the position; u, v and w for the velocity; roll (ϕ), pitch (θ), and yaw (ψ) for the attitude; gyro acceleration p, q and r for the gyro rate; a_x, a_y and a_z for the acceleration; v_a, v_g for the air speed and ground speed, respectively; α for the angle of attack and β for the slide-slip angle.

The control inputs of a small fixed-wing UAV generally include: aileron (δ_a), elevator (δ_e), rudder (δ_r), and throttle (δ_t). Many flying wing aircrafts have elevons which combine the functions of the aileron and the elevator, and some small delta wing fixed-wing UAVs just have elevator, aileron and throttle without a rudder.

The nonlinear equation set below (20.1), (20.2) and (20.3) can be used to model the six degrees of freedom small fixed-wing UAV dynamics [7], [24],

$$\dot{x} = f(x, u), \tag{20.1}$$

$$x = [p_x \ p_y \ p_z \ u \ v \ w \ \phi \ \theta \ \psi \ p \ q \ r]^T, \tag{20.2}$$

$$u = [\delta_a \ \delta_e \ \delta_r \ \delta_t]^T. \tag{20.3}$$

Following a preplanned 3-D trajectory with pre-specified orientations is the ultimate objective of small fixed-wing UAV flight control. Due to the hardware limitations, most current small fixed-wing UAVs can only achieve autonomous way-point navigation. For the flight controller design, the precise-model based nonlinear controller design and the in-flight tuning-based PID controller design are the most typically used methods. A precise dynamic model, which is normally very expensive to obtain, is required in the first approach. On the other hand, in flight control, the nonlinear dynamic system can be treated as a simple single input and single output (SISO) or multiple input and multiple output (MIMO) linear system by linearizing around certain trimming points, so the cascaded PID controller can be used for the small fixed-wing UAV flight control. Actually, the cascaded PID controllers are widely used in the commercial UAV autopilots for autonomous flight control [41]. The small fixed-wing UAV dynamics can be decoupled into two modes for the low level control: longitudinal mode for pitch loop and lateral mode for roll loop. After dividing the complicated flight control problem into several decoupling control loops, cascaded controllers can be designed to accomplish the loop control, and to achieve the small fixed-wing UAV flight control task.

In this chapter, the target flight control problem is focused on the decoupled roll-loop or lateral dynamics. The roll-loop of a small fixed-wing UAV can be treated as an SISO (roll-aileron) system around the equilibrium point. That is to say, the flight control system can be treated as a linear system around the point where it can achieve a steady state flight, which means all the force and moment components in the body coordinate frame are constant or zero [205].

20.2.2 The ChangE Small Fixed-Wing UAV Flight Control Platform

ChangE [43], an AggieAir small fixed-wing UAV [42] developed at CSOIS,[1] is used as the experimental platform for the flight controller designs and validations [62]. It is built from a delta wing airframe by the authors. The small fixed-wing UAV airborne system includes an open source Paparazzi [34] TWOG autopilot, navigation sensors (MicroStrain GX2 IMU and uBlox 5 GPS receiver), actuators (two servos for elevon and a throttle motor), a radio modem, a remote control (RC) receiver and Lithium Polymer batteries, as shown in Figure 20.1. The MicroStrain GX2 IMU can provide angle information (ϕ, θ, ψ) of the UAV in flight, at up to 100 Hz sampling frequency with a typical accuracy of $\pm 2°$ under dynamic conditions [213]. The major specifications of the ChangE small fixed-wing UAV are summarized in Table 20.1.

The ChangE small fixed-wing UAV has both manual mode and autonomous mode. The two modes can be switched on either through the ground station or through the RC transmitter by the safety pilot. ChangE UAV communicates with the Paparazzi

---

[1]Center for Self-Organizing and Intelligent Systems, Department of Electrical and Computer Engineering, Utah State University, Utah, USA.

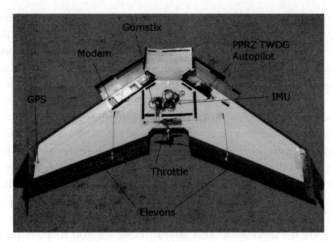

Figure 20.1 ChangE UAV Platform

ground control station (GCS) using a 900 MHz serial radio modem. The Paparazzi GCS software provides on-line parameter changing and plotting functions [34], which can easily be modified for in-flight tuning of controller parameters.

20.2.3 Closed-Loop System Identification

The low-level controllers including the longitudinal and lateral channels have a huge impact on the final flight performance since these controllers deal directly with the control inputs. There are basically two ways to identify the system models: closed-loop methods and open-loop methods. The open-loop analysis is used for the system response with no controls, such as the aileron-roll system model or the aileron-roll rate system model. However, this method can only be used for small reference (as small as 0.02 rad [165]) and the UAV can easily go out of control during flight tests. The closed-loop system identification is employed to identify the system with the loop closed. The only prior condition is that the parameters need to be tuned

Table 20.1 ChangE small fixed-wing UAV major specifications

ChangE UAV	Specifications
Weight	about 5.5 lbs
Wingspan	60 inches
Control Inputs	elevon & throttle
Flight Time	$\leqslant$ 1 hour
Cruise Speed	15 m/s
Take-off	bungee

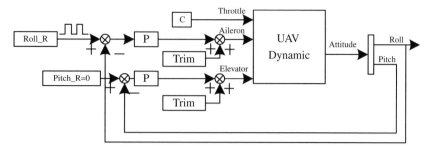

Figure 20.2 UAV flight controller design block diagram

to obtain a roughly working controller before the system identification experiment (for example, P-controller). An example implementation of the inner loop is shown in Figure 20.2.

The UAV needs to be tuned first so that the closed-loop model (ϕ_r-ϕ) can be identified. The idea here is to find the model around the point that the steady flight can be achieved. For simplicity, an ARX model [126] is used, which is defined as the following.

$$\frac{Y(z)}{R(z)} = \frac{a_0 + a_1 z^{-1} + \cdots + a_m z^{-m}}{b_0 + b_1 z^{-1} + \cdots + b_n z^{-n}}, \tag{20.4}$$

where $Y(z)$ is the system output, for example, the roll angle output, and $R(z)$ is the reference signal, for example the roll angle reference. The excitation signals for the system identification can be step, square wave, or a combination of several basic signals. The step and square wave references are commonly used in the identification experiments [126].

In real flight tests for the system identification, the ChangE UAV is manually tuned first to achieve a steady state flight with zero trims on the elevons at the nominal throttle (70% full throttle for ChangE UAV). The Paparazzi flight controller is replaced by the user-designed flight controller running at 60 Hz, following the implementation shown in Figure 20.2. The inner roll and pitch channel controllers only have the proportional part. The inner P gain K_0 for roll-loop is chosen as 10038 count/rad. The aileron inputs are limited within [-9600, 9600] counts. The square wave reference signals with different amplitudes ([$-10°$, $10°$] and [$-20°$, $20°$]) are generated and logged for the system identification. The reference pitch angle is set as zero all the time during the flight experiment. Typical system response (roll) and the reference roll angle are shown in Figure 20.3 and Figure 20.4 during one of our field tests.

The 5th-order ARX model of $Roll\_R - Roll$ loop is calculated based on the flight log (20 Hz) using least squares algorithm. The square wave responses based on the identified model are simulated and plotted together with the real system response in Figure 20.3. Thus, "ref" means the reference signal; "$roll$" is for the roll channel

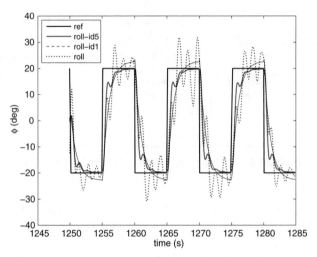

Figure 20.3 Roll-channel system identification with $(-20°, 20°)$ square wave reference

output signal; "$roll - id5$" is for the response with identified 5th-order ARX model and "$roll - id1$" stands for the response with the first order one.

For our controller designs, the following FOPTD model is used, which is a simplified model via frequency-domain fitting [239] from the high order ARX model (20.4) obtained from the flight test data.

$$P(s) = \frac{Y(s)}{R(s)} = \frac{K}{Ts + 1}e^{-Ls}. \qquad (20.5)$$

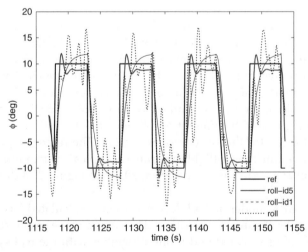

Figure 20.4 Roll-channel system identification with $(-10°, 10°)$ square wave reference

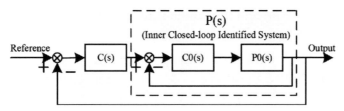

Figure 20.5   Identified inner closed-loop and outer closed-loop system diagram

The continuous-time FOPTD model can be calculated from the above 5th-order ARX model by using `getfoptd` function [239]:

$$P(s) = \frac{0.9912}{0.3414s + 1}e^{-0.2793s}. \tag{20.6}$$

It is worth mentioning here that the identified models are almost the same for excitations with different amplitudes.

20.3 Integer/Fractional Order Controller Designs

Once the system model is derived, another cascaded outer loop controller $C(s)$ can be designed based on the modified Ziegler-Nichols tuning algorithm or fractional order controller design algorithm. The overall control architecture is shown in Figure 20.5. Both the integer order and the fractional order controllers can be designed based on the identified model.

20.3.1 Integer/Fractional Controllers Considered and Design Rules

20.3.1.1 Integer/Fractional Controllers Considered

The first order plus time delay system discussed in this chapter has the following form of transfer function:

$$P(s) = \frac{1}{Ts + 1}e^{-Ls}. \tag{20.7}$$

Note that, the plant gain in (20.7) is normalized to 1 without loss of generality since the proportional factor in the transfer function (20.7) can be incorporated in the proportional part of the controller [119].

The integer order modified Ziegler-Nichols PI controller and integer order PID controller have the following transfer functions in (20.8) and (20.9) below,

$$C_1(s) = K_{p1}\left(1 + \frac{K_{i1}}{s}\right), \tag{20.8}$$

$$C_2(s) = K_{p2}\left(1 + \frac{K_{i2}}{s} + K_{d2}s\right). \tag{20.9}$$

The fractional order PI^λ and $(PI)^\lambda$ controllers have the following forms of transfer function, respectively,

$$C_3(s) = K_{p3}\left(1 + \frac{K_{i3}}{s^\gamma}\right), \tag{20.10}$$

$$C_4(s) = \left(K_{p4} + \frac{K_{i4}}{s}\right)^\lambda, \tag{20.11}$$

where $\gamma \in (0, 2)$ and $\lambda \in (0, 2)$.

20.3.1.2 Design Rules

(1) Modified Ziegler-Nichols design rule

As one of the most popular PID controller tuning rules, the Modified Ziegler-Nichols PI (MZNPI) tuning rule is chosen to make a comparison with the designed PI^λ controllers. The MZNPI tuning method [239] divides the tuning problem into several cases based on different system dynamics:

(i) Lag-dominated dynamics ($L < 0.1T$)

$$K_p = 0.3T/(KL),\ K_i = 1/(8L);$$

(ii) Balanced dynamics ($0.1T < L < 2T$)

$$K_p = 0.3T/(KL),\ K_i = 1/(0.8T);$$

(iii) Delay-dominated dynamics ($L > 2T$)

$$K_p = 0.15/K,\ K_i = 1/(0.4L).$$

(2) The proposed specifications for controllers design

Assuming that the gain crossover frequency is ω_c and phase margin is ϕ_m, the three specifications in Section 2.3 of Chapter 6 are applied to design the integer/fractional order controllers.

20.4 Modified Ziegler-Nichols PI Controller Design

The lateral directional (roll-channel) flight control model of the 60-inch UAV is identified as the first order plus time delay system in (20.6), with $L = 0.2793$s and $T = 0.3414$s. Similarly, an integer order PI controller can be designed using the modified Ziegler-Nichols tuning rule based upon the identified first order plus time delay model (20.6),

$$K_p = \frac{0.3T}{KL} = 0.37,\quad K_i = 0.8T = 3.66.$$

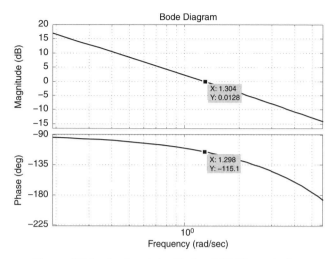

Figure 20.6   Bode plot with the MZNPI controller

So, the MZNPI controller in (20.8) can be tuned as follows

$$C_1(s) = 0.37 \left(1 + \frac{3.66}{s}\right). \tag{20.12}$$

The Bode plot of the open-loop transfer function with the modified Ziegler-Nichols PI controller is presented in Figure 20.6. We can see that the gain crossover frequency is $\omega_c = 1.3 \ rad/s$, and the phase margin is $\phi_m = 65°$.

20.5 Fractional Order $(PI)^\lambda$ Controller Design

The open-loop transfer function $G_4(s)$ of the fractional order $(PI)^\lambda$ controller with the first order plus delay time system is that,

$$G_4(s) = C_4(s)P(s).$$

From (20.7) and (20.11), the phase and gain of the open-loop frequency response can be obtained as follows,

$$Arg[G_4(j\omega)] = -\lambda \arctan \frac{K_{i4}}{1 + K_{p4}\omega} - \arctan(\omega T) - L\omega, \tag{20.13}$$

$$|G_4(j\omega)| = \frac{\left(K_{p4}^2 + \left(\frac{K_{i4}}{\omega}\right)^2\right)^{\frac{\lambda}{2}}}{\sqrt{1 + (\omega T)^2}}. \tag{20.14}$$

From specification (i) of the proposed design rule in Section 20.3.1.2, for the phase margin of $G_3(j\omega)$,

$$Arg(G_4(j\omega))|_{\omega=\omega_c} = -\pi + \phi_m,$$

then, it can be obtained that,

$$K_{i4} = K_{p4}\omega_c \tan\frac{\pi - \arctan(\omega_c T) - \phi_m - L\omega_c}{\lambda}. \tag{20.15}$$

From specification (ii) of the proposed design rule, an equation can be established as below,

$$|G_4(j\omega_c)|_{\omega=\omega_c} = \frac{\left(K_{p4}^2 + \left(\frac{K_{i4}}{\omega_c}\right)^2\right)^{\frac{\lambda}{2}}}{\sqrt{1+(\omega_c T)^2}} = 1,$$

so, we can obtain,

$$K_{i4}^2 = \omega_c^2 \left(1 + T^2\omega_c^2\right)^{\frac{1}{\lambda}} - (K_{p4}\omega_c)^2. \tag{20.16}$$

From specification (iii) of the proposed design rule, for the robustness to gain variations in the plant,

$$K_{i4} = \frac{(T + L\sqrt{1 + T^2\omega_c^2})K_{p4}^2\omega_c^2}{(\lambda K_{p4} - L)\sqrt{1 + T^2\omega_c^2} - T}. \tag{20.17}$$

Clearly, we can solve the three above equations (20.15), (20.17) and (20.16), to obtain the three parameters λ, K_{i4} and K_{p4}. However, as the complexity of three equations is set, the analytical resolution for the parameters of the $(PI)^\lambda$ controller cannot be obtained easily. So, a digital graphical method is used to find the solution of the designed controller as shown below.

According to the identified lateral directional flight control model (20.6) of the 60-inch UAV, the design procedure of the fractional order PI controller is summarized below:

(1) Given $T = 0.3414s$, $L = 0.2793s$, $\omega_c = 1.3 \ rad/s$, $\phi_m = 65°$.
(2) Plot curve 1, K_{i4} with respect to λ and plot curve 2, K_{i4} with respect to λ, then find the values of λ and K_i from the intersection point on the above two curves, which reads $\lambda = 1.4741$, $K_i = 1.0214$.
(3) Calculate $K_{p4} = 0.7158$.
(4) Then we can obtain the designed $(PI)^\lambda$ controller,

$$C_4(s) = \left(0.7158 + \frac{1.0214}{s}\right)^{1.4741}. \tag{20.18}$$

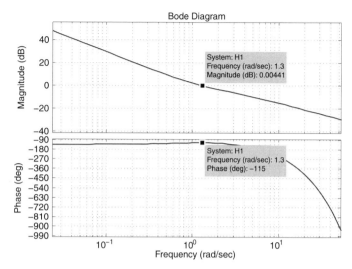

Figure 20.7   Bode plot with the $(PI)^\lambda$ controller

The Bode plot of the open-loop system with the designed $(PI)^\lambda$ controller is shown in Figure 20.7. It can be seen that the phase is flat around the gain crossover frequency. All of the three specifications in the proposed tuning rule in Section 20.3.1.2 for the controller design are satisfied.

20.6 Fractional Order PI Controller Design

The open-loop transfer function $G_3(s)$ of the fractional order PI controller with the first order plus delay time system is that,

$$G_3(s) = C_3(s)P(s).$$

Following the similar deduction and procedure for the $(PI)^\lambda$ controller design in Section 20.5, with the proposed tuning rule for controllers design in Section 20.3.1.2, three parameters λ, K_{i3} and K_{p3} of the PI^λ controller can be solved with three equations concerned in theory. However, as the complexity of the equations is set, it is not easy to find the parameters' resolution of the PI^λ controller. So, the digital graphical method is also used to find the solution of the designed PI^λ controller as follows:

(1) Given $T = 0.3414$s, $L = 0.2793$s, $\omega_c = 1.3$ rad/s, $\phi_m = 65°$.
(2) Plot curve 1, K_{i3} with respect to γ and plot curve 2, K_{i3} with respect to γ. Find the values of γ and K_{i3} from the intersection point on the above two curves, which reads $\gamma = 1.2029$, $K_{i3} = 2.0965$.

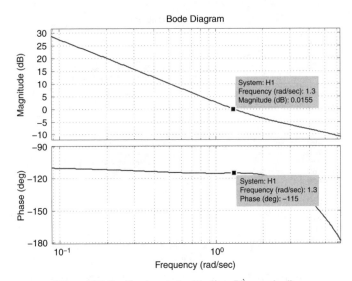

Figure 20.8 Bode plot with the PI^λ controller

(3) Calculate $K_{p3} = 0.7092$.
(4) Then the designed PI^λ controller is obtained,

$$C_{31}(s) = 0.7092 \left(1 + 2.0965 \frac{1}{s^{1.2029}} \right). \tag{20.19}$$

The Bode plot of the open-loop system with the designed PI^λ controller is presented in Figure 20.8. It can be seen that the three specifications in the proposed tuning rule in Section 20.3.1.2 for the controller design are all satisfied.

For the comparison, the PI^λ controller with 80° phase margin, and the same other setting $T = 0.3414$s, $L = 0.2793$s, $\omega_c = 1.3$ rad/s, are also designed as follows,

$$C_{32}(s)(80°) = 0.8461 \left(1 + 1.482 \frac{1}{s^{1.1546}} \right). \tag{20.20}$$

20.7 Integer Order PID Controller Design

The open-loop transfer function $G_2(s)$ of the integer order PID controller for the first order plus delay time system is

$$G_2(s) = C_2(s)P(s).$$

Following the similar deduction for the $(PI)^\lambda$ and PI^λ controllers design in Sections 20.5 and 20.6, respectively, three parameters K_{p2}, K_{i2} and K_{d2} of the IOPID

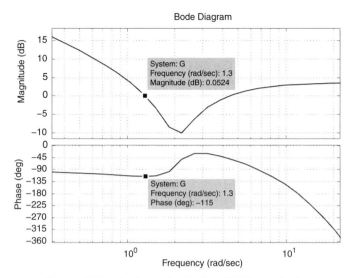

Figure 20.9 Bode plot with the IOPID controller

controller can be solved with three equations. The digital graphical method is also used to find the parameters solution of the designed IOPID controller as shown below.

(1) Given $T = 0.3414$s, $L = 0.2793$s, $\omega_c = 1.3 \ rad/s$, $\phi_m = 65°$.
(2) Plot curve 1, K_{i2} with respect to K_{d2} and plot curve 2, K_{i2} with respect to K_{d2}. Obtain the values of K_{i2} and K_{d2} from the intersection point on the above two curves, which can be read as $K_{i2} = 2.2148$, $K_{d2} = 0.5184$.
(3) Calculate the $K_{p2} = 0.3695$.
(4) Then we can obtain the designed IOPID controller,

$$C_2(s) = 0.3695 + 2.2148\frac{1}{s} + 0.5184s. \qquad (20.21)$$

The Bode plot of the open-loop system with the designed IOPID controller is shown in Figure 20.9. It also can be seen that the three specifications in the proposed tuning rule are all satisfied.

20.8 Simulation Illustration

20.8.1 Fractional Order Controllers Implementation

To implement the PI^λ and $(PI)^\lambda$ controllers, some approximation technique must be used since the fractional order operators have infinite dimensions. The Oustaloup Recursive Algorithm uses a band-pass filter to approximate the fractional order operator $1/s^\gamma$ based on the frequency domain response [183]. However, the Oustaloup approximation cannot be used in different kinds of fractional order operators, for example $(K_p + K_i/s)^\gamma$ for the $(PI)^\lambda$ controller. Fortunately, the fractional order operators for the PI^λ and $(PI)^\lambda$ controllers can both be realized by the impulse response invariant discretization (IRID) method in the time domain as introduced in Chapters 2 and 3 [51], [52], where a discrete-time finite dimensional (z) transfer function is computed to approximate the continuous irrational fractional order operator transfer function s^λ or $(K_p + K_i/s)^\gamma$, s is the Laplace transform variable, and λ is a real number in the range of (-1,1). s^λ is called a fractional order differentiator if $0 < \lambda < 1$ and a fractional order integrator if $-1 < \lambda < 0$. γ is a real number in the range of (0,1), K_p and K_i are constants. These approximations keep the impulse response invariant.

The $(PI)^\lambda$ controller $C_4(s) = (0.7158 + 1.0214/s)^{1.4741}$ which is designed in Section 20.5, can be approximated by a second-order discrete controller using IRID algorithm (sampling period $T_s = 0.0167s$) [51],

$$C_4(z) = \frac{0.3054z^2 + 0.0113z - 0.2941}{z^2 - 0.9975z - 0.0025}. \tag{20.22}$$

The Bode plot of this approximated (z) transfer function $C_4(z)$ can be compared with the true Bode plot of the designed $(PI)^\lambda$ controller $C_4(s)$, as shown in Figure 20.10. It can be observed that the second-order discrete controller $C_4(z)$ can approximate the frequency response of $C_4(s)$ around the designed gain crossover frequency 1.3 rad/s.

For the designed PI^λ controller $C_{31}(s) = 0.7092(1 + 2.0965/s^{1.2029})$ in Section 20.6, with 65° phase margin in design, the fractional order operator $1/s^{0.2029}$ can be

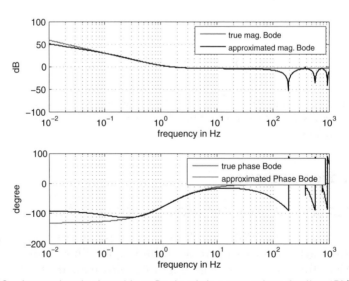

Figure 20.10 Approximated and true Bode plots comparison for the $(PI)^\lambda$ controller

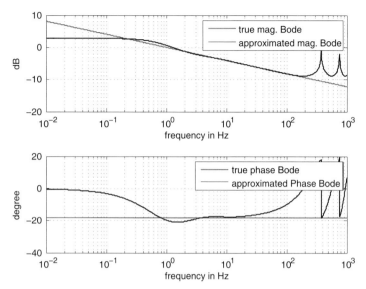

Figure 20.11 Approximated and true Bode plots comparison for the PI^λ controller operator

approximated by a third-order discrete (z) transfer function using the IRID algorithm (sampling period $T_s = 0.0167s$) [51],

$$C_{31}(z) = \frac{0.4243z^3 - 0.6963z^2 + 0.3036z - 0.0231}{z^3 - 1.8684z^2 + 1.0095z - 0.1342}. \tag{20.23}$$

The Bode plot of this approximated fractional order operator (z) transfer function $C_{31}(z)$ is compared with the true Bode plot of this designed PI^λ controller operator $1/s^{0.2029}$ in Figure 20.11. It can be seen that the third-order discrete controller $G_{31}(z)$ can approximate the frequency response of $1/s^{0.2029}$ around the gain crossover frequency 1.3 rad/s designed.

For the other designed PI^λ controller $C_{32}(s) = 0.8461(1 + 1.4821/s^{1.1546})$ in Section 20.6, with an 80° phase margin, the fractional order operator $1/s^{0.1546}$ can also be approximated by a third-order discrete (z) transfer function using the IRID algorithm (sampling period $T_s = 0.0167s$) [51],

$$C_{32}(z) = \frac{0.5203z^3 - 0.8720z^2 + 0.3964z - 0.0352}{z^3 - 1.8454z^2 + 0.9801z - 0.1266}. \tag{20.24}$$

20.8.2 Simulation Results

In this section, all the designed controllers are tested in the simulation. The comparison results among the designed fractional order $PI^\lambda/(PI)^\lambda$, IOPID, and MZNPI controllers are presented in the following section.

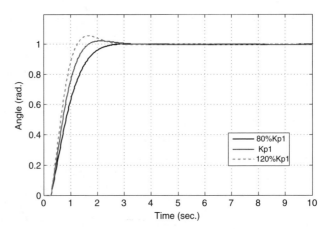

Figure 20.12 Simulation: Step responses using the designed MZNPI controller (20.12) with system gain variations

As the baseline, the MZNPI controller is tested first. According to the identified roll-channel flight control model (20.6), the MZNPI controller (20.12) in Section 20.4 is used in the simulation. As shown in Figure 20.12, the step responses using the MZNPI controller with the system gain changes ±20% are presented. It can be seen that the overshoots of the step responses vary remarkably when the system gain changes, that is to say, the MZNPI controller is not robust to the gain variations.

From the open-loop Bode plot with the MZNPI controller of Figure 20.6 in Section 20.4, it can be seen that the gain crossover frequency is 1.3 rad/s, and the phase margin is 65°. For the fairness of the comparisons, all the IOPID, PI^λ and $(PI)^\lambda$ controllers are designed with the condition of gain crossover frequency $\omega_c = 1.3$ rad/s and phase margin $\Phi_m = 65°$.

From Figure 20.13, we can see that the designed IOPID controller (20.21) is robust to the gain variations, as all of the three responses have no overshoot with the system gain changes ±20%. However, the performance with the designed IOPID controller is not as good, as the response is much slower compared to that using the MZNPI.

The step responses using the designed PI^λ controller (20.19) implemented as (20.23) with 65° phase margin in design, are presented in Figure 20.14 with system gain variations. It is clear that this PI^λ controller is robust to the gain changes, whereas the overshoots are bigger than that with the MZNPI controller in Figure 20.12. In this case, to show the potential advantage of the fractional order controller, the other PI^λ controller is designed as (20.20) with 80° phase margin and the same other settings in Section 20.6. It is implemented as (20.24) for comparison. From Figure 20.15, it can be seen that, not only gain variation robustness, but also smaller overshoots than that using MZNPI, are presented when using the PI^λ with 80° phase margin in design.

Furthermore, Figure 20.16 shows the step responses using the designed $(PI)^\lambda$ controller (20.18) implemented as (20.22), with the gain variations ±20%. The character is obvious that the response is rapid until reaching the point around 90% of the reference

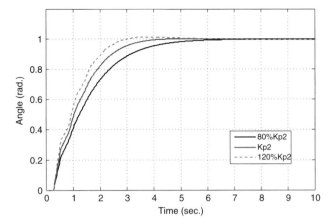

Figure 20.13 Simulation: Step responses using the designed IOPID controller (20.21) with system gain variations

in a short time, after that, the response slows down and approaches the setting line without overshoot and static state error. The design specification for the robustness to the system gain variations is also satisfied from Figure 20.16.

In order to compare all the designed stabilizing controllers in this chapter clearly, Figure 20.17 is illustrated with the labels for different responses. It can be seen obviously that, compared to the step response using the MZNPI controller as a baseline shown in the black curve, IOPID cannot achieve an overshoot, but has a slower response in the dashed line; PI^λ with 65° phase margin design is faster, but has a higher overshoot response in the dark line at the top; with a larger phase margin as 80° design, the other PI^λ can have a faster response with smaller overshoot in dashed

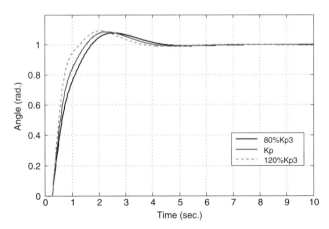

Figure 20.14 Simulation: Step responses using the designed $(PI)^\lambda$ (65°) controller (20.23) with system gain variations

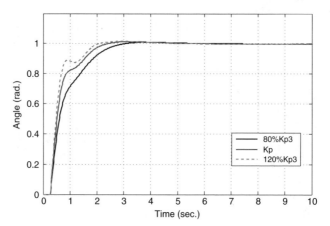

Figure 20.15 Simulation: Step responses using the designed $(PI)^\lambda$ (80°) controller (20.24) with system gain variations (For a color version of this figure, see Plate 24)

black line; when the designed $(PI)^\lambda$ controller is used in the simulation test, the best step response is obtained in the solid red line with the fastest response and without overshoot.

20.9 Flight Experiments

In this section, the designed stabilizing fractional order controllers following the proposed specifications are tested and compared with the designed stabilizing MZNPI and IOPID controllers in the real-time flight experiments, using the ChangE UAV and the flight control platform introduced in Section 20.2.

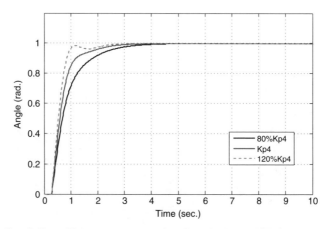

Figure 20.16 Simulation: Step responses using the designed $(PI)^\lambda$ controller (20.22) with system gain variations (For a color version of this figure, see Plate 25)

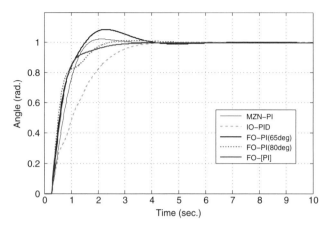

Figure 20.17 Simulation: Step responses comparison using the designed stabilizing fractional/integer order controllers (For a color version of this figure, see Plate 26)

The designed MZNPI controller (20.12) in Section 20.4, the PI^λ controller (20.19)/ (20.20) with the 65°/80° design phase margin in Section 20.6 and the $(PI)^\lambda$ controller (20.18) in Section 20.5 used in the simulation, are all implemented in the real-time flight experiments.

However, the designed IOPID controller (20.21) in Section 20.7 cannot be used in the real-time flight test directly as the simple derivative item in this controller, which can amplify the noise significantly. So the derivative item with a low pass filter $s/(T_l s + 1)$ is used to perform the designed IOPID controller. Thus, the implemented IOPID controller in the flight experiment is shown as below,

$$C_{2-lpf}(s) = 0.3695 \left(1 + 2.2148\frac{1}{s} + 0.5184\frac{s}{T_l s + 1} \right). \tag{20.25}$$

As the gain crossover frequency ω_c is set as 1.3 rad/s, and the bandwidth of the system (20.6) is 2.93 rad/s, so the cutoff frequency of the low pass filter is designed as 5 rad/s and the parameters T_l is chosen as 1.256s.

The anti-windup part is added for the designed MZNPI ($\lambda = 1$), IOPID ($\lambda = 1$) and PI^λ ($\lambda \in (0, 2)$) controllers as shown in Figure 20.18, where k_t is chosen as $2k_i$. The

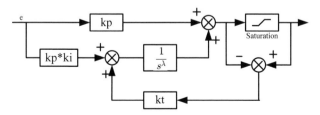

Figure 20.18 Anti-windup for the MZNPI and PI^λ controllers

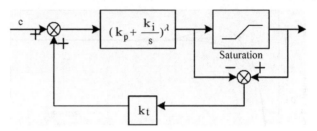

Figure 20.19 Anti-windup for the $(PI)^\lambda$ controller

anti-windup part for $(PI)^\lambda$ controllers is shown in Figure 20.19, where k_t is chosen as 2 following the same design idea in Figure 20.18.

As the baseline, the flight experimental step responses using the MZNPI controller with the proportional gain K_{p1} changes $\pm 20\%$ which is equal to the system gain changes, are presented in Figure 20.20. It can be seen that the overshoots of the step responses vary remarkably, that is to say, the MZNPI controller is not robust to the plant gain variations.

From Figure 20.21, the designed IOPID controller (20.25) is robust to the system gain variations. However, the responses are much slower than that using the MZNPI. The performance with the designed IOPID controller is not satisfied.

In Figure 20.22, the experimental step responses using the designed PI^λ controller (20.19) implemented as (20.23) with a $65°$ design phase margin are presented with system gain variations. It is clear that this PI^λ controller is robust to the gain changes. But, the overshoots are almost as big as that with the MZNPI controller. In this case, the PI^λ controller with the $80°$ phase margin design is also tested in the experiment,

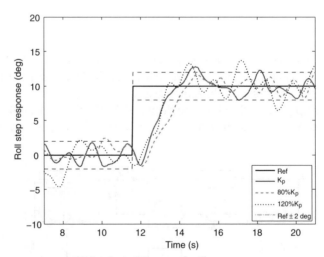

Figure 20.20 Experiment: Step responses using the designed MZNPI controller (20.12) with system gain variations

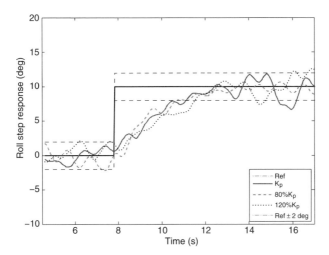

Figure 20.21 Experiment: Step responses using the designed IOPID controller (20.25) with system gain variations

and compared with the MZNPI and $PI^\lambda(65°)$ in Figure 20.23. It can be seen that the response using the $PI^\lambda(80°)$ (20.24) obtains a much smaller overshoot compared to that using the MZNPI and $PI^\lambda(65°)$.

Figure 20.24 shows the experimental step responses using the designed $(PI)^\lambda$ controller (20.18) implemented as (20.22), with the system gain variations ±20%. One can see that the design specification for the robustness to the system gain variations is also satisfied, and the overshoots are very small. At the same time, the designed $(PI)^\lambda$

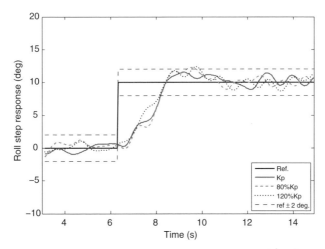

Figure 20.22 Experiment: Step responses using the designed PI^λ (65°) controller (20.23) with system gain variations

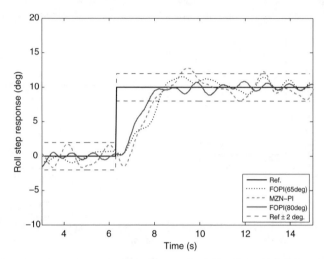

Figure 20.23 Experiment: Step responses comparison using MZNPI (20.12) and two designed PI^λ (20.23) (20.24) controllers

is also compared with the better PI^λ with the $80°$ phase margin design. From Figure 20.25, it is obvious that the designed $(PI)^\lambda$ performs better. Actually, it performs best among the four designed IO/FO controllers in this chapter.

Comparing the simulation with the experiments, one can conclude that the flight experimental results are consistent with the simulation results; the simulation performances are validated in the flight experimental tests.

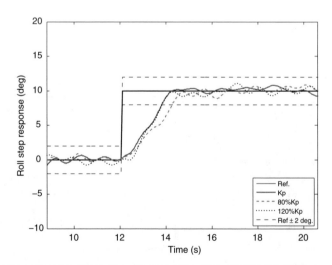

Figure 20.24 Experiment: Step responses using the designed $(PI)^\lambda$ controller (20.22) with system gain variations (For a color version of this figure, see Plate 27)

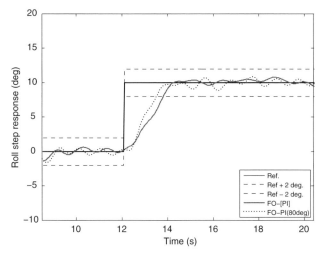

Figure 20.25 Experiment: Step responses comparison using the designed PI^λ (80°) (20.24) and $(PI)^\lambda$ (20.22) controllers (For a color version of this figure, see Plate 28)

20.10 Chapter Summary

In this chapter, the PI^λ and the $(PI)^\lambda$ controllers are developed and applied, to improve the flight control performance of a small unmanned aerial vehicle. The lateral directional (roll-channel) inner closed-loop of a 60-inch UAV is approximately decoupled and identified as a first order plus time delay model. According to this first order plus time delay model, an integer order PI controller is designed following the traditional modified Ziegler-Nichols tuning rule; and according to three design specifications, the PI^λ and $(PI)^\lambda$ controllers are designed, respectively, with the same gain crossover frequency and phase margin settings for fair comparisons. From the simulation illustration and flight experiment verification, it can been seen that the designed stabilizing fractional order controllers outperform the designed stabilizing MZNPI and IOPID controllers. Furthermore, the designed $(PI)^\lambda$ controller can achieve even better performance than the designed PI^λ controller. Our fractional order flight controller design technique is used in the roll-channel control in this chapter, and it can also be applied to other channels such as pitch and yaw for small UAVs, so as to achieve more desired flight control performance.

21

Fractional Order PD Controller Synthesis and Implementation for an HDD Servo System

21.1 Introduction

Hard-disk-drives (HDD) have been widely used as data-storage medium not only for computers but also many other data-processing devices [44], [149]. HDD servomechanism is a high accuracy control system, which plays a critical role in the increasing high density and high performance requirements of HDD [60], [214]. In the HDD servo system, a voice coil motor (VCM) as an actuator is mechanically connected to the heads for reading and writing data. The VCM horizontally activates the heads to the target track on the surface of the disks as the recording medium, where the data reading and writing can be implemented [149]. All the hardware components in the HDD servo system exist offsets and gain variations, for example, the VCM motor and driver, combo circuits, different heads and disks, and also different track widths on the disks. For example, the length of the magnetic fringing field and the magnetic flux density are not uniform in VCM, therefore, the force from the magnetic interaction between the moving coil magnetic field and the permanent magnets varies [104]. Meanwhile, the flying height of the head, the external disturbances and the temperature changes also affect the system gain significantly [79], [227]. Consequently, the loop gain of the HDD servo system is far from a constant value, the tracking control performance varies easily, and even the stability margins are not reliable with significant loop gain variations. Therefore, the effect of the loop gain changes has to be suppressed in the HDD servo system, to achieve consistent tracking performance over all the track locations.

Fractional Order Motion Controls, First Edition. Ying Luo and YangQuan Chen.
© 2013 John Wiley & Sons, Ltd. Published 2013 by John Wiley & Sons, Ltd.

In order to minimize the loop gain variation effect, many algorithms have been proposed, which can be classified into two categories. One is concerned with calibration methods [20], [101], [104], the automatic gain control methods have been widely applied for the loop gain calibration. However, after the servo gain calibration is fixed, the gain variations due to environmental changes cannot be handled. In [227], a model reference adaptation scheme is proposed for the servo loop gain calibration in the HDD servo system. The other way is to design robust controllers to reject the loop gain variations [61], [64], [110]. In [110], a method is investigated to adjust the controller gain automatically to maintain the open-loop gain in the presence of the plant gain variation, and to enhance the robustness of the feedback loop. In [33], a novel robust structure of the model reference adaptive control controller is presented for field-oriented-controlled drives with significant loop gain variations when the plant inertia is large.

In this chapter, based on a frequency response data model, a fractional order (FO) proportional derivative (PD) controller synthesis and implementation are presented for an HDD servo system. The open-loop system using the designed FO controller is given with a unique feature of "flat phase", for example, "iso-damping property" [48], to obtain the robustness on the system loop gain variations. Thus, the control performance can be more consistent than that with the optimized traditional integer order controller when the loop gain changes. Furthermore, the systematic design scheme of the FO controller can adjust the loop shape for the desired control achievement for users with smaller efforts than the traditional optimization methods. This FO controller is designed for the track-following control in the seek tracking control course of HDD servo. First, the basic methodology of the FO controller design synthesis is introduced. Thereafter, the implementation details for the real-time HDD servo system are presented, which are important for the real applications of the FO controllers in practice. From the experimental validation, the proposed FO proportional derivative (FOPD) controller and the design scheme are efficient to improve the track-following control performance for the HDD servo system.

21.2 Fractional Order Controller Design with "Flat Phase"

In this section, the methodology of the fractional order PI/D controller systematic design is presented. The open-loop with the designed FO controller is shown with the "flat phase" feature for the iso-damping property. Therefore, the servo system can be more robust to the loop gain variations, and obtain a more consistent servo performance from track to track and from head to head, than that with the traditional integer order controller in the HDD servomechanism. The form of the proposed FO controller is,

$$C(s) = K_p(1 + K_d s^r), \tag{21.1}$$

where, if $r \in (0, 2)$, the designed controller is FOPD one; and if $r \in (-2, 0)$, the designed controller is FOPI one.

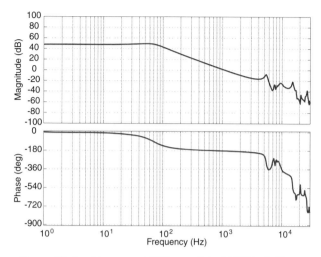

Figure 21.1 Measured FRD model of HDD platform

The plant model of HDD servo system is presented as $P(s)$, which can be a frequency response data (FRD) model, as shown in Figure 21.1, directly measured from the real hard-disk-drive, without fitting or approximation. The FO controller (21.1) is designed following three tuning specifications in Section 2.3.

The initial design procedures of the FO controller can be summarized as follows, without loss of generality:

(i)   Given the parametric model or the non-parametric model $P(s)$ of the plant.
(ii)   Given the desired gain crossover frequency ω_g.
(iii)   Given the desired phase margin ϕ_m.
(iv)   According to the plant model phase $\mathrm{Arg}(P(j\omega_g))$, and the FO controller phase $\mathrm{Arg}(C(j\omega_g))$ at the gain crossover frequency ω_g, find the first relationship in terms of the K_d and r from specification (ii) on phase margin.
(v)   According to the specification (iii) on "flat-phase", the phase of the open-loop system is flat around the gain crossover frequency ω_g, obtain the second relationship between K_d and r.
(vi)   From the two relationships of the K_d and r in (iv) and (v), the two parameters K_d and r of the FO controller can be solved in theory; although the analytical solution is difficult to calculate, the numerical method can be used to find the parameter values [136].
(vii)   With the plant model gain $|P(j\omega_g)|$, and the FO controller gain $|C(j\omega_g)|$ at the gain crossover frequency ω_g, obtain the equation in terms of the K_d, r, and K_p according to specification (i) on gain at crossover frequency.

$$K_p = \frac{1}{|P(j\omega_g)|\sqrt{1 + K_d^2\omega_g^{2r} + 2K_d\omega_g^r \cos(r\pi/2)}};$$

using the solved values of K_d and r in (vi), the third parameter K_p can be calculated. Therefore, the three parameters can be fixed and the FO controller is designed, with the desired "flat phase" feature for the open-loop system.

21.3 Implementation of the Fractional Order Controller

With the given plant model, the FO controller can be designed following three specifications, satisfying the desired gain crossover frequency, phase margin, and "flat phase". The implementation of the designed FO controller in the real-time servo system is the critical issue to achieve the expected control performance benefits. The key point of the FO controller application is the approximation implementation of the FO operator s^r, where $r \in (-2, 2)$. In this section, the fraction order operator s^r is implemented by the impulse response invariant discretization (IRID) method [51], which escapes the constraint of the frequency range.

21.3.1 Phase Loss from the Sampling Delay

According to the Bode plot of the implemented fractional order operator s^r, the desired gain and phase can be satisfied with the continuous approximation by the high-order transfer function. However, in the course of discretization which is necessary for the controller implementation in the real system, the phase of the designed controller is lost, especially, in the high frequency range closing to the Nyquist frequency. This phase loss is unavoidable due to the theory of the discretization, because of the sampling time delay.

In this chapter, the following FO controller design example on HDD servo system is illustrated to make the statement clear. In order to compare the FO controller with the original integer order controller, the gain crossover frequency is set as $\omega_g = 1400\,Hz$, and phase margin is set as, $\phi_m = 35°$. According to the procedures in Section 21.2, the FO controller can be designed as,

$$C(s) = 0.90814(1 + 9.5398 \times 10^{-5} s^{1.0611}),\qquad(21.2)$$

with the fraction order $r = 1.0611 \in (0, 2)$, a FO proportional derivative controller is obtained.

The true open-loop Bode plot is shown in Figure 21.2, with the designed FOPD controller in continuous form. It can be seen from Figure 21.2 that, the gain crossover frequency, phase margin, and the plat phase specifications are all satisfied.

In the approximation implementation of the FO operator $s^{1.0611}$, $8 - th$ order transfer function is used following the impulse response invariant implementation method [51], and the detailed transfer function is,

$$G_{s^{1.0611}} = \frac{A_{sr}}{B_{sr}},$$

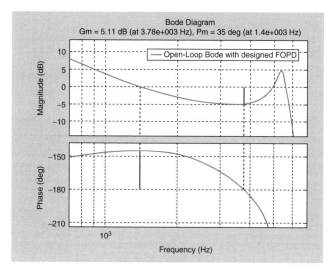

Figure 21.2 Open-loop Bode plot with the designed FO controller in continuous time domain

where,

$$A_{sr} = z^8 - 5.26z^7 + 11.63z^6 - 13.99z^5 + 9.872z^4$$
$$-4.097z^3 + 0.9427z^2 - 0.1027z + 0.003426, \tag{21.3}$$
$$B_{sr} = 1.961 \times 10^{-5}z^8 - 7.981 \times 10^{-5}z^7 + 12.97 \times 10^{-5}z^6$$
$$-10.6 \times 10^{-5}z^5 + 4.442 \times 10^{-5}z^4$$
$$-8.129 \times 10^{-6}z^3 + 1.123 \times 10^{-7}z^2 + 9.75 \times 10^{-8}z - 3.156 \times 10^{-9}. \tag{21.4}$$

After the discretization of the designed FO controller, the sampling delay is induced into the system, and the phase of the implemented FO controller is delayed in sampling. The sampling frequency is $23.914\,kHz$, namely, the sampling period is $T_s = 0.0418ms$. The signal period with the gain crossover frequency is $T_p = 0.7143ms$, and the lost phase of the FO operator (s^r) from discretization can be calculated as follows,

$$\frac{T_s}{T_p} \times \frac{2\pi}{2} = \frac{0.0418 \times 10^{-3}}{0.7143 \times 10^{-3}} \times \frac{360°}{2} = 10.5336°. \tag{21.5}$$

According to the designed FO controller (21.2), the Bode plots of FO operator $s^{1.0611}$ are presented and compared in Figure 21.3. It can be seen that the phase loss between the continuous (the dashed line in Figure 21.3(a)) and the discretized (the solid line in Figure 21.3(a)) FO operator at $1.4k\,Hz$ is around $8.1°$. This phase loss is smaller than the

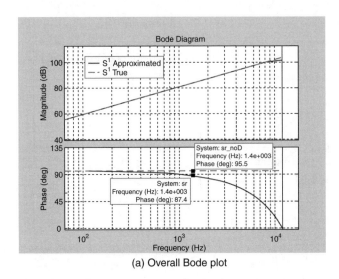

(a) Overall Bode plot

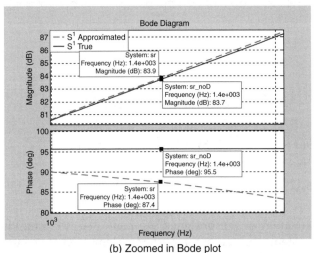

(b) Zoomed in Bode plot

Figure 21.3 Bode plot comparison of the continuous true and discretization approximated FO operator $s^{1.0611}$

calculated $10.5336°$ above. The reason is that the discretized gain of $s^{1.0611}$ is slightly higher than that in the continuous form around the gain cross over frequency, which can be seen in Figure 21.3 as well. The phase delay is reduced by about $2.5°$. Thus, the exact phase loss of the implemented FO operator is around $8°$ at the gain crossover frequency, which can be validated in Figure 21.3.

In the frequency range $\omega \in (1k, 2k)Hz$ of Figure 21.3, the gain difference between the continuous and discretized FO operators is almost constant around $0.2dB$. So, $2.5°$ phase difference can be obtained in this frequency range which is the chosen gain crossover frequency. Therefore, the exact phase loss of the implemented FO operator at the frequency ω can be calculated following the formula below,

$$\phi_{loss} = \omega T_s \times \frac{360°}{2} - 2.5°,$$

where, T_s is the sampling period.

Then, the phase delay θ_d of the implemented FO controller at the designed gain crossover frequency ω_g can be calculated according to the phase loss of the implemented FO operator. The frequency response expression of the continuous FOPD controller $C_1(s)$ is,

$$C_1(j\omega) = K_p(1 + K_d A e^{j\alpha}),\tag{21.6}$$

where $A = \omega^r$, $\alpha = \pi r/2$. The phase of C_1 is,

$$\theta_1 = \arctan\left(\frac{K_d A \sin\alpha}{1 + K_d A \cos\alpha}\right) \times \frac{180°}{\pi}.\tag{21.7}$$

The frequency response expression of the implemented FOPD controller $C_2(s)$ with phase loss of the FO operator is,

$$C_2(j\omega) = K_p\left(1 + K_d A e^{j(\alpha-\delta)}\right),\tag{21.8}$$

where $A = \omega^r$, $\alpha = \pi r/2$, $\delta = \phi_{lost}$. The phase of C_2 is,

$$\theta_2 = \arctan\left(\frac{K_d A \sin(\alpha-\delta)}{1 + K_d A \cos(\alpha-\delta)}\right) \times \frac{180°}{\pi}.\tag{21.9}$$

Then, the phase delay θ_d can be calculated,

$$\theta_d = \theta_1 - \theta_2.\tag{21.10}$$

When $\omega = 1400 \times 2\pi$ rad/s, the phase delay of the implemented FO controller is $\theta_d = 59.4149° - 53.9271° = 5.4877°$. The simulated Bode plots in Figure 21.4(a) and Figure 21.4(b) can validate this calculation result.

Because of the phase loss from the discretization, the gain crossover frequency, phase margin, and flat phase specifications cannot be satisfied in the open-loop Bode plot as shown in Figure 21.4(a) and Figure 21.4(b).

21.3.2 Gain Boosting from Discretization

In the course of the discretization of the FO operator s^r, not only the phase delay is induced, but also the magnitude is slightly boosted. The increase is around $0.2dB$

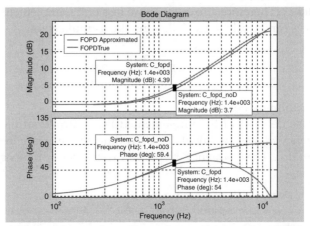

(a) Bode plot comparison of the continuous and descretized FOPD controllers

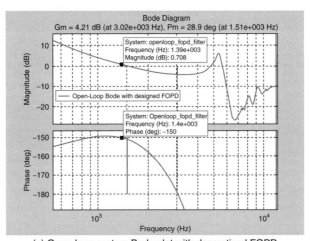

(a) Open-loop system Bode plot with descretized FOPD

Figure 21.4   Bode plot comparison of FOPD controller and open-loop system

in the frequency range $(1k, 2k)Hz$, where the gain crossover frequency is chosen. So, the gain change k_d of the implemented FOPD controller from the discretization can be calculated. The frequency response expression of the continuous FOPD controller $C_1(s)$ is,

$$C_1(j\omega) = K_p(1 + K_d A e^{j\alpha}),\tag{21.11}$$

where $\alpha = \pi r/2$, $A = \omega^r$. The gain of $C_1(s)$ is,

$$k_1 = 20Log_{10}(K_p\sqrt{(1 + K_d A\cos(\alpha))^2 + (K_d A\sin(\alpha))^2}),\tag{21.12}$$

The frequency response expression of the implemented FOPD controller $C_2(s)$ with phase loss of the FO operator is,

$$C_2(j\omega) = K_p(1 + K_d A e^{j\beta}),\tag{21.13}$$

where $\alpha = \pi r/2$, $\beta = \alpha - \phi_{lost}$, $A = \omega^r$. The gain of $C_2(s)$ is,

$$k_2 = 20 Log_{10}(K_p\sqrt{(1 + K_d A_2 \cos(\beta))^2 + (K_d A_2 \sin(\beta))^2}),\tag{21.14}$$

where $\alpha = \pi r/2$, $\beta = \alpha - \phi_{lost}$, $k_{srd} = 0.20$, $A_2 = \omega^r \times 10^{k_{srd}/20}$.
Then, the gain change k_d can be obtained as,

$$k_d = k_1 - k_2.\tag{21.15}$$

When, $\omega = 1400 \times 2\pi \; rad/s$, the gain change of the implemented FOPD controller is $k_d = 3.7211 - 4.4363 = -0.7152$. This calculation result can be validated in the simulation Bode plot of Figure 21.4.

21.4 Adjustment of the Designed FOPD Controller

According to the presentation in Section 21.3 for the implementation of the designed FOPD controller, the initial designed and implemented FOPD controller after discretization cannot satisfy the proposed three specifications in Section 2.3. Therefore, the designed FOPD controller needs to be adjusted following predicting and considering the phase loss and gain boosting in advance, to satisfy the desired specifications after the discretization.

21.4.1 Phase Margin Adjustment with Phase Loss Prediction

As presented in Section 21.3.1, the phase delay of the implemented FOPD controller or the open-loop system can be calculated according to the phase loss of the implemented FO operator s^r. So, this phase loss can be considered in the phase margin setting for the FOPD controller design. The desired phase margin can be adjusted as,

$$\phi'_m = \phi_m + \theta_d.\tag{21.16}$$

If the final phase margin is desired as $\phi_m = 35°$, and the phase loss can be calculated in advance as $\theta_d = 5.4877°$. Therefore, the phase margin needs to be set as,

$$\phi'_m = 35° + 5.4877° = 40.4877°.\tag{21.17}$$

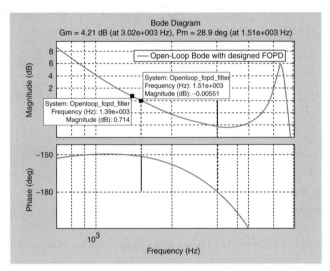

Figure 21.5 The slope of the open-loop magnitude in frequency range $(1k, 2k)$

21.4.2 Gain Crossover Frequency Adjustment with Gain Boosting Prediction

The gain change of the implemented FOPD controller can be calculated as $k_d = -0.7152$ which is mentioned in Section 21.3.2. Meanwhile, according to the observation from Figure 21.5, the slope of the open-loop magnitude in the frequency range $(1k, 2k)Hz$ is around, $0.0063 dB/Hz$. Then, the frequency offset ω_{os} can be calculated as,

$$\omega_{os} = -0.7152/0.0063 = -113.5238 Hz. \tag{21.18}$$

So, the gain crossover frequency setting for the FOPD controller design needs to be adjusted as,

$$\omega'_g = \omega_g + \omega_{os} = 1400 Hz - 113.5 Hz = 1286.5 Hz. \tag{21.19}$$

21.4.3 Phase Slope Adjustment with the Phase Loss Slope Prediction

Since the phase delay can be calculated as in Section 21.3.1, the phase slope change can also be calculated according to the derivative of the phase delay with respect to the frequency ω,

$$\frac{d(\phi_{lost})}{d\omega} = \frac{d(\omega T_s \times \frac{360°}{2} - 2.5°)}{d(\omega)} = T_s \times \frac{360°}{2}(°/Hz) = \frac{T_s}{2}(s). \tag{21.20}$$

So, the phase derivative with respect to the frequency at the gain crossover frequency point needs to be set as $-T_s \times \frac{360°}{2}(°/Hz)$ other than zero, to compensate for the phase loss in the discretization. This compensation can guarantee the "flat phase" specification in the implemented control system with the discretized FOPD controller.

According to the adjustments for the phase margin, the gain crossover frequency, and the phase slope, the FOPD controller can be re-designed as,

$$C_{re}(s) = 0.86878(1 + 4.7552 \times 10^{-5}s^{1.1411}). \tag{21.21}$$

For the approximation implementation of the fractional order operator $s^{1.1411}$, $8 - th$ order transfer function is also used following the impulse response invariant implementation method [51], and the detailed transfer function is,

$$G_{s^{1.1411}} = \frac{A_{srr}}{B_{srr}},$$

where,

$$A_{srr} = z^8 - 5.299z^7 + 11.82z^6 - 14.35z^5 + 10.23z^4$$
$$- 4.3z^3 + 1.005z^2 - 0.1117z + 0.003846, \tag{21.22}$$
$$B_{srr} = 8.653 \times 10^{-6}z^8 - 3.509 \times 10^{-5}z^7$$
$$+ 5.707 \times 10^{-5}z^6 - 4.726 \times 10^{-5}z^5 + 2.085 \times 10^{-5}z^4$$
$$- 4.69 \times 10^{-6}z^3 + 4.966 \times 10^{-7}z^2 - 3.556 \times 10^{-8}z$$
$$+ 2.752 \times 10^{-9}. \tag{21.23}$$

Using this discretized FOPD controller, the open-loop Bode plot can be drawn as shown in Figure 21.6. It can be seen that the desired phase margin, gain crossover

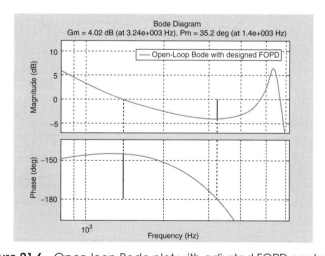

Figure 21.6   Open-loop Bode plot with adjusted FOPD controller

frequency, and "flat phase" are all satisfied with the discretization implementation of the designed FOPD controller.

21.4.4 FO Controller Design and Implementation Procedure Summary

After the discussion for the implementation details of the designed FO controller, the final design and implementation procedures of the FO controller can be summarized as follows:

(1) Given the parametric model or the non-parametric model of the plant.
(2) Given gain crossover frequency.
(3) Given phase margin.
(4) Gain crossover frequency adjustment with gain boosting prediction.
(5) Phase margin adjustment with phase loss prediction.
(6) Phase slope adjustment with phase slope change prediction.
(7) From specification (ii) on phase margin, find one relationship between K_d and r, one curve in terms of these two parameters can be drawn with a numerical method as the red line shown in Figure 21.7.
(8) From specification (iii) on "flat-phase" obtain the other relationship between K_d and r, and the other two curves can be drawn in Figure 21.7.
(9) According to the intersection of two curves in Figure 21.7 from steps (7) and (8), the parameters K_d and r can be fixed.
(10) From specification (i) on gain at crossover frequency, the solution of K_d and r in (9), the third parameter K_p can be determined.

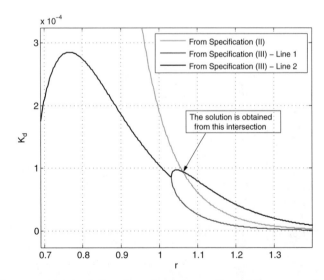

Figure 21.7   The numerical solution from the intersection of the curves in terms of r and K_d

Therefore, the FOPD controller is implemented satisfying the preselected gain crossover frequency, phase margin, and the "flat phase" requirement in the real HDD servo system.

21.5 Experiment

With the designed FOPD controller following the details presented in this chapter, the open-loop Bode plot comparison of the designed FO controller and original integer order controller can be seen in Figures 21.8(a). The dashed line stands for the

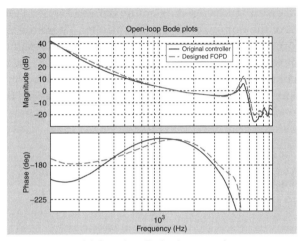

(a) Open-loop Bode plot comparison

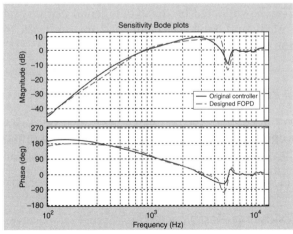

(b) Error sensitivity Bode plot comparison

Figure 21.8 Open-loop and error sensitivity Bode plot comparisons between the designed FOPD and original controllers

open-loop Bode plot with the designed FOPD controller, and the solid line shows that with the original integer order controller. This original controller used for the comparison with the designed FOPD is optimized by the loop shaping according to the performance requirements from the customers.

The "flat phase" feature of the open-loop system with the designed FOPD controller can be seen clearly in Figure 21.8(a). This preferable open-loop shape can be obtained following the presented systematic design scheme in this chapter.

Normally, the dominant external disturbances in HDD servo are in the low frequency range, for example, [10–1000Hz]. From the error sensitivity function (ESF) comparison in Figure 21.8(b), it can be seen that the attenuation in the low frequency range [150–800Hz] with FOPD is stronger than that with original controller, which means the external disturbances resisting ability of the designed FOPD is expected to be better than that of the optimized original controller. Therefore, the position error signal (PES) performance in low frequency range with the designed FOPD should be better than that with the optimized original controller. Because of the "waterbed effect", the ESF in high frequency range [3500–5000Hz] is boosted obviously in Figure 21.8(b). But the disturbance in this range [3500–5000Hz] is small in HDD servo system. So, this trade-off on ESF is positive for the overall PES performance improvement.

Remark 21.5.1 *However, this ESF boosting case in high frequency range can work and achieve benefit is because the high frequency modes are still phase stable in HDD servo system. In general, this ESF boosting in high frequency range should be applied carefully or avoided since it can make the system become sensitive to environment disturbance and measurement noise.*

The advantage of the designed FOPD controller observed from the open-loop Bode plot comparison, can be validated by the experimental demonstration presented in Figures 21.9 and 21.10, and Tables 21.1 and 21.2 for the track following and the throughput performances, respectively.

21.5.1 Original Integer Order Controller Design

In order to compare with this designed FOPD controller fairly, the original integer order controller is optimized according to the Random Neighborhood Search method [30], [80], where the gain crossover frequency and phase margin are set as the same values as those for the FO controller design, $\omega_g = 1400Hz$ and $\phi_m = 35°$.

21.5.2 Track Following Performance

For track following control, the track mis-registrations (TMR) are measured using the designed FOPD and original controllers with loop gain variations in Figure 21.9. In order to verify the robustness to loop gain variations with the designed FOPD

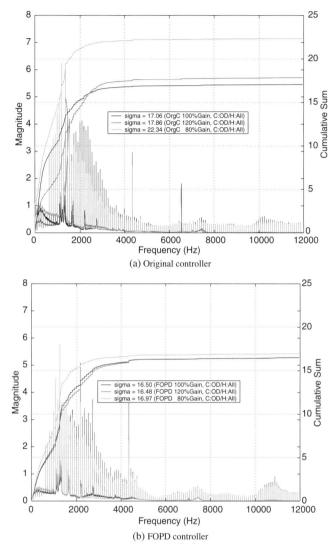

Figure 21.9 TMR comparison with original controller and designed FOPD controller (For a color version of this figure, see Plate 29)

controller clearly, and compare with the original controller fairly, the loop gain is artificially tuned through changing the proportional coefficient of the controller, for example, the K_p in (21.17). The blue line stands for the TMR without loop gain variation. The green and red lines represent the TMR with -20% and $+20\%$ system loop gain variations, respectively. It is clearly, the three TMR lines in Figure 21.9(b) are much more concentrative than that in Figure 21.9(a), namely, the track following performance using the designed FOPD controller is more robust to the loop gain variations than that using the original controller. The detailed comparison for the

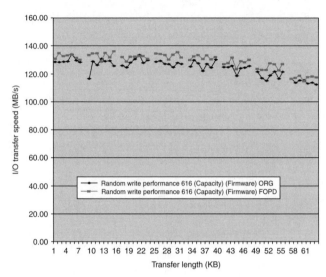

Figure 21.10 IO-Transfer performance comparison with original controller and designed FOPD controller

track following performance is shown in Table 21.1. From the loop shaping function of the systematic design, the overall TMRs with the designed FOPD is smaller than that with the original controller.

21.5.3 Throughput Performance

As shown in Figure 21.9, the TMRs with loop gain variations are compared using two controllers. However, the loop gain variations are produced artificially for simply and directly verifying the robustness of the system, which are not the real loop gain variations in the operations of HDD. Actually, the loop gain variations are severe in HDD with the operations from head to head and from track to track. In order to validate the real benefit of the designed FOPD controller, the I/O throughput performance is tested with the real loop gain variations in the read/write operations in HDD. In Figure 21.10, the black dot-line stands for the Input/Output (I/O) transfer efficiency with original controller, and the gray dot-line represents that with the designed FOPD controller. From Table 21.2, the I/O transfer performance has about 4.5% improvement. The throughput performance with FOPD is obviously better than that with original one.

Table 21.1 TMR performance improvement

	100%Gain	120%Gain	80%Gain
PES (Original)	13.53	12.59	18.60
PES (FOPD)	11.49	11.40	12.66
Improvement	15.05%	9.45%	31.94%

Table 21.2 Seek settle performance improvement

	Original	FOPD	Improvement
IO-Transfer Average Value	124.5 IOPS	130.1 IOPS	5.6 IOPS (4.5%)

21.6 Chapter Summary

This chapter provides a design synthesis for FO proportional derivative controller to achieve two pre-selections (phase margin and gain crossover frequency) and "flat phase" requirement for the HDD servo system. This designed FO controller is more robust to the loop gain variations, and makes the servo performance more consistent in HDD tracking system over the original optimized integer order controller. Meanwhile, as the systematic loop shaping function of the presented FO controller design synthesis, the sensitivity can be easily shaped based on the disturbance distribution to achieve desired control performance. Drive level test shows that both TMR and IO-Transfer performance are improved with the designed FOPD controller in the HDD servo system.

References

[1] J. Ackermann and D. Kaesbauer. Stable polyhedra in parameter space. *Automatica*, 39(5):937–943, 2003.

[2] O. P. Agrawal. A general formulation and solution scheme for fractional optimal control problems. *Nonlinear Dynamics*, 38(1-4):323–337, 2004.

[3] O. P. Agrawal and D. Baleanu. A Hamiltonian formulation and a direct numerical scheme for fractional optimal control problems. *Journal of Vibration and Control*, 13(9-10):1269–1281, 2007.

[4] Hyo-Sung Ahn and YangQuan Chen. State-periodic adaptive friction compensation. In *Proceedings of the 16th IFAC World Congress*, Prague, Czech Republic, July 2005.

[5] Hyo-Sung Ahn, YangQuan Chen, and Huifang Dou. State-periodic adaptive compensation of cogging and coulomb friction in permanent-magnet linear motors. *IEEE Transactions on Magnetics*, 41(1):90–98, 2005.

[6] Hyo-Sung Ahn, YangQuan Chen, and Zhongmin Wang. State-dependent disturbance compensation in low-cost wheeled mobile robots using periodic adaptation. In *Proceedings of IEEE/RSJ International Conference on Intelligent Robots and Systems*, pages 729–734, 2-6 Aug. 2005.

[7] E. P. Anderson, R. W. Beard, and T. W. McLain. Real-time dynamic trajectory smoothing for unmanned air vehicles. *IEEE Transactions on Control Systems Technology*, 13(3):471–477, 2005.

[8] B. Armstrong-Helouvry. *Control of Machines with Friction*. Kluwer Academic Publishers, Norwell, MA, 1991.

[9] B. Armstrong-Hilouvry, P. Dupont, and C. C. de Wit. A survey of models, analysis tools and compensation methods for the control of machines with friction. *Automatica*, 30(7):1083–1138, 1994.

[10] K. J. Astrom and T. Hagglund. Automatic Tuning of PID Controllers. *Instrument Society of America*, 1988.

[11] K. J. Astrom and T. Hagglund. Automatic tuning of simple regulators with specifications on phase and amplitude margins. *Automatica*, 20(5):645–651, 1984.

[12] K. J. Astrom and T. Hagglund. PID Controllers: *Theory, Design and Tuning*. Instrument Society of America, Research Triangle Park, NC, 1995.

[13] K. J. Astrom and T. Hagglund. *Advanced PID Control.* Research Triangle Park, NC: ISA, 2005.

[14] K. J. Astrom and C. C. Hang. Toward intelligent PID control. *Automatica*, 28(1):1–9, 1991.

[15] M. Axtell and E. M. Bise. Fractional calculus applications in control systems. In *Proceedings of the IEEE 1990 National Aerospace and Electronics Conference*, pages 563–566, New York, USA, 1990.

[16] M. Valenti, B. Bethke, and J. How. Cooperative vision based estimation and tracking using multiple UAVs. *Advances in Cooperative Control and Optimization*, Springer, Proceedings of the 7th International Conference on Cooperative Control and Optimization, Series: Lecture Notes in Control and Information Sciences, Vol. 369. M.J. Hirsch, P. Pardalos, R. Murphey, D. Grundel (Eds.). pages 179–189, 2007.

[17] R. L. Bagley and R. A. Calico. Fractional order state equations for the control of viscoelastic damped structures. *Journal of Guidance, Control and Dynamics*, 14(2):304–311, 1991.

[18] R. L. Bagley and P. Torvik. On the appearance of the fractional derivative in the behavior of real materials. *Journal of Applied Mechanics*, 51:294–298, 1984.

[19] D. Baleanu, O. Defterli, and O. P. Agrawal. A central difference numerical scheme for fractional optimal control problems. *Journal of Vibration and Control*, 15(4):583–597, 2009.

[20] E. Banta. Analysis of an automatic gain control (AGC). *IEEE Transactions on Automatic Control*, 9(2):181–182, 1964.

[21] R. Barbosa, J. Tenreiro, and I. Ferreira. PID controller tuning using fractional calculus concepts. *Fractional Calculus and Applied Analysis*, 7(2):119–134, 2004.

[22] R. Barbosa, J. A. Tenreiro, and I. M. Ferreira. Tuning of PID controllers based on Bode's ideal transfer function. *Nonlinear Dynamics*, 38:305–321, 2004.

[23] R. Barbosa, J. Tenreiro Machado, and Alexandra M. Galhano. Performance of fractional PID algorithms controlling nonlinear systems with saturation and backlash phenomena. *Journal of Vibration and Control*, 13(9-10):1407–1418, 2007.

[24] R. Beard, D. Kingston, M. Quigley, D. Snyder, R. Christiansen, W. Johnson, T. McLain, and M. Goodrich. Autonomous vehicle technologies for small fixed wing UAVs. *AIAA Journal of Aerospace Computing, Information, and Communication*, 2(1):92–108, 2005.

[25] Orham Beker, C. V. Hollot, Y. Chait, and H. Han. Fundamental properties of reset control systems. *Automatica*, 40:905–915, 2004.

[26] R. E. Bellman and K. L. Cooke. *Differential-Difference Equations.* New York: Academic Press, 1963.

[27] W. L. Bialkowski. Dreams versus reality: A view from both sides of the gap. *Pulp and Paper Canada*, 11:19–27, 1994.

[28] H. W. Bode. *Network Analysis and Feedback Amplifier Design.* Van Nostrand, New York, 1945.

[29] M. Bodson. Effect of the choice of error equation on the robustness properties of adaptive control schemes. *International Journal of Adaptive Contr. Signal Processing*, 2(1):249–257, 1988.

[30] M. Bodson and S. C. Douglas. Adaptive algorithm for the rejection of periodic disturbances with unknown frequency. *Automatica*, 33(12):2213–2221, 1997.

[31] M. Bodson, A. Sacks, and P. Khosla. Harmonic generation in adaptive feedforward cancellation schemes. *IEEE Transactions on Automatic Control*, 39(9):1939–1944, 1994.

[32] I. Boiko, L. Fridman. Analysis of chattering in continuous sliding-mode controllers. *IEEE Transactions on Automatic Control*, 50(9):1442–1446, 2005.

[33] S.R. Bowes and J. Li. New robust adaptive control algorithm for high-performance AC drives. *IEEE Transactions on Industrial Electronics*, 47(2):325–336, 2000.

[34] P. Brisset, A. Drouin, M. Gorraz, P. S. Huard, and J. Tyler. The Paparazzi solution. In *Proceedings of the MAV*, Sandestin, Florida, USA, 2006.

[35] H. Van Brussel, C.-H. Chen, and J. Swevers. Accurate motion controller design based on an extended pole placement method and a disturbance observer. *Annals of CIRP*, 43(1):367–772, 1994.

[36] G. Cai, B. M. Chen, T. H. Lee, and M. Dong. Design and implementation of a hardware-in-the-loop simulation system for small-scale UAV helicopters. *Mechatronics*, 19(7):1057–1066, 2009.

[37] M. Caputo. *Elasticita e Dissipacione*. Bologna: Zanichelli, 1969.

[38] M. Caputo and F. Mainardi. A new dissipation model based on memory mechanism. *Pure and Applied Geophysics*, 91(8):134–147, 1971.

[39] C. H. Chang and K. W. Han. Gain margins and phase margins for control systems with adjustable parameters. *Journal of Guidance, Control, Dynamics*, 13(3):404–408, 1990.

[40] H. Chao, M. Baumann, A. M. Jensen, Y. Q. Chen, Y. C. Cao, W. Ren, and M. McKee. Band-reconfigurable multi-UAV-based cooperative remote sensing for real-time water management and distributed irrigation control. In *Proceedings of the IFAC World Congress*, Seoul, Korea, July 2008.

[41] H. Chao, Y. C. Cao, and Y. Q. Chen. Autopilots for small unmanned air vehicles: A survey. *International Journal of Control, Automation and Systems*, 8(1):36–44, 2010.

[42] H. Chao, A. M. Jensen, Y. D. Han, Y. Q. Chen and M. McKee. Aggieair: towards low-cost cooperative multispectral remote sensing using small unmanned aircraft systems. *Advances in Geoscience and Remote Sensing, Gary Jedlovec*, Ed. Vukovar, Croatia, pages 463–490, IN-TECH, 2009.

[43] H. Chao, Y. Luo, L. Di, and Y. Q. Chen. Roll-channel fractional order controller design for a small fixed-wing unmanned aerial vehicle. *Control Engineering Practice*, 18(7):761–772, 2010.

[44] B. M. Chen, T. H. Lee, K. Peng, and V. Venkataramanan. *Hard Disk Drive Servo Systems*. Springer, Berlin, 2006.

[45] D. Chen and B. Paden. Nonlinear adaptive torque-ripple cancellation for step motors. In *Proceedings of the IEEE Conference on Decision and Control*, pages 3319–3324, Honolulu, HI, 1990.

[46] Qian Chen, Yossi Chait, and C. V. Hollot. Analysis of reset control systems consisting of a FORE and second-order loop. *ASME Journal of Dynamic Systems, Measurement and Control*, 123:279–283, 2001.

[47] Y. Q. Chen and Moore K. L. Analytical stability bound for a class of delayed fractional order dynamic systems. *Nonlinear Dynamics*, 29:191–200, 2002.

[48] Y. Q. Chen and K. L. Moore. Relay feedback tuning of robust PID controllers with iso-damping property. *IEEE Transactions on Systems, Man, and Cybernetics, Part B: Cybernetics*, 35(1):23–31, 2005.

[49] Y. Q. Chen, I. Petras, and D. Y. Xue. Fractional order control – a tutorial. In *Proceedings of the 2009 American Control Conference*, pages 1397–1411, Hyatt Regency Riverfront, St. Louis, MO, USA, 2009.

[50] YangQuan Chen. Ubiquitous fractional order controls? In *Proceedings of the Second IFAC Workshop on Fractional Derivatives and Applications*, Porto, Portugal, 19–21 July, 2006.

[51] YangQuan Chen. Impulse response invariant discretization of fractional order integrators/differentiators compute a discrete-time finite dimensional (z) transfer function to approximate s^r with r a real number. *Category: filter design and analysis, MATLAB Central,* http://www.mathworks.com/matlabcentral/fileexchange/loadFile. do objectId=21342 objectType=FILE, 2008.

[52] YangQuan Chen. Impulse response invariant discretization of fractional order lowpass filters discretize $[1/(\tau s + 1)]^r$ with r a real number. *Category: Filter Design and Analysis, MATLAB Central,* http://www.mathworks.com/matlabcentral/fileexchange/loadFile.do objectId=21365 objectType=FILE, 2008.

[53] YangQuan Chen and Kevin L. Moore. Discretization schemes for fractional-order differentiators and integrators. *IEEE Transactions on Circuits and Systems-I: Fundamental Theory and Applications,* 49(3):363–367, March 2002.

[54] YangQuan Chen, KianKeong Ooi, MingZhong Ding, LeeLing Tan, and KokTong Soh. An efficient sensorless rotational vibration and shock compensator (RVSC) for hard disk drives with higher TPI. *US PTO Published Patent Applications,* US20010036026, 2001.

[55] YangQuan Chen, Blas M. Vinagre, Igor Podlubny. Fractional order disturbance observer for robust vibration suppression. *Nonlinear Dynamics,* 38(1-2): 355–367, 2004.

[56] YangQuan Chen, Dingyu Xue, and Huifang Dou. Fractional calculus and biomimetic control. In *Proceedings of the IEEE International Conference on Robotics and Biomimetics (RoBio04),* pages (PDF–robio2004–347), Shenyang, China, August 22–25, 2004.

[57] R. K. Chhotaray and A. K. Mohanty. Delay controller: an improvement over PID regulator. *Journal of the Institution of Electronics and Telecommunication Engineers,* 28(8):403–407, 1982.

[58] J. C. Clegg. A nonlinear integrator for servomechanism. *Transactions of the American Institute of Electrical Engineers (A.I.E.E.) Part II,* 77:41–42, 1958.

[59] G. H. Cohen and G. A. Coon. Theoretical consideration of retarded control. *Transactions of the American Society of Mechanical Engineers,* 75:827–834, 1953.

[60] R. Conway, J. Choi, R. Nagamune, and R. Horowitz. Robust track-following controller design in hard disk drives based on parameter dependent Lyapunov functions. *IEEE Transactions on Magnetics,* 46(4):1060–1068, 2010.

[61] J. R. Corrado and W. M. Haddad. Static output feedback controllers for systems with parametric uncertainty and controller gain variation. In *Proceedings of American Control Conference,* pages 915–919, 1999.

[62] CSOIS. OSAM UAV website. Online, URL – http://www.engr.usu.edu/wiki/index.php/OSAM, 2008.

[63] P. Dahl. A solid friction model. *Technical Report TOR-0158(3107-18)-1,* The Aerospace Corporation, El Segundo, CA, 1968.

[64] P. K. Dash, S. Morris, and S. Mishra. Design of a nonlinear variable-gain fuzzy controller for FACTS devices. *IEEE Transactions on Control Systems Technology,* 12(3):428–438, 2004.

[65] Aniruddha Datta, Ming-Tzu Ho, and Shankar P. Bhattacharyya. *Structure and Synthesis of PID Controllers.* Springer-Verlag, London, 2000.

[66] C. C. de Wit and L. Praly. Adaptive eccentricity compensation. *IEEE Transactions on Control Systems Technology,* 8(5):157–766, 2000.

[67] C. Canudas de Wit, H. Olsson, and K. J. Astrom. A new model for control of systems with friction. *IEEE Transactions on Automatic Control,* 40(3):419–425, 1995.

[68] Richard C. Dorf and Robert H. Bishop. *Modern Control Systems*. Pearson Prentice Hall, Pearson Education, Upper Saddle River, NJ, pages 270–278, 2005.

[69] L. Dorčák. Numerical models for simulation the fractional order control systems. *UEF-04-94, The Academy of Science, Institute of Experimental Physics, (Kosice, Slovak Republic)*, pages 1–12, 1994.

[70] H. L. Du and S. S. Nair. Low velocity friction compensation. *IEEE Control Systems Magazine*, 18(2):61–69, 1998.

[71] Chang Duan, Guoxiao Guo, Chunling Du, and Tow Chong Chong. Robust compensation of periodic disturbances by multirate control. *IEEE Transactions on Magnetics*, 44(3):413–418, 2008.

[72] F. Duarte and J. T. Machado. Describing function of two masses with backlash. *Nonlinear Dynamics*, 56(4):409–413, 2009.

[73] F. Duarte and J. T. Machado. Fractional describing function of systems with nonlinear friction. *Intelligent Engineering Systems and Computational Cybernetics*, Springer, The Netherlands, pages 257–266, 2009.

[74] M. O. Efe. Fractional fuzzy adaptive sliding-mode control of a 2-DOF direct-drive robot arm. *IEEE Transactions on Systems, Man, and Cybernetics, Part B: Cybernetics*, 38(6):1561–1570, 2008.

[75] S. Endo, H. Kobayashi, C. Kempf, S. Kobayashi, M. Tomizuka, and Y. Hori. Robust digital tracking controller design for high-speed positioning systems. *Contr. Engineering Practice*, 4(4):527–535, 1996.

[76] D. Erdos and S. E. Watkins. UAV autopilot integration and testing. In *Proceedings of the IEEE Region 5 Conference*, pages 1–6, 17-20 April 2008.

[77] M. Fliess, R. Marquez, and H. Mounier. An extension of predictive control, PID regulators and Smith predictors to some linear delay systems. *International Journal of Control*, 75(10):728–743, 2002.

[78] B. Francis and W. Wonham. The internal model principle of control theory. *Automatica*, 12(5):457–465, 1976.

[79] G. F. Franklin, J. D. Powell, and M. L. Workman. *Digital Control of Dynamic Systems*. Addison-Wesley, Reading, MA, 1990.

[80] G. F. Franklin, J. D. Powell, and M. L. Workman. *Digital Control of Dynamic Systems (3rd Edition)*. Addison-Wesley, Reading, MA, 1998.

[81] Ch. Friedrich. Relaxation and retardation functions of the Maxwell model with fractional derivatives. *Rheol. Acta.*, 30:151–158, 1991.

[82] Arthur Gelb and Wallace E. Vander Velde. *Multiple-Input Describing Functions and Nonlinear System Design*. McGraw-Hill, New York, 1968.

[83] I. Goychuk and P. Hanggi. Fractional diffusion modeling of ion channel gating. *Phys. Rev. E 70, 051915*, 2004.

[84] M. R. Graham and R. A. de Callafon. An iterative learning design for repeatable runout cancellation in disk drives. *IEEE Transactions on Control Systems Technology*, 14(3):474–482, 2006.

[85] T. Hagglund and K. J. Astrom. PID controllers: Theory, design, and tuning. *ISA – The Instrumentation, Systems, and Automation Society (2nd edition)*, 1995.

[86] S. E. Hamamci and N. Tan. Design of PI controllers for achieving time and frequency domain specifications simultaneously. *ISA Transactions*, 45(4):529–543, 2006.

[87] S. E. Hamamci. An algorithm for stabilization of fractional order time delay systems using fractional order PID controllers. *IEEE Transactions on Automatic Control*, 52(10):1964–1969, Oct. 2007.

[88] S. H. Han, Y. H. Kim, and I. J. Ha. Iterative identification of state-dependent disturbance torque for high-precision velocity control of servo motors. *IEEE Transactions of Automatic Controls*, 43(5):724–729, 1998.

[89] C. C. Hang, K. J. Aström, and W. K. Ho. Refinements of the Ziegler-Nichols tuning formula. *IEE Proceedings Part D*, 138(2):111–118, 1991.

[90] T. T. Hartley, C. F. Lorenzo, and H. K. Qammar. Chaos in a fractional order Chua's system. *IEEE Transactions Circuits and Systems I*, 42(8):485–490, 1995.

[91] R. Hilfer. *Applications of Fractional Calculus in Physics*. World Scientific, 2000.

[92] W. K. Ho, O. P. Gan, E. B. Tay, and E. L. Ang. Performance and gain and phase margins of well-known PID tuning formulas. *IEEE Transactions on Control Systems Technology*, 4:473–477, 1996.

[93] N. Hohenbichler and J. Ackermann. Synthesis of robust PID controllers for time delay systems. In *Proceedings of European Control Conference*, page [CD ROM], Cambridge, U.K., 2003.

[94] Norbert Hohenbichler and Dirk Abel. Robust PID-controller design meeting pole location and gain/phase margin requirements for time delay systems. *Automatisierungstechnik*, 54:495–501, 2006.

[95] P. J. Hor, Z. Q. Zhu, D. Howe, and J. Rees-Jones. Minimization of cogging force in a linear permanent magnet motor. *IEEE Transactions on Magnetics*, 34(5):3544–3547, 1998.

[96] H. Hu, Y. Zheng, Y. Chait, and C.V. Hollot. On the zero-input stability of control systems with Clegg integrators. In *Proceedings of the American Control Conference*, pages 408–410, Albuquerque, New Mexico, USA, June 1997.

[97] C. Hwang, J. H. Hwang, and L. F. Hwang. Design of a PID-deadtime control for time-delay systems using the coefficient diagram method. *Journal of the Chinese Institute of Chemical Engineers*, 33(6):565–571, 2002.

[98] MathWorks Inc. Model predictive control toolbox user's guide. *[Online]:* http://www.mathworks.com/access/helpdesk/help/toolbox/mpc/.

[99] C. M. Ionescu and R. De Keyser. Time domain validation of a fractional order model for human respiratory system. In *Proceedings of The 14th IEEE Mediterranean Electrotechnical Conference (MELECON 2008)*, pages 89–95, 5–7 May, 2008.

[100] T. M. Jahns and W. L. Soong. Pulsating torque minimization techniques for permanent-magnet AC motor drives – a review. *IEEE Transactions Ind. Electronics*, 43(2):321–330, 1996.

[101] Q. W. Jia and G. Mathew. A novel AGC scheme for DFE read channels. *IEEE Transactions on Magnetics*, 32:2210–2212, 2000.

[102] K. H. Johansson, J. Ligeros, and S. Sastry. Modeling of hybrid systems. *Systems and Control Letters*, 38(5):141–150, 1999.

[103] T. H. Lee, K. K. Tan and Q. G. Wang. Enhanced automatic tuning procedure for process control of PI/PID controllers. *AIChE Journal*, 42(9):2555–2562, 1996.

[104] C. I. Kang and M. Abed. Servo loop gain identification and compensation in hard disk head-positioning servo. *IEEE Transactions on Magnetics*, 34(4):1889–1891, 1998.

[105] A. Karimi, D. Garcia, and R. Longcham. PID controller design using Bode's integrals. In *Proceedings of the American Control Conference*, pages 5007–5012, Anchorage, AK, USA, 2002.

[106] A. Karimi, D. Garcia, and R. Longchamp. Iterative controller tuning using Bode's integrals. In *Proceedings of the 41st IEEE Conference on Decision and Control*, pages 4227–4232, Las Vegas, Nevada, USA, 2002.

[107] L. H. Keel and S. P. Bhattacharyya. Robust, fragile or optimal? *IEEE Transactions on Automatic Control*, 42:1098–1105, 1997.

[108] Carl J. Kempf and Seiichi Kobayashi. Disturbance observer and feedforward design for a high-speed direct-drive positioning table. *IEEE Transactions on Control Systems Technology*, 7(5):513–526, September 1999.

[109] H. K. Khalil. *Nonlinear Systems* (third edition). Prentice Hall, Upper Saddle River, New Jersey, 2002.

[110] K. S. Kim and K. H. Rew. Enhancing robustness by feedback loop gain adjustment. In *Proceedings of International Conference on Control, Automation and Systems*, pages 2876–2879, 2007.

[111] C. S. Koh and J. S. Seol. New cogging torque reduction method for brushless permanent-magnet motors. *IEEE Transactions on Magnets*, 39(6):3503–3506, 2003.

[112] P. Krishnamurthy and F. Khorrami. Adaptive control of stepper motors without current measurements. In *Proceedings of American Control Conference*, pages 1563–1568, Arlington, VA, Jun. 25-27 2001.

[113] B. Kristiansson and B. Lennartsson. Optimal PID controllers including roll off and Smith predictor structure. In *Proceedings of IFAC 14th World Congress*, pages 297–302, Beijing, P. R. China, 1999.

[114] S. Ladaci and A. Charef. On fractional adaptive control. *Nonlinear Dynamics*, 43(4):365–378, 2006.

[115] P. Lanusse, V. Pommier, and A. Oustaloup. Fractional control system design for a hydraulic actuator. In *Proceedings of the First IFAC Conference on Mechatronics Systems*, Darmstadt, Germany, 2000.

[116] J. F. Leu, S. Y. Tsay, and C. Hwang. Design of optimal fractional order PID controllers. *Journal of the Chinese Institute of Chemical Engineers*, 33(2):193–202, 2002.

[117] A. Leva. PID autotuning algorithm based on relay feedback. *IEEE Proceedings Part-D*, 140(5):32–338, 1993.

[118] A. Leva and A. M. Colombo. On the IMC based synthesis of the feedback block of ISA PID regulators. *Transactions of the Institute of Measurement and Control*, 26(5):417–440, 2004.

[119] H. S. Li, Y. Luo, and Y. Q. Chen. A fractional order proportional and derivative (FOPD) motion controller: tuning rule and experiments. *IEEE Transactions on Control Systems Technology*, 18(2):516–520, 2010.

[120] Hongsheng Li, Jiacai Huang, Di Liu, Jianhua Zhang, and Fulin Teng. Design of fractional order iterative learning control on frequency domain. In *Proceedings of the International Conference on Mechatronics and Automation*, pages 2056–2060, 2011.

[121] Wen Li and Yoichi Hori. Vibration suppression using single neuron-based PI fuzzy controller and fractional order disturbance observer. *IEEE Transactions on Industrial Electronics*, 54(1):117–126, 2007.

[122] Yan Li, YangQuan Chen, and Hyo-Sung Ahn. Fractional order iterative learning control. In *Proceedings of the ICCAS-SICE*, pages 3106–3110, 2009.

[123] P. Lischinsky, C. Canudas deWit, and G. Morel. Friction compensation for an industrial hydraulic robot. *IEEE Control Systems Magazine*, 19:25–30, 1999.

[124] Jing Liu. Comparative study of differentiation and integration techniques for feedback control systems. Master's Thesis, Cleveland State University, 2002.

[125] Jinghan Liu and Yucong Yan. Automatic Control System Nonlinear Regulation. *Mechanical Industry Publisher*, pages 41–42, 1992 *(in Chinese)*.

[126] L. Ljung. *System Identification: Theory for the User*. Prentice Hall, Englewood Cliffs, NJ, 1999.

[127] Jun-Guo Lu and Guanrong Chen. Robust stability and stabilization of fractional order interval systems: An LMI approach. *IEEE Transactions on Automatic Control*, 54(6):1294–1299, 2009.

[128] Jun-Guo Lu and YangQuan Chen. Robust stability and stabilization of fractional order interval systems with the fractional order α : The $0 \le \alpha \le 1$ case. *IEEE Transactions on Automatic Control*, 55(1):152–158, 2010.

[129] L. Lu, Z. Chen, B. Yao, and Q. Wang. Desired compensation adaptive robust control of a linear-motor-driven precision industrial gantry with improved cogging force compensation. *IEEE/ASME Transactions on Mechatronics*, 13(6):617–624, 2008.

[130] Ying Luo and YangQuan Chen. Stabilizing and Robust FOPI Controller Synthesis for First Order Plus Time Delay Systems. *Autocatica*, 48(9):2159–2167, 2012.

[131] Y. Luo, Y. Q. Chen, H. S. Ahn, and Y. G. Pi. Fractional order periodic adaptive learning compensation for cogging effect in PMSM position servo system. In *Proceedings of the American Control Conference*, pages 937–942, St. Louis, Missouri, June 10–12, 2009.

[132] Y. Luo, Y. Q. Chen, H. S. Ahn, and Y. G. Pi. Fractional order robust control for cogging effect compensation in PMSM position servo systems: Stability analysis and experiments. *IFAC Control Engineering Practice*, 18(9):1022–1036, 2010.

[133] Y. Luo, Y. Q. Chen, and Y. G. Pi. Cogging effect minimization in PMSM position servo system using dual-high-order periodic adaptive learning compensation. *ISA Transactions*, 49(4):479–488, 2010.

[134] Y. Luo, C. Y. Wang, Y. Q. Chen, and Y. G. Pi. Tuning fractional order proportional integral controllers for fractional order systems. *Journal of Process Control*, 20(7):823–831, 2010.

[135] Ying Luo, Haiyang Chao, Long Di, and YangQuan Chen. Lateral directional fractional order $(PI)^\alpha$ control of a small fixed-wing unmanned aerial vehicles: controller designs and flight tests. *IET Control Theory and Applications*, 5(18):2156–2167, 2011.

[136] Ying Luo and YangQuan Chen. Fractional order [proportional derivative] controller for a class of fractional order systems. *Automatica*, 45(10):2446–2450, 2009.

[137] Ying Luo and YangQuan Chen. Fractional order [proportional derivative] controller for robust motion control: Tuning procedure and validation. In *Proceedings of the American Control Conference*, pages 1412–1417, St. Louis, Missouri, June 10–12, 2009.

[138] Ying Luo, YangQuan Chen, and Hyo-Sung Ahn. Fractional order adaptive compensation for cogging effect in PMSM position servo systems. In *Proceedings of the IFAC Workshop on Fractional Differentiation and its Applications*, pages 2087–2092, Ankara, Turkey, 2008.

[139] Ying Luo, YangQuan Chen, Hyo-Sung Ahn, and YouGuo Pi. A high order periodic adaptive learning compensator for cogging effect in PMSM position servo system. In *Proceedings of the IEEE International Conference on Systems, Man, and Cybernetics (SMC08)*, pages 3582–3587, Singapore, 12–15 October, 2008.

[140] Ying Luo, YangQuan Chen, and YouGuo Pi. Authentic simulation studies of periodic adaptive learning compensation of cogging effect in PMSM position servo system. In *Proceedings of the Chinese Conference on Decision and Control (CCDC08)*, pages 4760–4765, Yantai, Shandong, China, 2–4 July, 2008.

[141] Ying Luo, YangQuan Chen, and Youguo Pi. Fractional order ultra low-speed position servo: Improved performance via describing function analysis. *ISA Transactions*, 50:53–60, 2011.

[142] Ying Luo, YangQuan Chen, Hyosung Ahn, and Youguo Pi. Fractional order periodic adaptive learning compensation for the state-dependent periodic disturbance. *IEEE Transactions on Control Systems Technology*, 20(2): 465–472, 2012.

[143] Ying Luo, HongSheng Li, and YangQuan Chen. Experimental study of fractional order proportional derivative controller synthesis for fractional order systems. *Mechatronics*, 21:204–214, 2011.

[144] B. J. Lurie. Three-parameter tunable tilt-integral-derivative (TID) controller. *US Patent US5371670*, December 6th, 1994.

[145] Z. F. Lv. Time-domain simulation and design of SISO feedback control systems. Doctoral Dissertation, National Cheng Kung University, 2004.

[146] C. Ma and Y. Hori. Fractional order control: Theory and applications in motion control [past and present]. *IEEE Industrial Electronics Magazine*, 1(4):6–16, 2007.

[147] R. L. Magin. *Fractional Calculus in Bioengineering*. Begell House Publishers Inc., 2006.

[148] A. Makroglou, R. K. Miller, and S. Skaar. Computational results for a feedback control for a rotating viscoelastic beam. *Journal of Guidance, Control and Dynamics*, 17(1):84–90, 1994.

[149] A. Al Mamun, G. Guo, and C. Bi. *Hard Disk Drive: Mechatronics and Control*. Boca Raton, FL: CRC, 2007.

[150] S. Manabe. The non-integer integral and its application to control systems. *ETJ of Japan*, 6(3-4):83–87, 1961.

[151] G. Martelli. Comments on "new results on the synthesis of PID controller". *IEEE Transactions on Automatic Control*, 50(9):1468–1469, 2005.

[152] D. Matignon. Stability results for fractional differential equations with applications to control processing. In *Computational Engineering in Systems Applications*, pages 963–968, Lille, France, July 1996.

[153] D. Matignon. Generalized fractional differential and difference equations: Stability properties and modelling issues. In *Proceedings of Math. Theory of Networks and Systems Symposium*, Padova, Italy, 1998.

[154] A. Mees and A. Bergen. Describing functions revisited. *IEEE Transactions on Automatic Control*, 20(4):473–478, 1975.

[155] A. Le Mehaute and G. Crepy. Introduction to transfer and motion in fractal media: the geometry of kinetics. *Solid State Ionics*, 9(10):17–30, 1983.

[156] L. Merino, J. Wiklund, F. Caballero, A. Moe, J. R. M. De Dios, P. E. Forssen, K. Nordberg, and A. Ollero. Vision-based multi-UAV position estimation. *IEEE Robotics and Automation Magazine*, 13(3):53–62, 2006.

[157] Concepción Alicia Monje Micharet. Design methods of fractional order controllers for industrial applications. PhD thesis, University of Extremadura, Spain, 2006.

[158] K. S. Miller and B. Ross. *An Introduction to the Fractional Calculus and Fractional Differential Equations*. John Wiley and Sons, Inc., New York, USA, 1993.

[159] C. A. Monje, B. M. Vinagre, Y. Q. Chen, and V. Feliu. Proposals for fractional $PI^\lambda D^\mu$-tuning. In *Proceedings of The First IFAC Workshop on Fractional Differentiation and its Applications (FDA04)*, Bordeaux, France, 2004.

[160] C. A. Monje, B. M. Vinagre, V. Feliu, and Y. Q. Chen. Tuning and auto-tuning of fractional order controllers for industry applications. *Control Engineering Practice*, 16:798–812, 2008.

[161] Concepción A. Monje, YangQuan Chen, Blas M. Vinagre, Dingyu Xue, and Vicente Feliu. *Fractional Order Systems and Controls: Fundamentals and Applications (Advances in Industrial Control)*. London: Springer-Verlag, 2010.

[162] Concepción A. Monje, Antonio J. Calderon, Blas M. Vinagre, Vicente Feliu, and YangQuan Chen. On fractional PI^λ controllers: Some tuning rules for robustness to plant uncertainties. *Nonlinear Dynamics*, 38:369–381, 2004.

[163] M. Nakagawa and K. Sorimachi. Basic characteristics of a fractance device. *IEICE Transactions on Fundamentals*, E75-A(12):1814–1819, 1992.

[164] A. Nassirharand and S. R. M. Firdeh. Design of nonlinear lead and/or lag compensators. *International Journal of Control, Automation, and Systems*, 6(3):394–400, 2008.

[165] Jorge Nino, Flavius Mitrachea, Peter Cosynb, and Robin De Keyser. Model identification of a micro air vehicle. *Journal of Bionic Engineering*, 4(4):227–236, December 2007.

[166] T. F. Nonnenmacher and W. G. Glockle. A fractional model for mechanical stress relaxation. *Philosophical Magazine Letter*, 64(2):89–93, 1991.

[167] Aidan O'Dwyer. *Handbook of PI and PID Controller Tuning Rules* (3rd Edition). Imperial College Press, London, 2009.

[168] K. Ohnishi. A new servo method in mechatronics. *Transactions of Japanese Society of Electrical Engineers*, 107-D:83–86, 1987.

[169] K. B. Oldham and J. Spanier. *The Fractional Calculus*. Academic Press, New York, 1974.

[170] K. B. Oldham and C. G. Zoski. Analogue instrumentation for processing polarographic data. *Journal of Electroanal. Chem.*, 157:27–51, 1983.

[171] H. Olsson. Describing function analysis of a system with friction. In *Proceedings of the 4th IEEE Conference on Control Applications*, pages 310–315, 28-29 Sep. 1995.

[172] Manuel Duarte Ortigueira and J. A. Tenreiro Machado (Guest Editors). Special issue on fractional signal processing and applications. *Signal Processing*, 83(11):2285–2480, Nov. 2003.

[173] G. Otten, T. Vries, J. Amerongen, A. Rankers, and E. Gaal. Linear motor motion control using a learning feedforward controller. *IEEE/ASME Trans. Mechatronics*, 2(3):179–187, 1997.

[174] Alain Oustaloup, Jocelyn Sabatier, and Patrick Lanusse. From fractal robustness to the CRONE control. *Fractional Calculus and Applied Analysis*, 2(1):1–30, 1999.

[175] A. Oustaloup. Linear feedback control systems of fractional order between 1 and 2. In *Proceedings of the IEEE Symposium on Circuit and Systems*, Chicago, USA, 1981.

[176] A. Oustaloup. Nouveau systéme de suspension: La suspension CRONE. *INPI Patent 90 046 13*, 1990.

[177] A. Oustaloup. *La commande CRONE*. Edition Hermés, Paris, 1991.

[178] A. Oustaloup. *La dérivation non entiére: Théorie, synthése et applications*. Hermés, Paris, 1995.

[179] A. Oustaloup, F. Levron, F. Nanot, and B. Mathieu. Frequency band complex non integer differentiator: Characterization and synthesis. *IEEE Transactions on Circuits and Systems I*, 47:25–40, 2000.

[180] A. Oustaloup and B. Mathieu. *La commande CRONE: du scalaire au multivariable*. Hermés, Paris, 1999.

[181] A. Oustaloup, B. Mathieu, and P. Lanusse. The CRONE control of resonant plants: application to a flexible transmission. *European Journal of Control*, 1(2), 1995.

[182] A. Oustaloup, P. Melchoir, P. Lanusse, C. Cois, and F. Dancla. The CRONE toolbox for Matlab. In *Proceedings of the 11 th IEEE International Symposium on Computer Aided Control System Design – CACSD*, Anchorage, USA, 9 2000.

[183] A. Oustaloup, X. Moreau, and M. Nouillant. The CRONE suspension. *Control Engineering Practice*, 4(8):1101–1108, 1996.

[184] H. Panagopoulos, K. J. Astrom, and T. Hagglund. Design of PID controllers based on constrained optimization. In *Proceedings of the American Control Conference*, San Diego, CA, USA, 1999.

[185] I. Petras. The fractional order controllers: methods for their synthesis and application. *Journal of Electrical Engineering*, 50(9-10):284–288, 1999.

[186] V. Petrovic, R. Ortega, A. M. Stankovic, and G. Tadmor. Design and implementation of an adaptive controller for torque ripple minimization in PM synchronous motors. *IEEE Transactions on Power Electronics*, 15:871–880, 2000.

[187] I. Podlubny. The Laplace transform method for linear differential equations of the fractional order. *UEF-02-94, Slovak Acad. Sci., Kosice,* http://xxx.lanl.gov/abs/funct-an/9710005/, 1994.

[188] I. Podlubny. *Fractional Differential Equations*. Academic Press, New York, 1999.

[189] I. Podlubny. Fractional order systems and $PI^\lambda D^\mu$ controller. *IEEE Transactions Automatic Control*, 44(1):208–214, 1999.

[190] I. Podlubny, L. Dorcak, and J. Misanek. Application of fractional order derivatives to calculation of heat load intensity change in blast furnace walls. *Transactions of Technical University of Kosice*, 5(5):137–144, 1995.

[191] L. S. Pontryagin. On the zeros of some elementary transcendental function. *American Mathematical Society Translation*, 2:95–110, 1955.

[192] H.-F. Raynaud and A. Zergaïnoh. State-space representation for fractional order controllers. *Automatica*, 36:1017–1021, 2000.

[193] I. J. Rudas, J. K. Tar, and B. Patkai. Compensation of dynamic friction by a fractional order robust controller. In *Proceedings of the IEEE International Conference on Computational Cybernetics*, pages 1–6, 2006.

[194] J. Ryoo, T.-Y. Doh, and M. Chung. Robust disturbance obserer for the track-following control system of an optical disk drive. *Control Engineering Practice*, 12:577–585, 2004.

[195] S. Ladaci and A. Charef. MIT adaptive rule with fractional integration. In *Proceedings of the IMACS Multiconference Computational Engineering in Systems Applications* (CESA'03), pages 118–121, Lille, France, 2003.

[196] S. Ladaci and A. Charef. An adaptive fractional PID controller. In *Proceedings of TMCE International Symposium Series on Tools and Methods of Competitive Engineering*, pages 1533–1540, Ljubljana, Slovenia, 2006.

[197] Alexei Sacks, Marc Bodson, and William Messner. Advanced methods for repeatable runout compensation. *IEEE Transactions on Magnetics*, 31(2):1031–1036, 1995.

[198] S. G. Samko, A. A. Kilbas, and O. I. Marichev. *Fractional Integrals and Derivatives and Some of Their Applications*. Nauka i technika, Minsk, 1987.

[199] Lisa A. Sievers and Andreas H. von Flotow. Comparison and extensions of control methods for narrow-band disturbance rejection. *IEEE Transactions on Signal Processing*, 40(10):2377–2391, 1992.

[200] G. J. Silva, A. Datta, and S. P. Bhattacharyya. New results on the synthesis of PID controllers. *IEEE Transactions on Automatic Control*, 47:241–252, 2002.

[201] G. J. Silva, A. Datta, and S. P. Bhattacharyya. On the stability and controller robustness of some popular PID tuning rules. *IEEE Transactions on Automatic Control*, 48(9):1638–1641, 2003.

[202] C. A. Smith and A. B. Corripio. *Principles and Practice of Automatic Process Control*. John Wiley & Son, Inc., 1985.

[203] O. J. M. Smith. A controller to overcome dead time. *ISA Journal*, 6(2):28–33, 1959.

[204] M. T. Soylemez, N. Munro, and H. Baki. Fast calculation of stabilizing PID controllers. *Automatica*, 39(1):121–126, 2003.

[205] Brian L. Stevens and Frank L. Lewis. *Aircraft Control and Simulation, 2nd Edition*. John Wiley & Sons, Inc., New York, 2003.

[206] K. K. Tan, S. N. Huang, and T. H. Lee. Robust adaptive numerical compensation for friction and force ripple in permanent-magnet linear motors. *IEEE Transactions on Magnetics*, 38(1):221–228, 2002.

[207] Kok Kiong Tan, Qing-Guo Wang, Chang Chieh Hang, and Tore Hagglund. Advances in PID controllers. *Advances in Industrial Control*. Springer-Verlag, London, 2000.

[208] Nusret Tan. Robust phase margin, robust gain margin and Nyquist envelope of an interval plant family. *Computers & Electrical Engineering*, 30:153–165, 2004.

[209] Huajin Tang, Larry Weng, Zhao Yang Dong, and Rui Yan. Adaptive and learning control for SI engine model with uncertainties. *IEEE/ASME Transactions on Mechatronics*, 14(1):93–104, 2009.

[210] Y. Tarte, YangQuan Chen, Wei Ren, and K. Moore. Fractional horsepower dynamometer: a general purpose hardware-in-the-loop real-time simulation platform for nonlinear control research and education. In *Proceedings of IEEE Conference on Decision and Control*, pages 3912–3917, 13-15 Dec. 2006.

[211] M. S. Tavazoei and M. Haeri. Describing function based methods for predicting chaos in a class of fractional order differential equations. *Nonlinear Dynamics*, 57(3):363–373, 2008.

[212] M. S. Tavazoei, M. Haeri, S. Jafari, S. Bolouki, and M. Siami. Some applications of fractional calculus in suppression of chaotic oscillations. *IEEE Transactions on Industrial Electronics*, 55(11):4094–4101, 2008.

[213] The Microstrain Inc. GX2 IMU specifications. Online, http://www.microstrain.com/3dm-gx2.aspx, 2008.

[214] C. K. Thum, C. Du, B. M. Chen, E. H. Ong, and K. P. Tan. A unified control scheme for track seeking and following of a hard disk drive servo system. *IEEE Transactions on Control Systems Technology*, 18(4):294–306, 2010.

[215] M. Tomizuka. Zero-phase error tracking algorithm for digital control. *ASME Journal of Dynamic Syst., Measurement, Contr.*, 109:65–68, 1987.

[216] J. Tou and P. M. Schultheiss. Static and sliding friction in feedback systems. *Journal of Applied Physics*, 24(9):1210–1217, 1953.

[217] T. Umeno and Y. Hori. Robust speed control of DC servomotors using modern two degrees-of-freedom controller design. *IEEE Transactions on Industrial Electronics*, 38:363–368, October 1991.

[218] D. Valerio and J. S. da Costa. Time-domain implementation of fractional order controllers. *IEE Proceedings Part-D: Control Theory and Applications*, 152(5):539–552, 2005.

[219] B. M. Vinagre, I. Petras, I. Podlubny, and Y. Q. Chen. Using fractional order adjustment rules and fractional order reference models in model reference adaptive control. *Nonlinear Dynamics*, 29:269–279, 2002.

[220] B. M. Vinagre, I. Podlubny, A. Hernandez, and V. Feliu. On realization of fractional order controllers. In *Proceedings of the Conference Internationale Francophone d'Automatique*, pages 945–950, Lille, July 5–8, 2000.

[221] Blas M. Vinagre and YangQuan Chen. Lecture notes on fractional calculus applications in automatic control and robotics. In *the 41st IEEE CDC Tutorial Workshop #2 (Blas M. Vinagre and YangQuan Chen, Eds.)*, pages 1–310, [Online] http://mechatronics.ece.usu.edu/foc/cdc02_tw2_ln.pdf, Las Vegas, Nevada, USA, 2002.

[222] Blas M. Vinagre, Concepción A. Monje, and Ines Tejado. Reset and fractional integrators in control. In *Proceedings of the International Carpathian Control Conference (ICCC)*, pages 754–757, Strbske Pleso, Slovak Republic, May 24-27 2007.

[223] A. Wallen, K. J. Astrom, and T. Hagglund. Loop-shaping design of PID controllers with constant t_i/t_d ratio. *Asian Journal of Control*, 4(4):403–409, 2002.

[224] ChunYang Wang, Ying Luo, and YangQuan Chen. An analytical design of fractional order proportional integral and [proportional integral] controllers for robust velocity servo. In *Proceedings of The 4th IEEE Conference on Industrial Electronics and Applications*, pages 3448–3453, Xi'an, China, 25-27 May 2009.

[225] ChunYang Wang, Ying Luo, and YangQuan Chen. Tuning fractional order proportional integral controllers for fractional order systems. In *Proceedings of the Chinese Control and Decision Conference*, pages 307–312, Guilin, China, 17-19 June 2009.

[226] D. J. Wang. Further results on the synthesis of PID controllers. *IEEE Transactions Automat. Contr.*, 52(6):1127–1132, 2007.

[227] FuCai Wang, QingWei Jia, ChunFeng Wang, and JianYi Wang. Servo loop gain calibration using model reference adaptation in HDD servo systems. In *Proceedings of Chinese Control and Decision Conference*, pages 3377–3381, 2009.

[228] J. C. Wang. Realizations of generalized Warburg impedance with RC ladder networks and transmission lines. *Journal of Electrochemical Society*, 134(8):1915–1920, 1987.

[229] Yong Wang, Zhengping Cao, and Guobo Xiang. An intelligent integrator. *Information and Control* (in Chinese), 6:19–20, 1990.

[230] A. Weinmann. Robust improvement of the phase and gain margin and crossover frequency. *e&i Elektrotechnik und Informationstechnik*, 125:17–24, 2008.

[231] B. J. West, M. Bologna, and P. Grigolini. *Physics of Fractal Operators*. New York: Springer, 2003.

[232] S. Westerlund. Capacitor theory. *IEEE Transactions of Dielectrics and Electrical Insulation*, 1(5):826–839, 1994.

[233] S. C. Woon. Analytic continuation of operators – operators acting complex s – time. applications: From number theory and group theory to quantum field and string theories. *Reviews in Mathematical Physics*, 11(4):463–501, 1999.

[234] H. Wu, D. Sun, and Z. Y. Zhou. Micro air vehicle: configuration, analysis, fabrication and test. *IEEE/ASME Transactions on Mechatronics*, 9(1):108–117, Mar. 2004.

[235] H. Wu, D. Sun, and Z. Y. Zhou. Model identification of a micro air vehicle in loitering flight based on attitude performance evaluation. *IEEE Transactions on Robotics*, 20(4):702–712, Aug. 2004.

[236] J.-X. Xu, S. K. Pands, Y.-J. Pan, and T. H. Lee. A modular control scheme for PMSM speed control with pulsating torque minimization. *IEEE Transactions on Industrial Electronics*, 51:526–536, 2004.

[237] D. Xue. Computer Aided Design of Control Systems – MATLAB Language and its Applications. *Beijing Tsinghua University Press*, 1995 (in Chinese).

[238] D. Xue, Chunna Zhao, and Y. Q. Chen. Fractional order PID control of a DC-motor with elastic shaft: A case study. In *Proceedings of American Control Conference (ACC)*, pages 3182–3187, Minnesota, USA, 2006.

[239] Dingyu Xue and YangQuan Chen. *Linear Feedback Control: Analysis and Design with MATLAB*. SIAM Press, 2007.

[240] S. Yamamoto and I. Hashimoto. Recent status and future needs: The view from Japanese industry. In *Proceedings of the 4th International Conference of Chemical Process Control*. CP-CIV, 1991.

[241] S. M. Yang, F. C. Lin, and M. T. Chen. Micro-stepping control of a two-phase linear stepping motor with three-phase VSI inverter for high speed applications. In *Proceedings of 38th IAS Annual Meeting*, pages 473–479, Oct. 2003.

[242] O. Yaniv and M. Nagurka. Design of PID controllers satisfying gain margin and sensitivity constraints on a set of plants. *Automatica*, 40:111–116, 2004.

[243] J. Yi, S. Chang, and Y. Shen. Disturbance-observer-based hysteresis compensation for piezoelectric actuators. *IEEE/ASME Transactions on Mechatronics*, 14(4):456–464, 2009.

[244] Cheng-Ching Yu. *Autotuning of PID Controllers: Relay Feedback Approach. Advances in Industrial Control Series*. Springer-Verlag, London, 1999.

[245] Luca Zaccarian, Dragan Nesic, and Andrew R. Teel. First order reset elements and the Clegg integrator revisited. In *Proceedings of the American Control Conference*, pages 563–568, Portland, USA, June 8-10, 2005.

[246] P. Zavada. Operator of fractional derivative in the complex plane. *Communications in Mathematical Physics*, 192(2):261–285, 1998.

[247] Chunna Zhao, D. Xue, and Y. Q. Chen. A fractional order PID tuning algorithm for a class of fractional order plants. In *Proceedings of the IEEE International Conference on Mechatronics and Automation* (ICMA), pages 216–221, Niagara, Canada, 2005.

[248] S. Zhao and K. Tan. Adaptive feedforward compensation of force ripples in linear motors. *Control Engineering Practice*, 13:1081–1092, 2005.

[249] J. G. Ziegler and N. B. Nichols. Optimum settings for automatic controllers. *Transactions of American Society of Mechanical Engineers*, 64:759–768, 1942.

[250] Chunyang Wang, Yongshun Jin, Yangquan Chen. Auto-tuning of FOPI and FO[PI] Controllers with Iso-damping Property. In Proceedings of the 48th IEEE Conference on Decision and Control and 28th Chinese Control Conference, pages 7309–7314, Shanghai, P.R. China, December 16-18, 2009.

[251] Yongshun Jin, YangQuan Chen, and Dingyu Xue. Time-constant robust analysis of a fractional order [proportional derivative] controller. *IET Control Theory and Applications*, 5(1):164–172,2011.

[252] Ying Luo, YangQuan Chen, Youguo Pi, Concepcin A. Monje, and Blas M. Vinagre. Optimized fractional order conditional integrator. *Journal of Process Control*, 21(6):960–966, 2011.

[253] Ying Luo, YangQuan Chen, and YouGuo Pi. Fractional order adaptive feed-forward cancellation for disturbances. *Asian Journal of Control*, 2012 (to appear).

[254] Ying Luo and YangQuan Chen. Synthesis of robust PID controllers design with complete information on pre-specifications for the FOPTD systems. In *Proceedings of 2011 American Control Conference*, pages 5013–5018, San Francisco, CA, June 29-July 1, 2011.

Index

Fractional Order Motion Controls, First Edition. Ying Luo and YangQuan Chen.
© 2013 John Wiley & Sons, Ltd. Published 2013 by John Wiley & Sons, Ltd.